Magnetic Communications

Magnetic Communications

From Theory to Practice

Edited by
Fei Hu

CRC Press
Taylor & Francis Group
Boca Raton London New York

CRC Press is an imprint of the
Taylor & Francis Group, an **informa** business

CRC Press
Taylor & Francis Group
6000 Broken Sound Parkway NW, Suite 300
Boca Raton, FL 33487-2742

Printed on acid-free paper

International Standard Book Number-13: 978-1-4987-9975-1 (Hardback)

Library of Congress Cataloging-in-Publication Data
LoC Data here

Visit the Taylor & Francis Web site at
http://www.taylorandfrancis.com

and the CRC Press Web site at
http://www.crcpress.com

To my parents …

谨以此书献给我亲爱的父亲胡祖德和母亲董金花。

Contents

SECTION II ACOUSTIC COMMUNICATIONS

Preface

Today, wireless communication systems are mostly based on radio frequency (RF) bands, which typically vary from 1 MHz to 60 GHz. For example, WiFi operates at 2.45 GHz, most cellular networks operate from 800 M to 3 GHz, milliwave uses 28 G–60 GHz, etc.

However, in many applications, RF communications may not be able to provide good signal quality. For example, in underwater sensor networks, RF signals can quickly fade away after 1 m of distance. In underground mine exploration systems, usually the communication devices cannot reach each other through RF channels if they are more than 10 m away.

In those applications, non-RF wireless media, such as magnetic or acoustic channels, may play more important roles due to their strong penetration capability in water/ground environments. For example, acoustic signals can travel over 1,000 m in the water without much signal energy loss. Acoustic channels typically operate below 100 kHz.

Magnetic induction (MI) has proved to be a reliable and high-speed wireless communication method in underwater environments. While acoustic signals propagate at an extremely slow speed (~1,500 m/s), MI waves propagate at a speed of 3.33×10^7 m/s. While acoustic data rate is ~kb/s, MI can reach ~Mb/s. In inland (e.g. lake) water, MI can easily reach >50 m transmission distance without much fading loss. Under the sea, it can also reach around 10 m of range by using general low-cost coil antennas. Multiple MI nodes can be used to relay the data for a long distance.

This is the first book that introduces the acoustic/magnetic communications from a science and engineering design perspective. We have invited some experts who work in the related fields to contribute their research findings to this book. Particularly, we focus on the following six topics:

1. *Magnetic communications for wireless data /energy transfer:* MI has proved to be an effective approach to wirelessly charge electronic devices such as cell phones, iPads, and even electrical vehicles. This avoids the clumsy cable-based charging. We have included two chapters covering the MI-based power

charge principles. Note that the MI can be used for both wireless power and data transfer. This is useful in some data-sharing applications.

2. *Magnetic communication models:* Here we focus on the magnetic energy transfer efficiency between two induction coils and the channel characteristics during the magnetic signal coupling between the sender and receiver. Such models are important to the understanding of high-speed MI data transmission conditions. In addition, we have also included a chapter to cover the MI communication capacity model, which explains why MI has an upper bound for its data transmission throughput.
3. *Acoustic network routing protocols:* In underwater acoustic networks, it is necessary to establish an end-to-end routing path between any two nodes. For example, a sensor may need to deliver its collected undersea data to a buoy on the water surface. The acoustic links have long signal propagation delay (much longer than RF links). Therefore, the routing protocol must be able to tolerate such a long delay as well as the node mobility issues due to the impacts of water currents.
4. *Acoustic network media access control (MAC):* Due to the long propagation delay and slow data rate of acoustic channels, conventional 802.11-based MAC cannot be used in acoustic networks. A new schedule-based collision avoidance scheme may be more suitable to underwater acoustic links. This book will explain such a novel MAC scheme.
5. *Hybrid RF and acoustic communications:* In a complete underwater resource exploration system, the wireless network could consist of two parts: the acoustic links under the water and RF links above the water surface. The base station in the boat can use both acoustic modem and RF transceiver to communicate with all devices. Such a system has a complex network bandwidth management and protocol control. This book will explain those details.
6. *Acoustic signal propagation models:* We have also included a few chapters to explain the accurate math models to reflect the acoustic signal fading characteristics in free space and water. The corresponding interference alignment scheme is also described.

In some chapters, the authors have also pointed out the future research trends in those topics. Those trends are important to new product design, since they describe the unsolved issues and help to enhance existing products (such as acoustic modem and MI coils) by overcoming those issues.

Certainly, there could be some other topics to be covered, such as MI-based under vehicle swarming control, the acoustic localization problem, hybrid acoustic/magnetic networking, security and privacy issues in those systems, and so on. In our chapters, we have included many references that could help readers to further pursue those topics.

This book targets the following audiences: (1) *company R&D people*: when the company wants to manufacture the acoustic or magnetic communication systems, they must consider the earlier quoted topics. This book provides detailed technical discussions from an engineering design viewpoint. The R&D people can gain deep understanding of the non-RF networking schemes and use them to improve their existing products. (2) *Academia researchers*: Graduate students can use this book as the starting point when they want to pursue a topic in non-RF communications. Many research ideas are illustrated in the book. Other researchers can find new interesting issues to be pursued by reading our chapters. This book can also be used for teaching purpose for college engineering students, since it provides a solid foundation on acoustic/magnetic communications.

Finally, we thank all experts for their excellent work in chapter writing. Due to the time limitation, this book might have some issues. We have tried to give credits to each cited reference in the chapters. If you see any problems in this book, please let us know. Thank you for reading this book!

Editor

Dr. Fei Hu is currently a full professor in the Department of Electrical and Computer Engineering at the University of Alabama, Tuscaloosa, Alabama, USA. He obtained his Ph.D. degrees at Tongji University (Shanghai, China) in the field of signal processing (in 1999), and at Clarkson University (New York, USA) in electrical and computer engineering (in 2002). He has published more than 200 journal/conference papers and books. His research has been supported by U.S. National Science Foundation, Department of Energy, Department of Defense, Cisco, Sprint, and other agencies. His research expertise can be summarized as *3S: Security, Signals, Sensors:* (1) Security: This is about how to overcome different cyber attacks in a complex wireless or wired network. Recently, he focuses on cyberphysical system security and medical security issues. (2) Signals: This mainly refers to *intelligent signal processing*, that is, using machine learning algorithms to process sensing signals in a smart way to extract patterns (i.e., pattern recognition). (3) Sensors: This includes microsensor design and wireless sensor networking issues.

List of Contributors

Mohammad N. Abdallah
Department of Electrical Engineering and Computer Science
Syracuse University
Syracuse, New York

Naveed Ahmad
Department of Computer Science
University of Peshawar
Peshawar, Pakistan

Ian F. Akyildiz
Institute for Digital Communications
Friedrich-Alexander University Erlangen-Nuremberg
Erlangen, Germany

Muhammad Arshad
Department of Computer Science
Institute of Management Sciences
Peshawar, Pakistan

Clifton J. Barber
CJ Barber Consulting Services, Inc.
Jonesboro, Georgia

Mark Briggs
Underwriters Laboratories
Fremont, California

Yue Cao
Department of Computer & Information Sciences
Northumbria University
Newcastle upon Tyne, United Kingdom

Lucas S. Cerqueira
Computer Science Department
Universidade Federal de Juiz de Fora
Juiz de Fora, Minas Gerais, Brazil

Dong-Ho Cho
School of Electrical Engineering
Korea Advanced Institute of Science & Technology
Daejeon, Korea

Alexander D. de Sousa
Computer Science Department
Universidade Federal de Minas Gerais
Belo Horizonte, Minas Gerais, Brazil

Qiang Fu
Department of Computer and Electrical Engineering
The University of Alabama
Tuscaloosa, Alabama

Wolfgang H. Gerstacker
Institute for Digital Communications
Friedrich-Alexander University Erlangen-Nuremberg
Erlangen, Germany

Fei Hu
Department of Electrical and Computer Engineering
University of Alabama
Tuscaloosa, Alabama

Shengming Jiang
College of Information Engineering
Shanghai Maritime University
Shanghai, China

Shuchao Jiang
College of Information Engineering
Shanghai Maritime University
Shanghai, China

Muhammad Khalid
Department of Computer and Information Sciences
Northumbria University
Newcastle upon Tyne, United Kingdom

Waqar Khalid
Department of Computer Science
Institute of Management Sciences
Peshawar, Pakistan

Mohammad Reza Khosravi
Department of Electrical and Electronic Engineering
Shiraz University of Technology
Shiraz, Iran

Steven Kisseleff
Institute for Digital Communications
Friedrich-Alexander University Erlangen-Nuremberg
Erlangen, Germany

Kisong Lee
School of Information and Communication Engineering
Chungbuk National University
Cheongju, Korea

Conggai Li
College of Information Engineering
Shanghai Maritime University
Shanghai, China

Feng Liu
College of Information Engineering
Shanghai Maritime University
Shanghai, China

Yu Lu
Electrical and Computer Engineering
The University of Alabama
Tuscaloosa, Alabama

Rui Ma
Computer Science and Engineering
Southern University of Science and Technology
Shenzhen, China

Varun G. Menon
Department of Computer Science and Engineering
SCMS School of Engineering and Technology
Kochi, India

José Augusto M. Nacif
Instituto de Ciências Exatas e Tecnológicas
Universidade Federal de Viçosa
Florestal, Minas Gerais, Brazil

John Roman
Intel Corporation
Hillsboro, Oregon

Habib Rostami
Computer Engineering Department
Persian Gulf University
Bushehr, Iran

Magdalena Salazar-Palma
Department of Signal Theory and Communications
Universidad Carlos III de Madrid
Madrid, Spain

Seyedmohammad Salehi
Department of Computer and Information Sciences
University of Delaware
Newark, Delaware

Tapan K. Sarkar
Department of Electrical Engineering and Computer Science
Syracuse University
Syracuse, New York

Mehdi Shadloo-Jahromi
Information Technology Group
Petropars Operation and Maintenance Company
Bushehr, Iran

Chien-Chung Shen
Department of Computer and Information Sciences
University of Delaware
Newark, Delaware

Suraj Sindia
Intel Corporation
Hillsboro, Oregon

Aijun Song
Department of Computer and Electrical Engineering
The University of Alabama
Tuscaloosa, Alabama

Alex B. Vieira
Computer Science Department
Universidade Federal de Juiz de Fora
Juiz de Fora, Minas Gerais, Brazil

Luiz F. M. Vieira
Computer Science Department
Universidade Federal de Minas Gerais
Belo Horizonte, Minas Gerais, Brazil

Marcos A. M. Vieira
Computer Science Department
Universidade Federal de Minas Gerais
Belo Horizonte, Minas Gerais, Brazil

Zhen Yao
Intel Corporation
Santa Clara, california

Zhijing Ye
Department of Computer and
Electrical Engineering
The University of Alabama
Tuscaloosa, Alabama

Yuehai Zhou
Department of Computer and
Electrical Engineering
The University of Alabama
Tuscaloosa, Alabama

I

MAGNETIC COMMUNICATIONS

Chapter 1

Wireless Power Transfer (WPT) by Magnetic Induction

John Roman, Suraj Sindia, and Zhen Yao
Intel Corporation

Mark Briggs
Underwriters Laboratories

Clifton Barber
CJ Barber Consulting Services, Inc.

Contents

1.1 Basic Technology and Taxonomy of Wireless Power Transfer Mechanisms

Inductive wireless charging, while not considered a "communications" technology, is included here because of its use of magnetic fields and importance to the communications industry. As illustrated in Figure 1.1, wireless charging technologies can be broadly classified into nonradiative coupling-based charging and radiative based charging. The former consists of three techniques: inductive coupling, magnetic resonance coupling, and capacitive coupling, while the latter can be further sorted into directive Radiofrequency (RF) power beamforming, nondirective RF power transfer, and non-RF transfer (one example is ultrasound power transfer). In capacitive coupling, the achievable amount of coupling capacitance is dependent on the available area of the device. However, for a typical-size portable electronic device, it is hard to generate sufficient power density for charging, which imposes a challenging design limitation. For directive RF power beamforming, a limitation exists given the charger needs to know an exact location of the energy receiver.

Due to the limitation of the capacitive and beamforming mechanisms, wireless charging is usually realized through three techniques, i.e., magnetic inductive coupling, magnetic resonance coupling, and nondirective RF-based charging. The magnetic inductive and magnetic resonance coupling work in the near field, where the generated electromagnetic field dominates the region close to the transmitter or scattering object. The near-field power is proportional to the reciprocal of the cube of the charging distance. RF-based charging works in the near field but also in far field at a greater distance but usually at lower power capacities. The far-field power decreases according to the square of the reciprocal of the charging distance. Moreover, for the far-field technique, the absorption of electromagnetic energy does

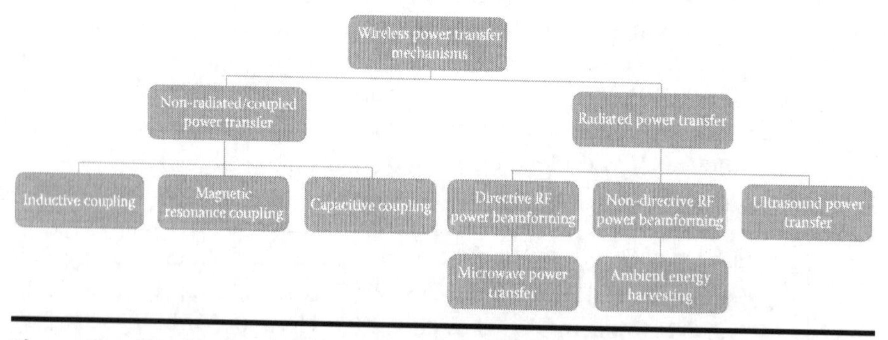

Figure 1.1 Taxonomy of wireless charging technologies [1].

not affect the transmitter. By contrast, for the near-field techniques, the absorption of energy influences the load on the transmitter. In general, far-field radiated power transfer techniques are of utility when the power to be transferred is low and the user experiences considerations mandating an easier to use interface; however, given the low-power application, the radiated field technology has not yet become generally available for use in consumer devices. On the other hand, near-field/coupled transfer techniques are useful for consumer devices, and for higher power transfer, for example for charging vehicles, and manufacturers are expanding creative ways of integration into current consumer electronics.

A comprehensive survey of wireless power transfer (WPT) mechanisms with a rather large bibliography delving further into each of these techniques is provided [1].

The examples of various power transfer techniques in action are shown in Figure 1.2.

1.2 Wireless Power Transfer Usage and Application

WPT eliminates wires and cables, where energy is directly transferred via air and harvested by receiving devices to power a device or charge a battery. It is desirable that all electronic devices are powered wirelessly, since WPT has many benefits compared with conventional charging. First, WPT helps to lessen the cord clutter

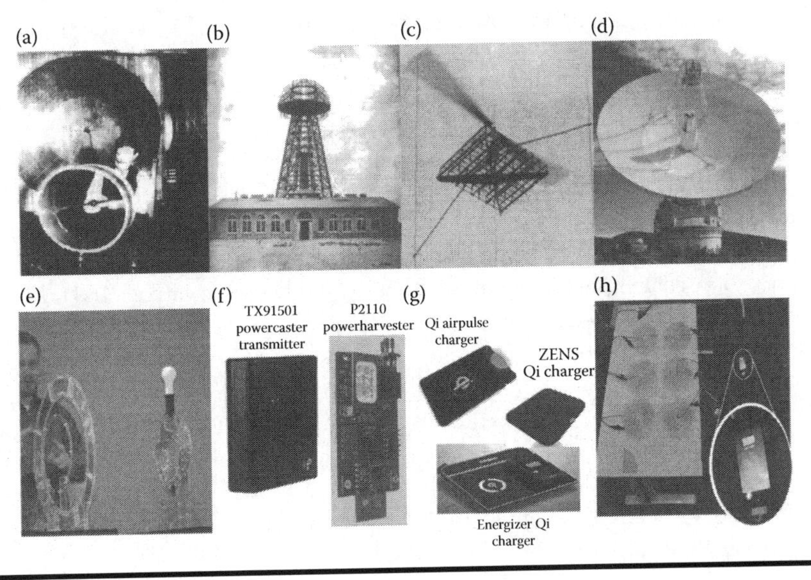

Figure 1.2 Illustrations of wireless power transmission systems. (a) Tesla coil, (b) Wardenclyffe Tower, (c) Microwave-powered airplane, (d) JPLs Goldstone Facility, (e) WiTricity system, (f) Powercaster transmitter and harvester, (g) Qi charging pads, (h) Magnetic MIMO system. (© 2018 IEEE. Reprinted with permission from *IEEE Communications Surveys & Tutorials*, Vol. 18, No. 2, Second Quarter 2016.)

problem. Most devices have their own power cable, and it is extremely annoying when those cords are tangled, or users have to search for a specific power cord to charge a particular device. Second, WPT allows devices to be completely sealed so that it can be waterproof, and such features are useful for devices such as an electric tooth brush, cellphone, etc. Another benefit is durability, where the physical connection required for the power cord inevitably leads to wearout due to the friction. WPT doesn't have this issue since no physical contact is required. WPT has been broadly applied in different fields, and this section briefly introduces some of its major applications.

1.2.1 Portable Electronics and Devices

The mainstream adoption for WPT so far is in the field of consumer electronics supporting devices such as wearables, cellphones, tablets, and laptops. Various wireless chargers compatible for different standards [2,3] and devices have been developed and are out in the market, e.g., Qi charging pad/stand and the desktop transmitter. Chargers can be designed as a lightweight and portable mat or stand, and they can also be built into furniture such as a wireless charging side table. They can also be built for automotive in-cabin applications, such as in commercially available hatchbacks and sedans, providing a convenient approach to charge one's cellphone. Another important marketing strategy is for public infrastructure deployment to enable wireless charging capability in areas such as airport lounges, conference rooms, hotels, restaurants, etc., which provides unprecedented user experience and convenience.

1.2.2 Transportation

Electric vehicles (EVs) are increasingly popular due to people's increasing concern with environmental impacts. However, one major drawback of EVs is the battery; currently, plug-in charging is required, which takes many hours to fully charge the vehicle. For example, some contemporary EVs need 13 hours to fully charge with a standard 110-V plug [4]. In the last few years, research has increased to enable wireless charging for EVs. Stevens et al. [5] proposed charging EVs using inductive coupling. Typically for inductive-based wireless charging, a small charging distance is required (4–10 mm). Recently, magnetic resonance based charging systems for EVs have been developed, which allow larger charging distance and higher charging efficiency [6]. Several pilot projects have been conducted to provide wireless charging to EVs [7]. In addition, WPT for public transportation is also being undertaken and wireless charging for light rail vehicles and electric buses is being evaluated at certain site projects [8,9].

1.2.3 Medical Implants

Another major application of WPT is to power implanted medical devices. Battery life is a vital concern for most implanted medical devices, since the embedded battery cannot be easily replaced. When the battery depletes, the entire implanted device needs to be replaced by surgery. So wireless charging has great potential to elongate the service life of medical implants. However, there are some technical challenges for this application: some implants are deep inside the human body, e.g., a pacemaker, which requires a larger charging distance. Furthermore, the small size of certain types of implants limits the receiver size and subsequently the corresponding efficiency. Both inductive-based and resonance-based wireless charging have been researched for several years. For example, in [10,11], different kinds of medical implants are powered with inductive wireless charging, where the charging distance is limited. In contrast, magnetic resonance based charging for biomedical implants [12,13] exhibits better properties for deeper penetration.

In this section, we briefly discuss the various advantages of wireless charging and its major applications in different areas. In the next chapter, we will give an introduction to industry standards of wireless charging technology.

1.3 Wireless Power Industry Standards

There are currently two prevailing wireless charging industry standards: one is called Qi (pronounced Chee) owned by the Wireless Power Consortium (WPC) and the other, which covers both inductive and resonant charging methods, is published by the AirFuel Alliance, which was formed in 2016 by the merger of two standards, namely the Alliance for Wireless Power and the Power Matters Alliance. Table 1.1 shows a brief comparison on different aspects between these standards.

The Qi standard was published in 2008, and it is primarily based on inductive technology (Qi specification v1.0 & v1.1); however in recent years, it started to embrace resonance-based technology (Qi specification v1.2) to target broader applications. Qi is backed by more than 200 companies and has more than 1000 products in the different markets.

Table 1.1 Comparison between WPT Standards

		AirFuel Alliance	
	WPC Qi	*Inductive*	*Resonant*
Members	~200	~150	
Products	>1000	>60	
Maximum power rate	15 W	70 W	
Power frequency band	100–200 kHz	235–275 kHz	6.78 MHz
Technology	Inductive	Inductive	Resonant

The former Power Matters Alliance standard, now the AirFuel Alliance Inductive standard, is also based on inductive coupling technology, and includes key features such as digital transceiver communication and cloud-based power management.

The former Alliance for Wireless Power Rezence standard, now called AirFuel Alliance Resonant Charging, differentiates itself from the Qi and AirFuel Alliance inductive standards by offering much more spatial freedom for wireless power [14]. To achieve such benefits, the charging system uses a technology where the transmitter and receiver are tuned to a certain resonant frequency to boost the electromagnetic field and coupling. This technology does not require a precise alignment, and the separation between a charger and charging devices can be much larger than that of inductive-based charging. Another advantage of the resonant technology is that power control communication is conducted by a separate, out-of-band Bluetooth signal. This allows support for charging multiple devices that have different power and charging requirements and demands more complex communication between the power transmitter unit (PTU—the device that provides the charging signal) and power receiving unit (PRU—the device that uses the charging signal).

1.3.1 Inductive-Based Technology

Inductive coupling is based on Faraday's law of induction. A transmitter coil generates a time-varying magnetic field across a receiver coil within a distance typically much less than a wavelength. The oscillating magnetic field then induces voltage and current across the receiver coil to power a device or charge a battery. The operating frequency of inductive coupling is typically in the kilohertz range. The quality factor is usually designed in small values. The effective charging distance is typically within a few centimeters. Despite the limited transmission range, the effective charging power can be very high [15] (e.g., kilowatt level for EV charging).

1.3.2 Magnetic Resonance Based Technology

Magnetic resonance coupling is also based on Faraday's law of induction, where energy is transferred by time-varying magnetic fields. The major difference with inductive-based coupling is that PTUs and PRUs are operating at the same resonant frequency, which is typically in the range of megahertz. The quality factor for the resonators is usually high, which compensates for the weak coupling when the distance between PTU and PRU is large [16]. Resonance-based technology also supports multiple device charging [17]. Resonant charging systems proposed by the AirFuel Alliance could support power transfer of 50 W at distances up to 5 cm, with the potential for up to eight devices to be charged from one charging pad. Table 1.2 is a comparison between inductive- and resonance-based technology [18].

Table 1.2 Comparison between Inductive and Resonant Technology

Wireless Charging Technology	*Pros*	*Cons*	*Effective Charging Distance*
Inductive	Simple implementation	Short charging distance, heating effect, need tight alignment between charger and charging devices	From a few millimeters to a few centimeters
Magnetic resonance	Loose alignment between chargers and charging devices, charging multiple devices simultaneously on different power.	Complex implementation	From a few centimeters to a few meters

1.4 Regulatory Standards of Wireless Power Transfer Systems

1.4.1 Introduction

Internationally coordinated frequencies of operation for wireless technologies are defined in the International Telecommunications Union (ITU) radio regulations [19]. Governments, civil society, and technology developers, known as sector members, convene to discuss the coordination and proposed assignment of radio frequencies, and governments define their agreements at periodic world radio conferences, which occur every 3–4 years in Geneva, Switzerland [20]. The specific use, any additional restrictions, or considerations are defined by each countries' respective regulatory agency and are published as national regulations. For example, the United States defines the regulations for wireless devices by frequency in the Federal Communications Commission (FCC) Code of Federal Regulations (CFR) [21], title 47 (Telecommunications), parts 0–199. Similarly in Canada, the regulations are provided by the department of Innovation, Science, and Economic Development [22]. In general, wireless charging technology is treated as a nonradio device; however in some cases, a distinction is made whether the wireless power device uses a communication protocol in the form of modulation on the wireless charging frequency. Specific requirements for the different wireless power technologies in a selective list of countries are provided in the following sections.

1.4.2 Standards and Regulations by WPT Type

Magnetic wireless power technologies have focused operation on different frequency bands based on the technology type; for example, inductive WPT uses 85 kHz and 100–357 kHz, and magnetic resonance is utilizing 6.78 MHz. The 85-kHz band is specifically targeted for EV charging, while the remaining frequencies are targeted more broadly for consumer devices. In all cases, the technology operates in what is called unlicensed or license exempt spectrum, where a specific license is not required to operate a technology in the designated bands; however, the equipment must comply with the applicable technical requirements for safety and to protect from interference. Given the technical and regulatory requirements are different for each band and country, we must look at each one separately.

While technical regulations are established by the regulatory agencies in each country, they often rely on national and international technical standards to help establish critical regulatory parameters. In the United States, the FCC generally looks to the relevant Institute of Electrical and Electronics Engineers (IEEE) and International Electrotechnical Committee (IEC) standards, as well as industry, and uses the Knowledge Database (KDB) system [23] to develop and provide additional technical guidance. KDB's 680106 and 648474 currently provide guidance for wireless power operation and approval in the United States. Further work is being undertaken by the American National Standards Institute (ANSI) to develop test methods for WPT devices that will be published in ANSI C63.30.

In Europe, both the IEC and the Comité International Spécial des Perturbations Radioélectriques (CISPR) have developed standards and guidance that pertain to wireless power systems. For example, the CISPR 11 standard [24], Industrial, scientific, and medical (ISM) equipment—Radio-frequency disturbance characteristics—Limits and methods of measurement, provides guidance for evaluating wireless power technology operating in the ISM bands, namely resonant wireless power operating at 6.78 MHz. The European Telecommunications Standards Institute is addressing inductive WPT systems through the development of the EN 303 417 standard.

Additionally, several committees have developed technical reports to assist technologists and governments how to understand and regulate wireless power technology. The IEC TC100 committee, focused on multimedia equipment, has formed the TA15 working group for WPT technology, and in addition to several standards, have developed the technical report, IEC TR 62869, which provides an initial overview of WPT technology, standards, and applications. Similarly in Asia, discussions are taking place at the Asia Pacific telecommunity, Asia Wireless Group, which has also developed a technical report on WPT [25]. The ITU has also published a technical report for WPT [26] and is working on providing further guidance for WPT frequencies in the ITU Working Party 1A. Additionally, separate discussions have taken place for higher power applications, such as charging EVs. For example, the Society of Automotive engineers (SAE) and the IEC technical committee TC69—Electric Road Vehicles and Electric Industrial trucks, are developing standards and guidance for powering and charging EVs.

Included in the regulatory requirements for most countries, wireless technologies must also comply with applicable RF exposure requirements. In the United States, the limits are established in the aforementioned CFRs, namely part 1.1310. Countries across Europe, and most of the world, have adopted the guidelines from the International Commission on Non-Ionizing Radiation Protection (ICNIRP) [27]. To further assist approval methods for WPT devices, the IEC TC106 committee has formed the special committee SC9—Addressing methods for assessment of WPT related to human exposures to electric, magnetic, and electromagnetic fields.

1.4.2.1 *Inductive Wireless Power: 85 kHz, 110–357 kHz*

As previously mentioned, commercial inductive wireless power systems operate between 100 and 357 kHz. The automotive industry is targeting the use of 85 kHz for high-power vehicle charging. In the United States, inductive wireless power operating in the low kilohertz ranges is typically governed by the requirements for intentional radiators, even though the transmission is tightly coupled between the source and the load, with minimal leakage. Specifically, FCC part 15C for intentional radiators or FCC Part 18 for ISM equipment contains the requirements that must be met to place the device on the market [28,29]. For all other devices, the Part 15 Subpart C limits or Part 18 limits for non-ISM bands would apply. As previously mentioned, the United States also provides guidance in their KDB system, specifically in KDB's 680106 and 648474. In Canada, the regulations for operation are provided in RSS-216 "Wireless Power Transfer Devices" [30].

1.4.2.2 *Resonant Wireless Power: 6.78 MHz*

Resonant wireless power is a special case of inductive wireless power, and industry has selected 6.78 MHz for standardized operation. The 6.78-MHz power is defined in the ITU radio regulations as an ISM frequency or part of the Industrial, Scientific, and Medical band. These frequencies have special rules that enable higher power; however, in some countries, for example the United States, they are limited to "local use." In some countries, further limitations are considered where there is modulation on the fundamental charging frequency; however, the United States currently allows for modulation on the fundamental charging frequency as long as it is used only for controlling the load/charging function. The US rules are defined for resonant wireless power systems in the US CFR, part 18, Industrial, Scientific and Medical equipment [30], and in Canada again in RSS-216 [21].

Table 1.3 Wireless Power Regulation

	Inductive	*Resonant*
United States	CFR 47 Part 15C or Part 18	CFR 47, Part 18
Canada	RSS-216	RSS-216
China	Short-range device rule (refer to MIIT [2005] No. 423)	Short-range device rule (refer to MIIT [2005] No. 423)
Japan	Item 3, Article 45, Japan radio act enforcement regulations	[Item 3, Article 45, Japan Radio Act Enforcement Regulations and Technical requirements on 6-MHz band Magnetic Resonance Coupling Wireless Power Transmission System and 400-MHz band Electric Field Coupling Wireless Power Transmission System, The MIC Information Communication Council Report, January 2015.
Korea	RRA notice 2013-19, MSI P notice 2015-05	RRA Notice 2013-19, MSI P notice 2015-05
Europe	EN 300 330	EN 55011
	Draft EN 303 417	

Table 1.3 provides a list of regulations for select countries for both inductive and resonant wireless power technologies.

1.5 Testing for RF Exposure Regulatory Standards

Since the focus of the book is on magnetically coupled systems, we will restrict our discussion in this section to exposure regulations that are applicable for WPT systems that are magnetically coupled. In the United States, the FCC is the agency that administers regulatory oversight on all WPT devices. According to the FCC, "Wireless power transfer devices operating at frequencies above 9 kHz are intentional radiators and are subject to either Part 15 and/or Part 18 of the FCC rules. The specific applicable rule part depends on how the device operates, and if there is communication between the charger and device being charged." It should be noted that short signals used to control charging are not considered communications. Further, the FCC also requires that "Devices specifically intended for use for WPT, or inductive charging, require FCC guidance for frequency exposure review. This includes Part 18 devices."

Intentional radiators transmitting information must be certified under the appropriate FCC Part 15 rules and will almost always require an equipment certification, except for special types of devices meeting requirements under Section 15.201 that are subject to the Supplier's Declaration of Conformity (SDoC) approvals procedures. A charger may operate in two different modes: charging and communications. It is possible for the device to be approved under Part 18 for the charging mode and Part 15 for the communications mode, if it can be shown that (1) the device complies with the relevant rule parts and (2) the functions are independent. Part 18 consumer devices can be either certified or approved under the SDoC approval procedures, only after the required Specific Absorption Rate (SAR), or RF exposure guidance has been given.

In the European Union and the rest of the world where the ICNIRP guidelines are followed, the requirements for RF exposure compliance are centered on two kinds of requirements. One is called Basic Restriction that seeks to be the fundamental limit and is the highest allowed value of RF exposure at a given frequency. The second kind of RF exposure metric is called Reference Level, which tends to be more relaxed compared to Basic Restriction as well as usually easier to evaluate given this background, a typical device under test (DUT) is evaluated for RF exposure by carrying out a sequence of steps summarized in the flow later. The DUT is first tested for Reference Level compliance. If this is met, the DUT is considered compliant, otherwise the remaining steps of Basic Restriction is evaluated.

1.6 Challenges

One of the biggest challenges to WPT technology is efficiency, not only matching the performance compared with wired systems, but also for defining the methods used to characterize the power transfer performance. The efficiency of WPT can vary depending on the application and the distance from the charging pad. As with any new technology, there are several regulatory challenges for wireless power transfer (WPT). Regulators must typically catchup with the industry knowledge of the technical characteristics, and then as needed, define new methods for evaluation and approval. For magnetic resonance, commercial measurement systems and methods were initially unavailable to evaluate conformance to the RF exposure limits, and ergo numerical techniques were developed in the interim to accurately characterize the compliance and allow the technology to safely advance. In fact, these groundbreaking techniques have catapulted the numerical modeling state of the art for regulatory compliance to enable its use more broadly. Another challenge to WPT, as with all new wireless technologies, was the definition and international availability of frequencies of operation. In particular, for 6.78 MHz, while the frequency was defined in the radio regulations, several countries had decided not to follow the ITU and assign the band for ISM. Similarly, some countries have not recognized the entire 100- to 357-kHz band to allow for inductive WPT operation. This poses challenges to the equipment manufacturers, given they have to carry different versions of the product for different countries, increasing cost and time to market.

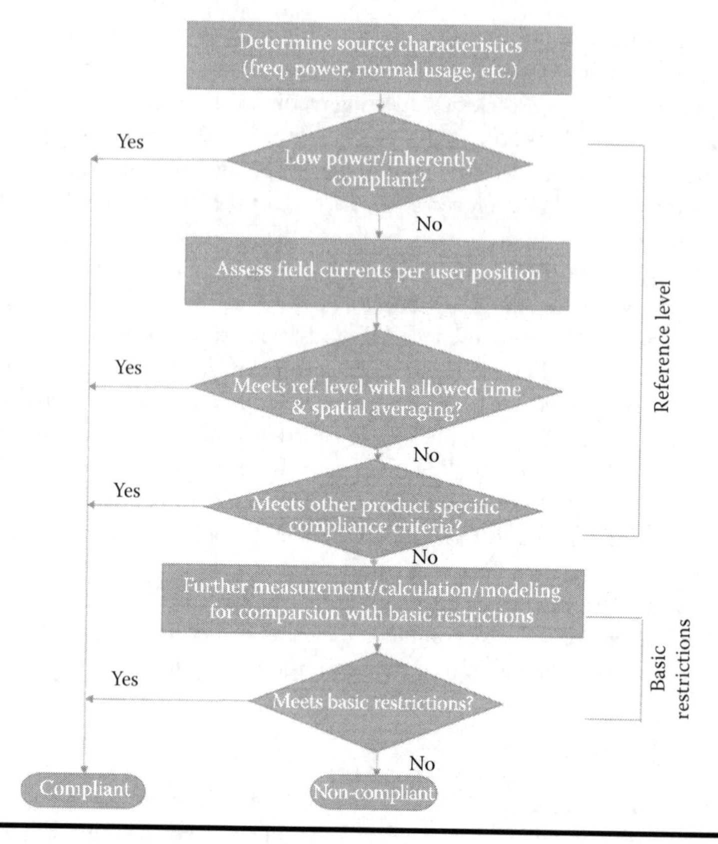

Figure 1.3 A typical flow for RF exposure compliance testing for a WPT device.

1.7 Conclusions

Wireless power technology is growing by leaps and bounds, and magnetic coupling is by far the most prevalent approach used today. As charging technology improves, we will continue to see more infrastructure growth as greater numbers of consumer devices, from laptops to watches and mopeds to busses, are enabled with the available technology.

References

1. X. Lu, P. Wang, D. Niyato, D. I. Kim and Z. Han, "Wireless charging technologies: Fundamentals, standards, and network applications," *IEEE Communications Surveys & Tutorials*, vol. 18, no. 2, pp. 1413–1452, 2016.
2. https://www.wirelesspowerconsortium.com/.
3. http://www.airfuel.org/.

4. GM-Volt, "Chevy Volt specs," Available: http://gm-volt.com/full-specifications/.
5. R. Stevens et al., "An ultra-compact transformer for a 100 W to 120 kW inductive coupler for electric vehicle battery charging," *Proc. of IEEE Applied Power Electronics Conference and Exposition*, 1996.
6. X. Wang et al., "Analysis on the efficiency of magnetic resonance coupling wireless charging for electric vehicles," *Proc. of IEEE Annual International Conference on Cyber Technology in Automation, Control and Intelligent Systems*, 2013.
7. https://www.hevopower.com/.
8. http://primove.bombardier.com/projects/europe.html.
9. http://spectrum.ieee.org/transportation/advanced-cars/the-allelectric-car-you-never-plug-in.
10. H. Jiang et al., "A low-frequency versatile wireless power transfer technology for biomedical implants," *IEEE Transactions on Biomedical Circuits and Systems*, 2013.
11. A.K. RamRakhyani et al., "On the design of efficient multi-coil telemetry system for biomedical implants," *IEEE Transactions on Biomedical Circuits and Systems*, 2013.
12. R. F. Xue et al., "High-efficiency wireless power transfer for biomedical implants by optimal resonant load transformation," *IEEE Transactions on Biomedical Circuits and Systems*, 2011.
13. G. Yilmaz et al., "An efficient wireless power link for implanted biomedical devices via resonant inductive coupling," *Proc. of IEEE Radio and Wireless Symposium*, 2012.
14. R. Tseng et al., "Introduction to the alliance for wireless power loosely-coupling wireless power transfer system specification version 1.0," *Proc. of IEEE Wireless Power Transfer (WPT)*, 2013.
15. H. H. Wu et al., "A high efficiency 5 kW inductive charger for EVs using dual side control," *IEEE Transactions on Industrial Informatics*, 2012.
16. A. Kurs et al., "Wireless power transfer via strongly coupled magnetic resonances," *Science*, 2007.
17. B. Cannon, "Magnetic resonant coupling as a potential means for wireless power transfer to multiple small receivers," *IEEE Trans. Power Electron*, 2009.
18. X. Lu et al., "Wireless charging technologies: Fundamentals, standards, and network applications," *IEEE Communications Surveys & Tutorials*, 2016.
19. International telecommunication Union (ITU), Radio Regulations, 2016 http://www.itu.int/pub/R-REG-RR.
20. International telecommunication Union (ITU), World Radio Conference Website http://www.itu.int/en/ITU-R/conferences/wrc/Pages/default.aspx.
21. United States Code of Federal Regulations, Title 47 Telecommunication, Volumes 1–5, parts 0–199, 2016 https://www.fcc.gov/general/rules-regulations-title-47.
22. Government of Canada, Innovation, Science, and Economic development Canada, Standards and Certification of Radio Apparatus and Electronic Equipment Used in Canada, 2016 http://www.ic.gc.ca/eic/site/smt-gst.nsf/eng/sf01698.html.
23. United States Federal Communications Commission (FCC) Office of Engineering and Technology Laboratory Division Knowledge Database (KDB), 2016 https://apps.fcc.gov/oetcf/kdb/index.cfm.
24. https://webstore.iec.ch/publication/22643.
25. http://www.aptsec.org/AWG-RECS-REPS.
26. https://www.itu.int/dms_pub/itu-r/opb/rep/R-REP-SM.2303-1-2015-PDF-E.pdf.
27. International Commission on Non Ionizing Radiation Protection (ICNIRP) http://www.icnirp.org/.

28. http://www.ecfr.gov/cgi-bin/text-idx?SID=ed9ffc6d5269b3879d4e33f5a73970ba&mc=true&node=sp47.1.15.b&rgn=div6 Briggs – suggest http://www.ecfr.gov/cgi-bin/text-idx?SID=19b2716413f9bc4a0f00139102a6e479&mc=true&node=pt47.1.15&rgn=div5 for part 15.
29. http://www.ecfr.gov/cgi-bin/text-idx?SID=19b2716413f9bc4a0f00139102a6e479&mc=true&node=pt47.1.18&rgn=div5 for Part 18.
30. https://www.ic.gc.ca/eic/site/smt-gst.nsf/vwapj/RSS-216-i2.pdf/$FILE/RSS-216-i2.pdf.

Chapter 2

Magnetic Induction Based Simultaneous Wireless Information and Power Transmission

Steven Kisseleff, Ian F. Akyildiz, and
Wolfgang H. Gerstacker
Friedrich-Alexander University Erlangen-Nuremberg

Contents

2.1 Introduction

The main issues in modern communications are resource allocation and utilization. The increasing demand on green energy has raised concerns about the correct utilization of electromagnetic radiation and invoked an even more thorough optimization in the area of communication networks and signal design. In particular, the possibility of utilizing the same signals for information and power transfer has been intensely studied in recent years [1–3]. This technique is commonly known as "Simultaneous Wireless Information and Power Transfer (SWIPT)" and has been shown to substantially improve the power utilization. Since the SWIPT technique enables the harvest of reasonable amounts of energy using the signals intended for data transmission, the design of such a system is related to the nontrivial tradeoff between the achievable data rate and the amount of harvested energy [3]. This design problem becomes even more tough with increasing system complexity involving multiple transceivers, antennas, resource blocks, etc. In this context, various aspects of signal design have been investigated. In [4], challenges for implementing SWIPT in practical communication systems have been investigated. A novel nonlinear energy harvesting model has been presented in [5] and incorporated into the global system optimization. Novel methods for the design of SWIPT with enabled physical layer security have been studied [6, 7]. Performance analysis for SWIPT with imperfect channel estimation has been provided [8]. In [9] and [10], signal transmission has been optimized for various multiuser SWIPT systems. Furthermore, relay-enhanced communication networks with enabled SWIPT have been investigated [11–13].

Assuming multiple transmit and receive antennas, the power efficiency of the resulting multiple-input multiple-output (MIMO) system is typically much higher compared with a single-input single-output (SISO) scenario, since the transmitted power can be steered in the preferred direction and also the antenna diversity can be exploited. This technique is commonly known as beamforming, and the corresponding spatial filter is called beamforming vector. While MIMO technique has been thoroughly studied for traditional radio frequency (RF) systems for the last few decades, the use of multiple magnetic antennas (coils) in magnetic communication systems has only seen some improvement in the recent years. One of the reasons is an extremely short transmission range of the near-field communication, such that there have not been many practical applications for a multicoil scheme in the past. On the contrary, in the modern times, the demand for higher data rates has led to a densification of communication networks, such that more and more practical scenarios for short distance transmissions have emerged. Hence, more effort has been put into the development of a comprehensive theory for magnetic MIMO systems with multiple transmitters [14], multiple receivers [15], and more general schemes [16–18].

In RF systems, the power efficiency optimization corresponds to the maximization of the receive power for a fixed L_2-norm of the beamforming vector, because the consumed power in the transmitting device depends only on the L_2-norm of the beamforming vector, but not on its particular coefficients. For Magnetic Induction (MI)-based communications, the power transfer efficiency and signal quality depend explicitly on the coupling between the transceivers, such that improving the coupling yields an increase of the system performance [19–21]. For short transmission distances, the coupling between magnetic devices is very strong and affects both the transmitters and the receivers. In particular, signals from multiple transmitting coils can overlap constructively or destructively at the transceivers, such that the transmit power depends on the choice of the beamforming coefficients. Correspondingly, a sole receive power maximization for a fixed L_2-norm of the beamforming vector becomes insufficient, cf. [18]. The resulting transmit power is nonconvex in general and can only be simplified for large transmission distances, i.e., for far-field communications. However, the wireless power transfer (WPT) is very inefficient in such cases.

In this chapter, we investigate the beamforming for magnetic SWIPT systems. A similar problem has been addressed in [22], where a special case of SWIPT with a single data stream and multiple power streams has been considered. We extend this scenario to a more general case with an arbitrary number of information and power receivers (PRs). In this context, the interference between the adjacent data streams is taken into account and the influence of the selected symbol alphabet on the transmit power is analyzed. We consider one transmitter with three orthogonal coils free of self-interference and multiple single antenna receivers randomly deployed in the near field of the transmitter. The transmitter can be viewed as an access point, which provides information and/or power upon request. To satisfy the needs of all users, we consider a multiobjective optimization problem with two objective functions. The first objective refers to the quality of service (QoS) for the data receivers (DRs). This QoS corresponds to the minimum achievable signal quality among all DRs. The second objective corresponds to the sum receive power of all PRs. The optimization problem turns out to be nonconvex, such that no globally optimum solution can be obtained with a practical approach. Hence, we provide an iterative approach, which is based on an alternating maximization of the first and the second objective function, respectively. In addition, the important role of a nonconvex transmit power constraint is revealed. To incorporate this constraint into the considered optimization problem, two strategies have been investigated. The corresponding nonconvex transmit power metric is replaced by either a squared L_2-norm of the beamforming vector (far-field approximation) or taken into account using an iterative convex approximation (proposed method), similar to [18] and [22]. In this context, substantial gains in terms of both sum receive power and QoS have been observed using the proposed solution compared with the traditional far-field approximation described in Section 2.3, which renders the considered magnetic SWIPT system power efficient and promising. Furthermore, additional insight is provided by analyzing various scenarios with different numbers of power and information receivers. Important aspects of channel estimation for magnetic

SWIPT, system coverage, and its application in larger communication systems are addressed in Section 2.4. Subsequently, Section 2.5 summarizes the contribution of this work and emphasizes some of the important observations.

2.2 Design of MI-SWIPT system

2.2.1 System Model

As mentioned earlier, we assume a system with one transmitter equipped with three orthogonally deployed coils (3D-coil) and multiple (K) receivers with one coil each, Figure 2.1. Furthermore, to establish a SWIPT scheme, we assume that K_{data} users receive information (we call these users DRs), whereas K_{power} users receive power (we call these users PRs). Every transmitter coil has inductivity L_t and is considered as part of a resonant circuit, which includes a capacitor with capacitance C_t and a resistor with resistance R_t (modeling the copper resistance of the coil). Similarly, the receiver circuit comprises a coil with inductivity L_r, a capacitor with capacitance C_r, and a resistor with resistance R_r. The capacitances C_t and C_r are selected to make the circuit resonant at the common resonance frequency f_0 using the relation $f_0 = \frac{1}{2\pi\sqrt{L_t C_t}} = \frac{1}{2\pi\sqrt{L_r C_r}}$.

Since the considered system can be viewed as a variation of the traditional Near Field Communication (NFC) and passive Radio Frequency Identification (RFID)-based systems, we select the resonance frequency f_0 and the bandwidth B according to the respective standard. For these values, the frequency selectivity of the transmission channel is typically very low. Such channels do not need any extensive equalization or frequency-selective prefiltering (precoding). However, in case of very strong couplings between the coils, the well-known effect of frequency splitting occurs [23], which can make the transmission channel frequency selective. Therefore, to provide a general approach for all types of channels (weak coupling and strong coupling), we must assume a nonvanishing frequency selectivity of the

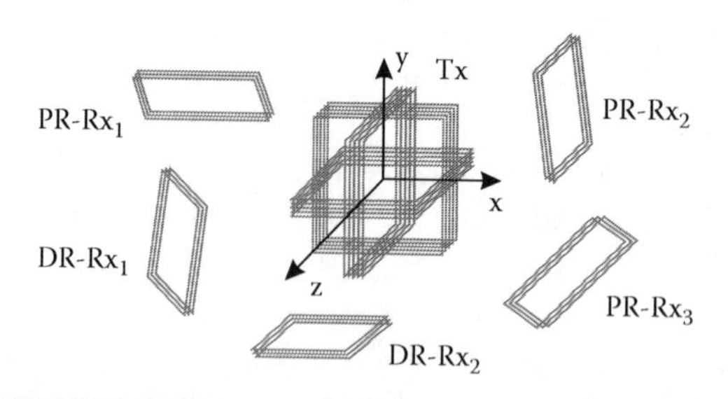

Figure 2.1 Example of a SWIPT system with 3D-coil based transmitter (Tx) and multiple ($K = 5$) single coil receivers (three PRs (PR–Rx$_1$, PR–Rx$_2$, and PR–Rx$_3$) and two DRs (DR–Rx$_1$ and DR–Rx$_2$).

transmission channels. Correspondingly, the DRs employ linear filters to remove the resulting intersymbol interference and equalize the channel. This filtering at the DRs seems to be sufficient, such that the transmit filter does not need to be frequency selective as well. Hence, a frequency-flat transmit filter is utilized in the following. Furthermore, the low-frequency selectivity justifies the use of a linear equalizer (LE) instead of a more advanced scheme, e.g., decision-feedback equalizer or maximum-likelihood sequence estimation, since the respective performance differences are not too heavy while the complexity of a LE is limited.

Each receiver circuit contains an additional real-valued load resistor Z_L. Hence, the inner impedance of each transmitter circuit can be given by

$$Z_{in,t}(f) = j2\pi fL_t + \frac{1}{j2\pi fC_t} + R_t, \tag{2.1}$$

and for the inner impedance of each receiver circuit

$$Z_{in,r}(f) = j2\pi fL_r + \frac{1}{j2\pi fC_r} + R_r + Z_L \tag{2.2}$$

is valid. The load impedance is optimized according to previous works, cf. [19]. The induced voltage is related to the coupling between the coils, which is determined by the mutual inductance M. The knowledge of mutual inductance is very important for the optimization of WPT systems [19] and can be acquired either by channel estimation or by distance estimation under the assumption that all other system parameters (coil dimensions, polarization, etc.) are known. In this work, we assume that the exact value of the mutual inductance between any pair of coils is available to the transmitter. Of course, imperfect channel state information (CSI) may lead to performance degradation; however, this problem is not new and has been studied in various previous works, especially for RF-based SWIPT [8, 9]. Hence, similar methods of robust optimization can be also applied to magnetic SWIPT. However, more suitable schemes might be developed in future With perfect knowledge of mutual inductance, the frequency selectivity of the MI channel is entirely known, such that not only the WPT can be established but also the information transmission [20].

The orientations and alignments of the coupled coils have a strong impact on the mutual inductance and correspondingly on the path loss, cf. [24]. Hence, we model the mutual inductance between coils m and l by

$$M_{m,l} = \bar{M}_{m,l} \cdot J_{m,l}, \tag{2.3}$$

$$J_{m,l} = 2\sin\theta_m \sin\theta_l + \cos\theta_m \cos\theta_l \cos\phi, \tag{2.4}$$

cf. [25], where θ_m and θ_l are the angles between the radial directions of the coils m and l, respectively, and the line connecting the two coil centers. ϕ is the angle difference between the coils' axes in the plane, which is orthogonal to the direction of transmission. $\bar{M}_{m,l}$ represents the (absolute) value of the mutual inductance for the case $J_{m,l} = 1$. For the subsequent derivations, we define $Z_{m,l}(f) = j2\pi f M_{m,l}, \forall m \neq l$.

In the following, $(\cdot)^T$, $(\cdot)^H$, and $(\cdot)^*$ stand for transpose, Hermitian transpose, and complex conjugation, respectively. In addition, $\in\{\cdot\}$ denotes the expectation operator. For simplicity, we assign the first three indices $m \in \{1,2,3\}$ to the transmitter circuits, indices $m \in \{4,\ldots,K_{\text{data}}+3\}$ to the DR circuits, and the remaining indices $m \in \{K_{\text{data}}+4,\ldots,K+3\}$ to the PR circuits, respectively.

For the signal transmission to the DR m, we assume a concatenation of a transmit filter with an impulse response $h(t)$ in time domain and a beamforming filter $\mathbf{b}_m = [b_{m,1}, b_{m,2}, b_{m,3}]$, $m \in \{4,\ldots,K_{\text{data}}+3\}$ in spatial domain. Thus, a sequence of statistically independent transmit symbols $s_m[n]$ spaced by symbol duration $T = 1/B$ (with bandwidth B) is fed into the transmit filter $h(t)$ and then mapped onto three transmit voltages using the beamforming filter $\mathbf{b}_m$, which generates the voltage signals $\mathbf{u}_m(t) = [u_{m,1}(t), u_{m,2}(t), u_{m,3}(t)]$ in all three resonance circuits. Furthermore, we assume that $\mathcal{E}\left\{\left|s_m[n]\right|^2\right\} = 1, \quad m \in \{4,\ldots,K_{\text{data}}+3\}$ and $\mathcal{E}\{s_m[n]^* s_\kappa[n]\} = 0, \quad \kappa \neq m$, i.e., the adjacent data streams are statistically independent.

In time domain, the complex-valued transmit voltage vector is given by

$$\mathbf{u}_m(t) = \mathbf{b}_m \sum_{n=-\infty}^{\infty} s_m[n] h(t - nT). \tag{2.5}$$

In frequency domain, we consider the complex-valued amplitudes $U_{m,l}(f) = \mathcal{F}\{u_{m,l}(t)\}$ of the voltages $u_{m,l}(t)$ of the lth circuit. Here, $\mathcal{F}\{\cdot\}$ denotes the Fourier transform operator. For simplicity, we assume a frequency-flat filter characteristic within the frequency band of the filter, which corresponds to a root-raised cosine impulse response $h(t)$ with a roll-off factor $\beta = 0$ and Fourier transform

$$H(f) = \mathcal{F}\{h(t)\} = \begin{cases} 1, & f_0 - 0.5B \leq f \leq f_0 + 0.5B, \\ 0, & \text{else.} \end{cases} \tag{2.6}$$

Correspondingly, the Fourier transform $\mathbf{U}_m(f)$ of a transmit voltage vector according to (2.5) is given by

$$\mathbf{U}_m(f) = \mathbf{b}_m \sum_{n=-\infty}^{\infty} s_m[n] e^{-j2\pi f n T}, \quad f_0 - 0.5B \leq f \leq f_0 + 0.5B. \tag{2.7}$$

Since we assume that all K_{data} transmissions are done in parallel, the resulting current $i_l(t)$ with the complex-valued amplitude $I_l(f) = \mathcal{F}\{i_l(t)\}$ in the circuit l depends on all voltage amplitudes $U_{m,l}(f), \quad m \in \{4,\ldots,K_{\text{data}}+3\}$ and on the current amplitudes $I_m(f), \quad \forall m \neq l$ in all surrounding circuits via the voltage equation

$$I_l(f)\cdot Z_{\text{in},t}(f)+\sum_{m\neq l}\left(I_m(f)\cdot Z_{m,l}(f)\right)=\sum_{m=4}^{K_{\text{data}}+3}U_{m,l}(f), \tag{2.8}$$

if l belongs to a transmitter or

$$I_l(f)\cdot Z_{\text{in},r}(f)+\sum_{m\neq l}\left[I_m(f)\cdot Z_{m,l}(f)\right]=0, \tag{2.9}$$

if l belongs to a receiver. In (2.9), $U_{m,l}(f)=0$ and correspondingly $U_{m,l}(t)\equiv 0$ holds, since only the transmitter is supposed to generate power. Hence, a set of voltage equations

$$\begin{bmatrix} \mathbf{Z}_{Tx}(f) & \mathbf{Z}_{Ch}(f) \\ \mathbf{Z}_{Ch}^{T}(f) & \mathbf{Z}_{Rx}(f) \end{bmatrix}\cdot\begin{bmatrix} \mathbf{I}_{Tx}(f) \\ \mathbf{I}_{Rx}(f) \end{bmatrix}=\begin{bmatrix} \sum_{m=4}^{K_{data}+3}\mathbf{U}_m(f) \\ \mathbf{0} \end{bmatrix} \tag{2.10}$$

needs to be solved to calculate the currents in all circuits of the coupled network. Here, 0 stands for an all-zero vector of dimension $K\times 1$. $\mathbf{I}_{Tx}(f)$ and $\mathbf{I}_{Rx}(f)$ denote vectors comprising the complex-valued currents of the transmitter and the receiver circuits, respectively. The matrices $\mathbf{Z}_{Tx}(f)$, $\mathbf{Z}_{Ch}(f)$, and $\mathbf{Z}_{Rx}(f)$ contain complex impedances and are defined in the following. In this work, we use a 3D-coil based transmitter, which means that all three transmitter coils' axes are orthogonal to each other, such that

$$\mathbf{Z}_{Tx}(f)=\begin{bmatrix} Z_{in,t}(f) & 0 & 0 \\ 0 & Z_{in,t}(f) & 0 \\ 0 & 0 & Z_{in,t}(f) \end{bmatrix}=Z_{in,t}(f)\cdot\mathbf{I}_3 \tag{2.11}$$

holds, where $\mathbf{I}_3$ stands for the identity matrix of dimension 3×3. The receiver coils are not necessarily orthogonal to each other, and we obtain

$$\mathbf{Z}_{Rx}(f)=\begin{bmatrix} Z_{in,r}(f) & \cdots & Z_{K+3,4}(f) \\ \vdots & \ddots & \vdots \\ Z_{4,K+3}(f) & \cdots & Z_{in,r}(f) \end{bmatrix}. \tag{2.12}$$

The purely imaginary matrices $\mathbf{Z}_{Ch}(f)$ and $\mathbf{Z}_{Ch}^{T}(f)$ in (2.10) stand for the influence of the receiver coils onto the transmitter coils and vice versa, respectively. Hence, $\mathbf{Z}_{Ch}(f)$ is defined by

$$\mathbf{Z}_{Ch}(f)=\begin{bmatrix} Z_{4,1}(f) & Z_{5,1}(f) & \cdots & Z_{K+3,1}(f) \\ Z_{4,2}(f) & Z_{5,2}(f) & \cdots & Z_{K+3,2}(f) \\ Z_{4,3}(f) & Z_{5,3}(f) & \cdots & Z_{K+3,3}(f) \end{bmatrix}. \tag{2.13}$$

By inverting the overall impedance matrix in (2.10) using [26], we obtain similar to [18]

$$\begin{aligned} \mathbf{I}_{Tx}(f) &= \left[\mathbf{Z}_{Tx}(f)-\mathbf{Z}_{Ch}(f)\mathbf{Z}_{Rx}^{-1}(f)\mathbf{Z}_{Ch}^{T}(f)\right]^{-1}\sum_{m=4}^{K_{data}+3}\mathbf{U}_m(f) \\ &= \mathbf{A}(f)\sum_{m=4}^{K_{data}+3}\mathbf{U}_m(f), \end{aligned} \tag{2.14}$$

$$\begin{aligned} \mathbf{I}_{Rx}(f) &= -\mathbf{Z}_{Rx}^{-1}(f)\mathbf{Z}_{Ch}^{T}(f)\mathbf{A}(f)\sum_{m=4}^{K_{data}+3}\mathbf{U}_m(f) \\ &= \mathbf{C}(f)\sum_{m=4}^{K_{data}+3}\mathbf{U}_m(f), \end{aligned} \tag{2.15}$$

with implicit definitions of $\mathbf{A}(f)$ and $\mathbf{C}(f)$. In addition, we define*

$$\mathbf{D}(f)=\left(\mathbf{Z}_{Rx}(f)-\mathbf{Z}_{Ch}^{T}(f)\mathbf{Z}_{Tx}^{-1}(f)\mathbf{Z}_{Ch}(f)\right)^{-1}, \tag{2.16}$$

which will be discussed in the context of noise power calculation.

As known from the fundamentals of electric power generation and transmission (e.g., [27]), to produce enough active power in electric circuits, the transmitter/generator needs to release also the reactive power, which corresponds to the imaginary part of the generated complex power. The reactive power is not absorbed by the load, but fluctuates between the power source and the load impedance, i.e., the magnetic field is generated and destroyed very fast. Furthermore, without the reactive power, the induction coils cannot be operated. However, a large amount of reactive power might limit the performance of communication system and impose additional hard constraints for the system design. This problem has been explicitly addressed in [28], where the reactive power for a magnetic WPT system has been minimized. Unfortunately, this approach does not necessarily maximize the power efficiency of the system. Similar to [18], we take both real and imaginary parts of

* Matrix $\mathbf{D}(f)$ is used to describe the influence of the noise voltages in the receiver circuits on the currents in the receiver circuits similar to the matrix $\mathbf{C}(f)$, which maps the signals from the transmitter circuits onto the currents in the receiver circuits.

the transmit power into account, such that the magnitude of the total generated complex power (the so-called apparent power [27]) is a better reference for the transmit power than the pure active [19] or reactive [28] power. The apparent transmit power spectral density in the lth transmitter circuit is given by

$$P_{t,l}(f)=\mathcal{E}\left\{\left|\sum_{m=4}^{K_{\text{data}}+3} U_{m,l}(f) I_l(f)\right|\right\}=\mathcal{E}\left\{\left|\sum_{m=4}^{K_{\text{data}}+3} U_{m,l}(f)\right|\left|I_l(f)\right|\right\}, \quad (2.17)$$

where $|\cdot|$ denotes the elementwise absolute value operator. Therefore, we define the total power provided by the transmitter within a given transmission band with center frequency f_0 and bandwidth B similar to [18] as

$$\begin{aligned} P_{t,\text{total}} &= \sum_{l=1}^{3} \int_{f_0-0.5B}^{f_0+0.5B} P_{t,l}(f)\,\mathrm{d}f \\ &= \int_{f_0-0.5B}^{f_0+0.5B} \mathcal{E}\left\{\left|\sum_{m=4}^{K_{\text{data}}+3} \mathbf{U}_m(f)\right|^T \left|\mathbf{A}(f) \sum_{m=4}^{K_{\text{data}}+3} \mathbf{U}_m(f)\right|\right\}\mathrm{d}f \\ &= \int_{f_0-0.5B}^{f_0+0.5B} \mathcal{E}\left\{\left|\sum_{m=4}^{K_{\text{data}}+3} s_m[n]\mathbf{b}_m\right|^T \left|\mathbf{A}(f) \sum_{m=4}^{K_{\text{data}}+3} s_m[n]\mathbf{b}_m\right|\right\}\mathrm{d}f, \end{aligned} \quad (2.18)$$

where (2.14) and (2.7) have been used. Obviously, the different data sequences $s_m[n],\ m \in \{4,\ldots,K_{\text{data}}\}$ cannot be separated in terms of transmit power, even despite their statistical independency. Hence, the respective beamforming vectors cannot be optimized separately from the modulation selection.

For weak couplings between coils (low mutual inductance), matrix $\mathbf{A}(f)$ is approximately* given by

$$\mathbf{A}(f) \approx \mathbf{Z}_{Tx}^{-1}(f) = Z_{\text{in},t}^{-1}(f) \cdot \mathbf{I}_3. \quad (2.19)$$

This yields

$$\begin{aligned} P_{t,\text{total}} &\approx \int_{f_0-0.5B}^{f_0+0.5B} \left|Z_{\text{in},t}^{-1}(f)\right| \mathcal{E}\left\{\left\|\sum_{m=4}^{K_{\text{data}}+3} s_m[n]\mathbf{b}_m\right\|_2^2\right\}\mathrm{d}f \\ &= \sum_{m=4}^{K_{\text{data}}+3} \mathbf{b}_m^H \mathbf{b}_m \int_{f_0-0.5B}^{f_0+0.5B} \left|Z_{\text{in},t}^{-1}(f)\right|\mathrm{d}f. \end{aligned} \quad (2.20)$$

* For $\left(2\pi f \max_{m,l}\{\bar{M}_{m,l}\} \ll \sqrt{R_t R_r}\right)$, $\mathbf{Z}_{Ch}(f)\mathbf{Z}_{Rx}^{-1}(f)\mathbf{Z}_{Ch}^T(f)$ in (2.14) is negligible.

This result motivates the use of a far-field approximation, where the transmit power is only related to the sum of squared L_2-norms of $\mathbf{b}_m$. A distinct advantage is that the data streams can be separated in this case and the modulation selection can be decoupled from the beamforming optimization. However, in case of strong couplings between coils (or imperfect coil deployment at the transmitter), this approximation is not valid, which may lead to a performance degradation.

For the received active power density at the load resistor Z_L of the receiver circuit $l \in \{4, \ldots, K+3\}$, we obtain

$$\begin{aligned} P_{r,l}(f) &= \mathcal{E}\left\{\left|I_l(f)\right|^2\right\} Z_L = \mathcal{E}\left\{\left|\mathbf{e}_{K,l-3}^H \mathbf{C}(f) \sum_{m=4}^{K_{\text{data}}+3} \mathbf{U}_m(f)\right|^2\right\} Z_L \\ &= \sum_{m=4}^{K_{\text{data}}+3} \mathbf{b}_m^H \mathbf{C}^H(f) \mathbf{e}_{K,l-3} \mathbf{e}_{K,l-3}^H \mathbf{C}(f) \mathbf{b}_m Z_L, \end{aligned} \tag{2.21}$$

where the $K \times 1$ vector $\mathbf{e}_{K,l}$ is defined as $\mathbf{e}_{K,l} = [0, \ldots, 0, 1, 0, \ldots, 0]^T$ with "1" at the lth position, and (2.15) has been used. In case of power transfer, the total receive power at circuit l is given by

$$P_{r,l,\text{total}} = \eta \sum_{m=4}^{K_{\text{data}}+3} \mathbf{b}_m^H \left(\int_{f_0-0.5B}^{f_0+0.5B} \mathbf{C}^H(f) \mathbf{e}_{K,l-3} Z_L \mathbf{e}_{K,l-3}^H \mathbf{C}(f) \, \mathrm{d}f \right) \mathbf{b}_m, \tag{2.22}$$

where η stands for the losses due to the conversion of the received high frequency signals into electric power. For simplicity, we assume $\eta = 1$ in this work.

In case of information transmission, the transmitted data may be corrupted by noise and interference, caused by parallel signal transmissions. We model the noise source in each resonant circuit as a voltage source, which provides an additive white Gaussian noise with the average power spectral density (PSD) N_0 [V^2s]. The uncorrelated noise signals may occur in all resonant circuits (including the receiver circuit). These signals influence not only the currents in the respective circuits but also the currents of the neighboring circuits via near-field coupling between the coils [29]. Hence, the additive noise power, as seen by the receiver m, can be calculated via summation of the active received noise powers from all noise sources of the system. Considering the Fourier transform $U_{N,m}(f)$ of the noise signal $u_{N,m}(t)$ in circuit m within a time window, we obtain $\mathcal{E}\left\{\left|U_{N,m}(f)\right|^2\right\} = N_0, \forall m$, see e.g., [30]. The current in the mth receiver can be expressed as

$$I_m(f) = \sum_{m=1}^{3} \mathbf{e}_{K,m-3}^H \mathbf{C}(f) \mathbf{e}_{3,m} U_{N,m}(f) + \sum_{m=4}^{K+3} \mathbf{e}_{K,m-3}^H \mathbf{D}(f) \mathbf{e}_{K,m-3} U_{N,m}(f). \tag{2.23}$$

The corresponding average received noise PSD $N_m(f) = \mathcal{E}\left\{\left|I_m(f)\right|^2\right\} Z_L$ can be therefore formulated as

$$\begin{aligned} N_m(f) &= \mathbf{e}_{K,m-3}^H \mathbf{C}(f) \mathcal{E}\left\{\sum_{m=1}^{3} \mathbf{e}_{3,m} \left|U_{N,m}(f)\right|^2 \mathbf{e}_{3,m}^H\right\} \mathbf{C}^H(f) \mathbf{e}_{K,m-3} Z_L \\ &+ \mathbf{e}_{K,m-3}^H \mathbf{D}(f) \mathcal{E}\left\{\sum_{m=4}^{K+3} \mathbf{e}_{K,m-3} \left|U_{N,m}(f)\right|^2 \mathbf{e}_{K,m-3}^H\right\} \mathbf{D}^H(f) \mathbf{e}_{K,m-3} Z_L \\ &= N_0 \mathbf{e}_{K,m-3}^H \left[\mathbf{C}(f)\mathbf{C}^H(f) + \mathbf{D}(f)\mathbf{D}^H(f)\right] \mathbf{e}_{K,m-3} Z_L, \end{aligned} \tag{2.24}$$

where $\sum_{m=1}^{K} \mathbf{e}_{K,m} \mathbf{e}_{K,m}^H = \mathbf{I}_K$ has been used. Obviously, the resulting noise PSD is not white anymore. Therefore, we employ a whitening filter $H_{w,m}(f) = \sqrt{\frac{1\,\mathrm{V}^2\mathrm{s}}{N_m(f) Z_L}}$ in the mth receiver,* which completely removes the frequency selectivity of the noise.

As shown in (2.21), the received power density is equal to the sum of power densities that correspond to multiple data streams. This indicates that in case of imperfect orthogonal beamforming vectors $\mathbf{b}_m$, the signal transmission might be corrupted by cochannel interference. Similar to (2.21), we express the interference power density for the DR m as

$$P_{I,m}(f) = \sum_{\kappa=4,\kappa\neq m}^{K_{\text{data}}+3} \mathbf{b}_\kappa^H \mathbf{C}^H(f) \mathbf{e}_{K,m-3} \mathbf{e}_{K,m-3}^H \mathbf{C}(f) \mathbf{b}_\kappa Z_L, \tag{2.25}$$

whereas the useful signal power density is

$$P_{U,m}(f) = \mathbf{b}_m^H \mathbf{C}^H(f) \mathbf{e}_{K,m-3} \mathbf{e}_{K,m-3}^H \mathbf{C}(f) \mathbf{b}_m Z_L. \tag{2.26}$$

Furthermore, a LE is utilized at the receiver, which minimizes the mean-squared error (MSE) of the equalized signal [minimum MSE (MMSE)]. As known from the literature (cf. [31]), the resulting signal-to-noise ratio (SNR) at the output of the unbiased MMSE-LE can be written as

$$\mathrm{SNR}_m = \left(\frac{1}{B} \int_{f_0-0.5B}^{f_0+0.5B} \frac{1}{\frac{P_{U,m}(f)}{P_{I,m}(f) + N_m(f)} + 1} \mathrm{d}f \right)^{-1} - 1, \tag{2.27}$$

* The noise PSD needs to be multiplied with Z_L, since the whitening filter is applied to a voltage signal. Hence, a unitless filter results.

where $P_{I,m}(f) + N_m(f)$ is the total disturbance of the signal transmission. For simplicity, we denote $\mathbf{W}_m(f) = \mathbf{C}^H(f)\mathbf{e}_{K,m-3}Z_L\mathbf{e}_{K,m-3}^H\mathbf{C}(f)$. Then, the substitution of (2.24–2.26) into (2.27) yields

$$\mathrm{SNR}_m = \left(\frac{1}{B} \int_{f_0-0.5B}^{f_0+0.5B} \frac{1}{\frac{\mathbf{b}_m^H \mathbf{W}_m(f)\mathbf{b}_m}{\sum_{\kappa=4,\kappa\neq m}^{K_{\text{data}}+3} \mathbf{b}_\kappa^H \mathbf{W}_m(f)\mathbf{b}_\kappa + N_m(f)} + 1} \mathrm{d}f \right)^{-1} - 1. \tag{2.28}$$

Hence, the receive power in terms of (2.22) needs to be maximized for the PRs and the signal quality in terms of (2.28) for the DRs. Therefore, a multiobjective optimization problem related to the design of a magnetic SWIPT system is discussed in the following.

2.2.2 Problem Formulation

Different optimization problems have been discussed in the literature in the context of SWIPT systems for traditional RF communication, cf. e.g., [1,32]. We consider a SWIPT system with multiple DRs that are supposed to receive an independent information stream with a certain QoS and multiple PRs that are intended to receive power. Hence, we select the minimum SNR among all DRs as the QoS metric to be maximized and the sum receive power of all PRs at the WPT metric. Correspondingly, an optimization problem with two objective functions can be formulated:

$$\begin{aligned} &1) \max_{\mathbf{b}_m, m\in\{4,\dots K_{\text{data}}+3\}} \quad \max_{m\in\{4,\dots K_{\text{data}}+3\}} \mathrm{SNR}_m \\ &2) \max_{\mathbf{b}_m, m\in\{4,\dots K_{\text{data}}+3\}} \quad \sum_{l=K_{\text{data}}+4}^{K+3} P_{r,l,\text{total}}, \\ &\qquad \text{s.t.:} \quad P_{t,\text{total}} \leq P, \end{aligned} \tag{2.29}$$

The first objective function is not convex, since it represents a generalization of the nonconvex problem discussed in [33]. In addition, the transmit power constraint is nonconvex in general, which can be shown by inspecting (2.18). Therefore, the well-known tools of convex optimization [34] are not applicable. Also, no closed-form solution can be provided.

This type of optimization problem potentially has an infinite number of optimal solutions, since we do not provide any priority for the two objectives yet. Typically, a pareto optimal solution is selected in such cases. Furthermore, as known from the previous works on SWIPT systems, e.g., cf. [1], the optimization of the WPT and the optimization of the data transmission oppose each other, i.e., the improvement with respect to the first objective would mean a deterioration with respect to the second objective and vice versa. This tradeoff can be resolved using the following strategy. If the signal transmission to the DRs shows a poor performance, i.e., the signal quality is low, we prefer information transmission over WPT and optimize the beamforming to the DRs. On the other hand, if the signal quality is high, we focus on the WPT. Through this, we avoid wasting valuable resources on data transmissions with too low or too high signal quality.

Except for the transmit power constraint, the two objective functions do not differ much from the traditional beamforming objectives known from the fundamentals of communications and MIMO systems (cf. [35]), where, however, the transmit power constraint is convex. In Section 2.2.3, the problem (2.29) is simplified via the approximation of the transmit power (2.18) by a squared L_2-norm of the beamforming vector. This approximation is valid for all types of far-field communications. Hence, the corresponding simplified optimization problem can be viewed as traditional beamforming problem and is related to a baseline scheme that we use for performance comparisons. However, the resulting optimized beamforming vectors strictly provide only suboptimum solutions to the original problem formulations (2.29), if the devices are in the near field of the transmitter. Hence, we utilize a successive convex approximation suggested in [18], which has shown substantial improvement of the power transfer efficiency even for near-field transmissions. This approximation is combined with the approach from Section 2.2.3 in each iteration of an iterative algorithm proposed in Section 2.2.5.

2.2.3 Far-Field Approximation

We consider an optimization problem, which is identical to (2.29), except for the transmit power constraint. For convenience, we stack all beamforming vectors $\mathbf{b}_m$ into one vector $\mathbf{y} = \left[\mathbf{b}_4^T, \dots, \mathbf{b}_{K_{\text{data}}+3}^T\right]^T$. Hence, the explicit problem formulation can be given by

$$\begin{aligned}
&1) \max_{\mathbf{y}} \quad \min_{m} \left(\int_{f_0-0.5B}^{f_0+0.5B} \frac{\mathrm{d}f}{\dfrac{\mathbf{y}^H kr\left\{\mathbf{E}_m, \mathbf{W}_m(f)\right\}\mathbf{y}}{\mathbf{y}^H kr\left\{\mathbf{I}_{K_{\text{data}}} - \mathbf{E}_m, \mathbf{W}_m(f)\right\}\mathbf{y} + N_m(f)} + 1} \right) \\
&2) \max_{\mathbf{y}} \quad \mathbf{y}^H kr\left\{\mathbf{I}_{K_{\text{data}}}, \sum_{l=K_{\text{data}}+4}^{K+3} \int_{f_0-0.5B}^{f_0+0.5B} \mathbf{W}_l(f)\,\mathrm{d}f \right\}\mathbf{y}, \\
&\text{s.t.:} \quad \mathbf{y}^H \mathbf{y} \le P_0
\end{aligned} \tag{2.30}$$

where P_0 is a properly chosen constant* related to the transmit power P, $kr\{\cdot,\cdot\}$ denotes the Kronecker operator, and $\mathbf{E}_m = \mathbf{e}_{K_{\text{data}},m-3}\mathbf{e}^H_{K_{\text{data}},m-3}$.

By separating the objectives, we obtain two subproblems that can be solved independently. The first subproblem corresponds to the sum power maximization, which represents an eigenvalue problem. The solution for this problem is given by the eigenvector of $kr\left\{\mathbf{I}_{K_{\text{data}}}, \sum_{l=K_{\text{data}}+4}^{K+3} \int_{f_0-0.5B}^{f_0+0.5B} \mathbf{W}_l(f)\,\mathrm{d}f\right\}$ that pertains to the maximum eigenvalue. The second subproblem corresponds to a max–min problem, which, however, can only be solved iteratively. For this, we provide a Lagrangian

$$\mathcal{L}(\mathbf{y}) = \min_m \text{SNR}_m + \lambda \mathbf{y}^H \mathbf{y}, \tag{2.31}$$

where λ denotes the Lagrange multiplier. The optimum vector $\mathbf{y}_{\text{opt}}$ has to fulfill $\frac{\partial \mathcal{L}(\mathbf{y})}{\partial \mathbf{y}^*} = \mathbf{0}$, which leads to

$$\frac{\partial \min_m \text{SNR}_m}{\partial \mathbf{y}^*} = \tilde{\lambda}\mathbf{y}. \tag{2.32}$$

Similar to [33], (2.32) can also be viewed as a nonlinear eigenvalue problem, where $\tilde{\lambda}$ is the eigenvalue. Correspondingly, the left-hand side of (2.32) can be viewed as a local gradient $\text{grad}\left(\text{SNR}_{m_{\min}}\right) = \frac{\partial \min_m \text{SNR}_m}{\partial \mathbf{y}^*}$ that maximizes the signal quality of DR $m_{\min}$ with $m_{\min} = \arg\min_m \text{SNR}_m$. Hence, we propose a gradient-based approach for solving the second subproblem, see Algorithm 1 given in pseudocode notation.

In each iteration i of the gradient search, the direction of the gradient is determined as

$$\begin{aligned}
\text{grad}_i\left\{\text{SNR}_{m_{\min}}\right\} &= \frac{\partial \text{SNR}_{m_{\min}}}{\partial \mathbf{y}_i^*} = \\
&\int_{f_0-0.5B}^{f_0+0.5B} \frac{\mathbf{y}_i^H kr\left\{\mathbf{I}_{K_{\text{data}}} - \mathbf{E}_{m_{\min}}, \mathbf{W}_{m_{\min}}(f)\right\}\mathbf{y}_i kr\left\{\mathbf{E}_{m_{\min}}, \mathbf{W}_{m_{\min}}(f)\right\}\mathbf{y}_i}{\left(1 + \mathbf{y}_i^H kr\left\{\mathbf{I}_{K_{\text{data}}}, \mathbf{W}_{m_{\min}}(f)\right\}\mathbf{y}_i\right)^2}\mathrm{d}f \\
&+ \int_{f_0-0.5B}^{f_0+0.5B} \frac{N_{m_{\min}}(f)\, kr\left\{\mathbf{E}_{m_{\min}}, \mathbf{W}_{m_{\min}}(f)\right\}\mathbf{y}_i}{\left(1 + \mathbf{y}_i^H kr\left\{\mathbf{I}_{K_{\text{data}}}, \mathbf{W}_{m_{\min}}(f)\right\}\mathbf{y}_i\right)^2}\mathrm{d}f \\
&- \int_{f_0-0.5B}^{f_0+0.5B} \frac{\mathbf{y}_i^H kr\left\{\mathbf{E}_{m_{\min}}, \mathbf{W}_{m_{\min}}(f)\right\}\mathbf{y} kr\left\{\mathbf{I}_{K_{\text{data}}} - \mathbf{E}_{m_{\min}}, \mathbf{W}_{m_{\min}}(f)\right\}\mathbf{y}_i}{\left(1 + \mathbf{y}_i^H kr\left\{\mathbf{I}_{K_{\text{data}}}, \mathbf{W}_{m_{\min}}(f)\right\}\mathbf{y}_i\right)^2}\mathrm{d}f
\end{aligned} \tag{2.33}$$

---

* The value for P_0 corresponds to $P \Big/ \left(\int_{f_0-0.5B}^{f_0+0.5B} \left|Z_{\text{in},t}^{-1}(f)\right| \mathrm{d}f\right)$, as can be deduced from (2.20).

where $\mathbf{y}_i$ denotes the state of the vector $\mathbf{y}$ in the beginning of the ith iteration. The iteration consists of an update step

$$\mathbf{y}_{i+1} = \mathbf{y}_i + \delta_i \cdot \text{grad}_i \left\{ \text{SNR}_{m_{\min}} \right\} \tag{2.34}$$

and an evaluation step, where the value of $\text{SNR}_{m_{\min}}$ is calculated. If $\mathbf{y}_i$ and $\mathbf{y}_{i-1}$ are similar (i.e., if $1 - \frac{\left|\mathbf{y}_{i-1}^H \mathbf{y}_i\right|}{\left|\mathbf{y}_{i-1}\right|^T \left|\mathbf{y}_i\right|} < \epsilon$ holds with a sufficiently small ϵ, e.g., $\epsilon = 10^{-3}$), we assume that a local maximum has been reached. Hence, the algorithm stops. Alternatively, the algorithm stops if the minimum SNR_m reaches a certain value SNR_{thr}, which is 10 dB in this work*. The choice of the step size δi is important for the system performance as well. Due to the mentioned similarity of the considered subproblem with the problem in [33], we follow the recommendations provided in [31] and select the step size accordingly.

Algorithm 1 Gradient-Based Improvement of the Minimum SNR_m

1: Input: $\mathbf{y}_1$

2: Calculate $\text{SNR}_m, \quad m \in \{4, \ldots, K_{\text{data}}\}$;

3: $i = 1$, $\mathbf{y}_0 = \mathbf{0}$;

4: **while** $\left(1 - \frac{\left|\mathbf{y}_{i-1}^H \mathbf{y}_i\right|}{\left|\mathbf{y}_{i-1}\right|^T \left|\mathbf{y}_i\right|} \geq \epsilon\right) \& \left(\min_m \text{SNR}_m < \text{SNR}_{\text{thr}}\right)$ **do**

5: $\quad m_{\min} = \arg\min_m \text{SNR}_m$;

6: $\quad$ Calculate $\text{grad}_i \left\{ \text{SNR}_{m_{\min}} \right\}$ from (2.33) using $\mathbf{y}_i$;

7: $\quad$ Update $\mathbf{y}_{i+1} = \mathbf{y}_i + \delta_i \cdot \text{grad}_i \left\{ \text{SNR}_{m_{\min}} \right\}$;

8: $\quad$ Normalize $\mathbf{y}_{i+1} \Rightarrow P_{t,\text{total}} = P$;

9: $\quad$ Calculate $\text{SNR}_m, \quad m \in \{4, \ldots, K_{\text{data}}\}$;

10: $\quad$ Update $i = i + 1$;

11: **end while**

12: Output: $\mathbf{y}_i$.

The respective solutions for the two subproblems are combined to solve the original problem (2.29). The solution for the first subproblem via eigenvalue

* The value of 10 dB seems to be sufficient for the QoS under the assumption of a Binary Phase Shift Keying (BPSK) modulation.

decomposition as discussed earlier provides the starting point for Algorithm 1. The algorithm is run till convergence, i.e., the local maximum of the second subproblem, which corresponds to the local maximum of the QoS, as mentioned earlier. This local optimum is typically the closest one to the starting point, such that the decrease of the sum power for the PRs is not too large. Since the starting point corresponds to the optimal solution for the first subproblem, this strategy seems promising. However, since only a far-field approximation has been used, this starting point may become suboptimal in case of strong couplings between magnetic devices.

2.2.4 Convex Transmit Power Approximation

The optimization problems and solutions discussed in Section 2.2.3 are mostly suitable for far-field communications. With increasing coupling strength between the devices (in the near field), the transmit power is more and more influenced by the presence of receivers, as can be deduced from (2.18). Hence, the power transfer efficiency can be improved by taking into account this influence in the optimization of the beamforming vector [18]. For this, an iterative algorithm has been proposed, where the total transmit power in (2.29) is successively approximated by a squared L_2-norm in each iteration. Employing this approximation, the optimal beamforming vector is calculated, which is used for improving the approximation and updating the solution in the next iteration.

Unfortunately, the method proposed in [18] cannot be directly applied to the SWIPT system with multiple data streams, since the transmitted symbols have a nonlinear influence on the transmit power, such that the data streams cannot be easily separated despite their statistical independence. For simplicity, we assume that all data streams utilize a BPSK modulation. Hence, there are exactly $2^{K_{\text{data}}}$ different combinations of symbols $s_m[n]$ that can be transmitted in each symbol interval n. We simplify the optimization by taking a closer look at $n \in \{0,\ldots,2^{K_{\text{data}}}-1\}$ symbol intervals with all different combinations of $s_m[n]$, $m \in \{4,\ldots,K_{\text{data}}+3\}$ symbols. The respective combination in each time interval n is represented by vector $\mathbf{s}_n = \left(s_4[n],\ldots,s_{K_{\text{data}}+3}[n]\right)^T$. Furthermore, similar to Section 2.2.3, we stack all beamforming vectors $\mathbf{b}_m$ in vector $\mathbf{y}$. At first, we rewrite the transmit power (2.18) using $\mathbf{y}$ and $\mathbf{s}_n$:

$$\begin{aligned} P_{t,\text{total}} &= \int_{f_0-0.5B}^{f_0+0.5B} \mathcal{E}\left\{\left|kr\left\{\mathbf{s}_n^T,\mathbf{I}_3\right\}\mathbf{y}\right|^T \left|\mathbf{A}(f)\,kr\left\{\mathbf{s}_n^T,\mathbf{I}_3\right\}\mathbf{y}\right|\right\}\mathrm{d}f \\ &= \frac{1}{2^{K_{\text{data}}}} \sum_{n=0}^{2^{K_{\text{data}}}-1} \int_{f_0-0.5B}^{f_0+0.5B} \left|kr\left\{\mathbf{s}_n^T,\mathbf{I}_3\right\}\mathbf{y}\right|^T \left|\mathbf{A}(f)\,kr\left\{\mathbf{s}_n^T,\mathbf{I}_3\right\}\mathbf{y}\right|\mathrm{d}f, \end{aligned} \tag{2.35}$$

We denote $\mathbf{y}_i$ the state of the vector $\mathbf{y}$ at the end of the ith iteration. For the approximation, we assume that in case of convergence of this algorithm,

$$|\mathbf{y}_i| \approx |\mathbf{y}_{i-1}| \tag{2.36}$$

holds. Then, we express $\left|kr\left\{\mathbf{s}_n^T,\mathbf{I}_3\right\}\mathbf{y}_{i-1}\right|^T$ as

$$\left|kr\left\{s_n^T,\mathbf{I}_3\right\}\mathbf{y}_{i-1}\right|^T = [1,1,1]\mathbf{V}_{n,i}, \tag{2.37}$$

using matrix $\mathbf{V}_{n,i}$ with all zero elements except for the main diagonal, which contains the elements of $\left|kr\left\{\mathbf{s}_n^T,\mathbf{I}_3\right\}\mathbf{y}_{i-1}\right|$. Hence, we obtain

$$\begin{aligned}\left|kr\left\{\mathbf{s}_n^T,\mathbf{I}_3\right\}\mathbf{y}_i\right|^T\left|\mathbf{A}(f)kr\left\{\mathbf{s}_n^T,\mathbf{I}_3\right\}\mathbf{y}_i\right| &\approx [1,1,1]\mathbf{V}_{n,i}\left|\mathbf{A}(f)kr\left\{\mathbf{s}_n^T,\mathbf{I}_3\right\}\mathbf{y}_i\right| \\ &\approx [1,1,1]\left|\mathbf{V}_{n,i}\mathbf{A}(f)kr\left\{\mathbf{s}_n^T,\mathbf{I}_3\right\}\mathbf{y}_i\right|.\end{aligned} \tag{2.38}$$

Next, we define $\mathbf{S}_{n,i}(f) = \mathbf{V}_{n,i}\mathbf{A}(f)kr\left\{\mathbf{s}_n^T,\mathbf{I}_3\right\}$ and approximate (2.38) similar to [18] by

$$\left|\mathbf{S}_{n,i}(f)\mathbf{y}_i\right| \approx \left\{\left|\mathbf{S}_{n,i}(f)\mathbf{y}_i \oslash \sqrt{\left|\mathbf{S}_{n,i}(f)\mathbf{y}_{i-1}\right|}\right|\right\}^2, \tag{2.39}$$

where $\oslash$ and $(\cdot)^2$ represent elementwise division and square operator, respectively. By multiplying (2.39) with a vector [1,1,1], (2.38) can be expressed as a squared L_2-norm:

$$\left|kr\left\{\mathbf{s}_n^T,\mathbf{I}_3\right\}\mathbf{y}_i\right|^T\left|\mathbf{A}(f)kr\left\{\mathbf{s}_n^T,\mathbf{I}_3\right\}\mathbf{y}_i\right| \approx \left|\mathbf{S}_{n,i}(f)\mathbf{y}_i \oslash \sqrt{\left\|\mathbf{S}_{n,i}(f)\mathbf{y}_{i-1}\right\|}\right|_2^2. \tag{2.40}$$

The elementwise division in (2.40) can be formulated as a multiplication with a matrix $\mathbf{Q}_{n,i}(f)$, where

$$\mathbf{Q}_{n,i}(f) = \begin{bmatrix} \frac{1}{\sqrt{\mathbf{e}_{3,1}^T\left|\mathbf{S}_{n,i}(f)\mathbf{y}_{i-1}\right|}} & 0 & 0 \\ 0 & \frac{1}{\sqrt{\mathbf{e}_{3,2}^T\left|\mathbf{S}_{n,i}(f)\mathbf{y}_{i-1}\right|}} & 0 \\ 0 & 0 & \frac{1}{\sqrt{\mathbf{e}_{3,3}^T\left|\mathbf{S}_{n,i}(f)\mathbf{y}_{i-1}\right|}} \end{bmatrix}. \tag{2.41}$$

Hence, we obtain

$$\mathbf{S}_{n,i}(f)\mathbf{y}_i \oslash \sqrt{|\mathbf{S}_{n,i}(f)\mathbf{y}_{i-1}|} = \mathbf{Q}_{n,i}(f)\mathbf{S}_{n,i}(f)\mathbf{y}_i. \tag{2.42}$$

Finally, by inserting (2.42) into (2.40) and (2.35), we obtain

$$\begin{aligned} P_{t,\text{total}} &\approx \frac{1}{2^{K_\text{data}}} \sum_{n=0}^{2^{K_\text{data}}-1} \int_{f_0-0.5B}^{f_0+0.5B} |\mathbf{Q}_{n,i}(f)\mathbf{S}_{n,i}(f)\mathbf{y}_i|_2^2 \, \mathrm{d}f \\ &= \mathbf{y}_i^H \mathbf{P}_i \mathbf{y}_i, \end{aligned} \tag{2.43}$$

where $\mathbf{P}_i$ is given by

$$\mathbf{P}_i = \frac{1}{2^{K_\text{data}}} \sum_{n=0}^{2^{K_\text{data}}-1} \int_{f_0-0.5B}^{f_0+0.5B} \left(\mathbf{Q}_{n,i}(f)\mathbf{S}_{n,i}(f)\right)^H \left(\mathbf{Q}_{n,i}(f)\mathbf{S}_{n,i}(f)\right) \mathrm{d}f. \tag{2.44}$$

In case of convergence, the approximations (2.36), (2.39), and correspondingly (2.43) are valid. In the corresponding iterative procedure of Section 2.2.5, the transmit power constraint of (2.29) becomes convex in each iteration step. As shown in Section 2.3, the proposed convex approximation of the transmit power dramatically improves the performance of the SWIPT scheme compared with the far-field approximation based SWIPT under near-field (strong coupling) conditions.

2.2.5 *Iterative Approach*

Using the convex transmit power approximation from Section 2.2.4, the optimization problem (2.29) can now be solved more accurately than with the approach of Section 2.2.3. For this, an iterative algorithm is proposed in the following and described via pseudocode notation in Algorithm 2.

Algorithm 2 Proposed Solution for the Near-Field SWIPT

1: Input: $\mathbf{y}_0$, N_iter;
2: **for** $i = 1$ **to** N_iter **do**
3: Calculate $\mathbf{P}_i$ using $\mathbf{y}_{i-1}$ and (2.44);
4: Solve generalized eigenvalue problem (2.45) $\rightarrow \mathbf{y}_i$;
5: Normalize $\mathbf{y}_i \Rightarrow P_{t,\text{total}} = P$;

6: Execute Algorithm 1;

7: **end for**

8: Output: $\mathbf{y}_{N_{\text{iter}}}$.

In each iteration, the matrix $\mathbf{P}_i$ is calculated using the beamforming solution $\mathbf{y}_{i-1}$ from the previous iteration. Then, a generalized eigenvalue problem is formulated by disregarding the first objective function:

$$\max_{\mathbf{y}_i} \quad \mathbf{y}_i^H kr \left\{ \mathbf{I}_{K_{\text{data}}}, \sum_{l=K_{\text{data}}+4}^{K+3} \int_{f_0-0.5B}^{f_0+0.5B} \mathbf{W}_l(f)\,\mathrm{d}f \right\} \mathbf{y}_i, \qquad \text{s.t.:} \quad \mathbf{y}_i^H \mathbf{P}_i \mathbf{y}_i \leq P. \tag{2.45}$$

This problem can be easily solved using known methods, e.g., simultaneous diagonalization [36]. The resulting beamforming vector is then used as a starting point for the gradient-based search according to Algorithm 1. Hereafter, $\mathbf{y}_i$ replaces $\mathbf{y}_{i-1}$.

Similar to [18], we observe that a low number of iterations is sufficient for a clear convergence to a local optimum, which provides a good performance in terms of sum receive power and minimum data rate. We use $N_{\text{iter}} = 5$ in this work.

2.3 Results

In this section, we present numerical results for the magnetic SWIPT. We assume that all coils have identical parameters, i.e., $L_t = L_r = L$, $C_t = C_r = C$, and $R_t = R_r = R$. Following the convention of the WPT community [37], we define a factor $F_{m,l} = \frac{2\pi f_0 \bar{M}_{m,l}}{R}$, which corresponds to the product of the quality factor $\frac{2\pi f_0 L}{R}$ and the coupling coefficient $\frac{\bar{M}_{m,l}}{L}$ between coils m and l, respectively. Similar to [18], we assume that all receivers are placed at the same distance d from the transmitter, such that the coupling coefficient is identical for all transmitter–receiver links, which means $\bar{M}_{m,l} = \bar{M}, \quad m \in \{1,2,3\}, \quad l \in \{4,\ldots,K+3\}$. Furthermore, we restrict ourselves to the case $F_{m,l} = F, \quad m \in \{1,2,3\}, \quad l \in \{4,\ldots,K+3\}$. Since the distances between the receivers may vary, the mutual inductance between them differs from $\bar{M}$. Defining the distance between two adjacent receivers with indices l_1 and l_2 as d_{l_1,l_2}, $\bar{M}_{l_1,l_2}$ can be expressed as $\bar{M}_{l_1,l_2} = \bar{M}\left(\frac{d}{d_{l_1,l_2}}\right)^3$, because mutual inductance scales with the third power of the transmission distance [38].

In this work, we assume a random distribution of receiver devices (on a circle around the transmitter) and that their orientation in the three-dimensional space is random. As mentioned earlier, we choose a practically relevant parameter set known from the passive RFID standards and NFC, i.e., a carrier frequency $f_0 = 13.56$ MHz and a bandwidth $B = 14$ kHz. Furthermore, we select $P = 1$ W. We assume that to satisfy the transmit power constraint, the transmitter adjusts the amplification of the beamforming vector based on the observation of the current in all three transmit circuits. In our simulations, we consider 200 scenarios for each set of system parameters. For each scenario, the beamforming solution is calculated using the far-field approximation (baseline scheme) and the proposed method. For the QoS performance analysis, we calculate the data rate according to the Shannon's channel capacity formula, $rate = B\log_2\left(1+\text{SNR}_{m_{\min}}\right)$, for the DR with index $m_{\min}$, which experiences the lowest SNR among all DRs. For the WPT, we calculate the sum receive power.

We start with a performance comparison of the proposed solution (Section 2.2.5) with the far-field approximation for two PRs and two DRs, see Figure 2.2. We observe that with decreasing noise variance (from $N_0 = 10^{-6}$ to $N_0 = 10^{-10}$ V^{2}s), the minimum data rate can be substantially increased, especially for $F < 10$. On the other hand, the proposed method shows an improvement compared with the far-field approximation only for $F > 10$. For $F > 0.2$, the sum receive power of the proposed solution is significantly larger than that of the baseline scheme. In case of very low noise variance, i.e., $N_0 = 10^{-10}$ V^{2}s, the sum receive power can be increased by up to 0.24 W (for $F = 40$).

For a better insight, we also investigate the system performance for a constant noise variance, constant number of DRs, and a variable number of PRs, see Figure 2.3. This scenario is a special case of magnetic SWIPT that has been

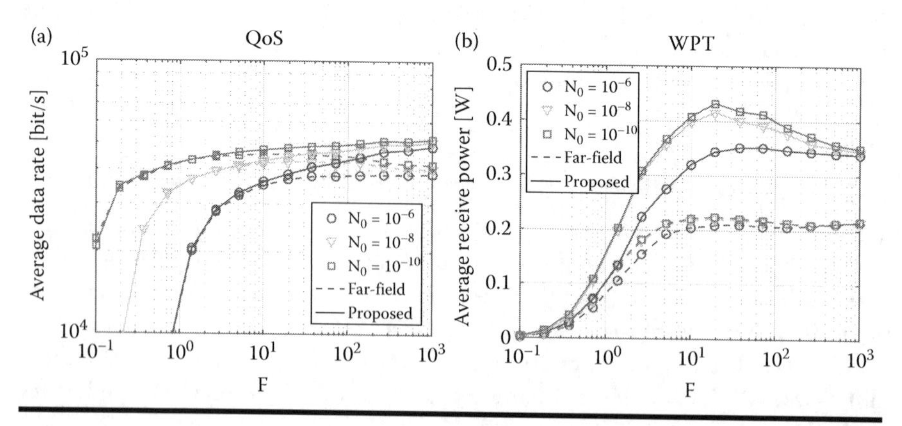

Figure 2.2 Average system performance from 200 simulations with random receiver locations, $K_{\text{data}} = 2$ and $K_{\text{power}} = 2$. (a) Minimum data rate; (b) sum receive power.

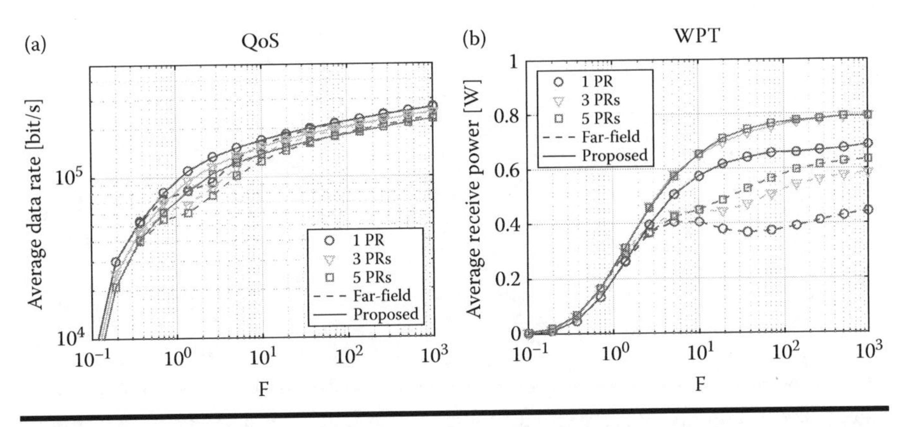

Figure 2.3 Average system performance from 200 simulations with random receiver locations, $K_{data} = 1$ and $N_0 = 10^{-8}$ V²s. (a) Minimum data rate; (b) sum receive power.

considered in [22]. Obviously, the average receive power for $K_{power} = 2$ in Figure 2.3 is equal to the average receive power obtained in [22]. From this, we deduce that the proposed solution contains the solution obtained in [22]. With increasing number of PRs, the minimum data rate slightly decreases. This is due to the fact that more and more PRs absorb the transmitted power, such that less power is left for the DR. Also, the proposed method shows a slightly worse performance than the far-field approximation in terms of minimum data rate for $F > 20$. The rate loss of the proposed scheme compared with the far-field approximation is bounded below 12%. This is due to the fact that the baseline scheme prioritizes information transmission over WPT, such that more effort is put into the maximization of the minimum SNR than into the maximization of the sum receive power. Hence, the resulting QoS is typically much higher than the assumed $SNR_{thr} = 10$ dB. On the other hand, the proposed method switches between QoS and WPT in an alternating manner, such that both objectives are maximized. Hence, by sacrificing a small part of the signal quality, a better solution for WPT is found. Interestingly, with increasing number of PRs, the WPT performance improves for both schemes. With three PRs and the proposed solution, an average sum receive power of up to 0.8 W can be reached for $F > 100$. This corresponds to 80% power efficiency, if only the WPT is considered*. This effect comes from the fact that with increasing number of devices, the degree of freedom increases (the power received at multiple locations is summed up), such that a better power efficiency results for a large number of receivers. Furthermore, it

* The actual power efficiency is of course somewhat lower due to the losses from the conversion of the received signal into electrical power. However, we assumed $\eta = 1$ in this work, such that these losses are not considered.

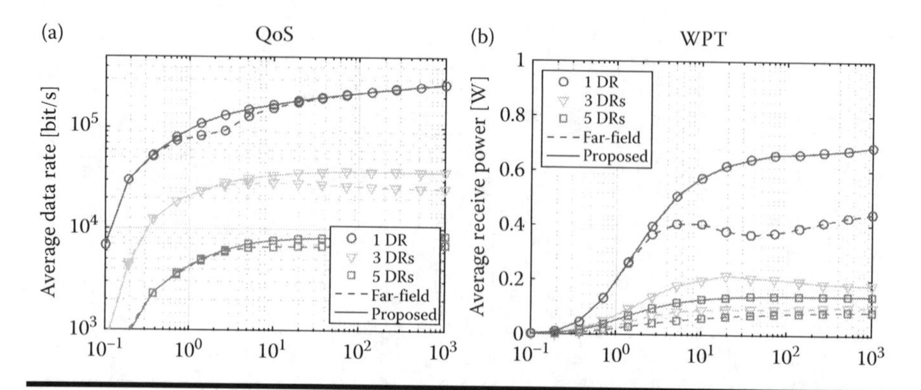

Figure 2.4 Average system performance from 200 simulations with random receiver locations, $K_{\text{power}} = 1$ and $N_0 = 10^{-8}$ V²s. (a) Minimum data rate; (b) sum receive power.

can be clearly seen that the baseline scheme and the proposed solution completely agree for $F < 1$, since the far-field approximation becomes valid, as explained earlier.

In addition, we show the system performance for the adjacent case, i.e., for a constant noise variance, constant number of PRs, and a variable number of DRs, see Figure 2.4. Since each additional data stream generates interference to all other streams, the minimum data rate decreases dramatically with increasing number of DRs. Correspondingly, the optimization focuses more and more on the interference mitigation, such that the receive power at the single PR reduces as well. The asymptotical value of the sum receives power decreases on average by 50% with every additional DR. Moreover, the proposed method still provides an undeniable advantage compared with the far-field approximation, especially for large values of F. Even for $K_{\text{data}} = 3$, we observe a difference of up to 0.14 W (for $F = 20$) between the two schemes.

2.4 Discussion

Using the described method, simultaneous information and power transmission to multiple devices can be established. The resulting system performance seems promising, especially for very strong couplings between magnetic devices, i.e., in the near-field region. This implies that the considered SWIPT system should preferably be used as an access point for short-distance transmissions. Such magnetic access points can not only be employed as stand-alone service providers but also integrated into larger communication systems like 5G, Internet of Things, and other applications typically in the area of consumer electronics. Moreover, the interconnection of multiple access points is of special interest, since a distributed beamforming can be applied both to information transmission and power transfer, such that the system performance might be further improved. The resulting magnetic SWIPT networks seem promising especially for indoor applications.

The coverage of the magnetic SWIPT systems is not very high. In practice, it may not be possible to achieve a very strong coupling beyond $F = 20$ for a distance between the transmitter and the receiver above 20 cm. In principle, the maximum transmission distance, for which $F = 1$ and $F = 10$, can be up to 1 m and 0.5 m, respectively, depending on the geometry of the employed coils. This implies, however, the use of very large coils at least in the transmitter. Furthermore, by selecting a different frequency band with higher carrier frequency, e.g., in the range of few hundred megahertz, the coupling between the devices can be substantially strengthened, such that the values of F beyond 100 can be reached. However, with increasing resonance frequency, more and more parasitic effects occur that are usually neglected in the low-frequency bands, e.g., proximity effect, skin effect, etc. These effects are typically nonlinear, which can make the system design even more challenging. Their influence on the magnetic beamforming remains an open issue for future investigation.

One of the issues of the presented scheme is the assumption of a perfectly known mutual inductance between any two devices of the network. Unfortunately, this perfect CSI cannot be obtained due to the limitations provided by the channel estimation theory, especially assuming a nonvanishing mobility of the users. Hence, a performance decrease due to the imperfect channel estimation is inevitable. Nevertheless, as known from the previous works on magnetic induction based communication, the pattern of the magnetic field lines can be well estimated, if the geometry, position, and orientation of the magnetic device are available. Hence, the estimation process can be further enhanced by taking into account the information from additional sensors, e.g., a gyroscope, Global Positioning System (GPS), etc. Furthermore, the recently proposed channel estimation inside the transmitter circuits, cf. [39], can be utilized and combined with the traditional methods of channel estimation, such that multiple observations of the same channel statistics may increase the estimation accuracy. However, the performance of the presented SWIPT scheme under imperfect CSI and the corresponding robust optimization are beyond the scope of this work.

Despite a very promising system performance of a 3D-coil based access point, it might not be possible to further increase the number of coils, since the coil deployment would not be orthogonal anymore. Correspondingly, the coupling between the transmitter coils may become much stronger than the coupling between the transmitter and the receiver. Typically, the energy exchange occurs mostly among the coils with the strongest coupling, such that almost no energy would reach the receiver. Hence, the performance of such a scheme can be even worse than using a 3D-coil. Furthermore, the angular diversity of the three-dimensional space is completely exploited in the presented scheme, such that the use of additional coils does not imply an increase of the degree of freedom. Correspondingly, the use of an access point with more than three coils cannot be justified.

2.5 Summary

In this work, a magnetic induction based SWIPT system has been proposed, which utilizes a transmitter with three coils and multiple single coil receivers. From the set of randomly distributed receiver devices, some users (DRs) are selected for information reception, whereas the remaining users (PRs) are intended to receive the power. Since all transmissions are supposed to be performed simultaneously, a nontrivial multiobjective optimization problem results. The considered problem refers to the maximization of the sum receive power of all PRs and the maximization of the signal quality for the worst DR. Furthermore, the near-field coupling between coils has been taken into account, which results in an additional nonconvex transmit power constraint. Correspondingly, the overall optimization problem turns out to be nonconvex, such that no globally optimum solution can be obtained. Also, the nonconvex transmit power constraint does not allow to apply the well-known beamforming techniques according to the literature. Hence, the problem has been split into an eigenvalue problem and a QoS problem. The QoS problem has been solved via a gradient-based iterative algorithm. In order to cope with the nonconvexity of the transmit power constraint, it has been replaced by an L_2-norm constraint (far-field approximation) for the baseline solution and using a more accurate iterative convex approximation for the proposed solution. In this context, significant performance gains have been observed for the proposed solution in most of the considered scenarios in terms of minimum achievable data rate and average sum power compared with the baseline scheme. The overall high system performance renders the presented magnetic SWIPT scheme efficient and promising. Moreover, in this work, the spectrum of possible applications, the coverage, the extendability, and the practical issue of imperfect CSI have been addressed. The corresponding discussion should help system designers to gain more insight into magnetic SWIPT and its applicability and potential.

References

1. R. Zhang and C.K. Ho (2013) MIMO broadcasting for simultaneous wireless information and power transfer. *IEEE Transactions on Wireless Communications*, **12**(5), 1989–2001.
2. L.R. Varshney (2008) Transporting information and energy simultaneously, in *Proceedings of IEEE International Symposium Information Theory (ISIT)*, pp. 1612–1616.
3. P. Grover and A. Sahai (2010) Shannon meets Tesla: Wireless information and power transfer, in *Proceedings of IEEE International Symposium Information Theory (ISIT)*, pp. 2363–2367.
4. I. Krikidis, S. Timotheou, S Nikolaou, G. Zheng, D.W.K. Ng, and R. Schober (2014) Simultaneous wireless information and power transfer in modern communication systems. *IEEE Communications Magazine*, **52**, 104–110.

5. E. Boshkovska, D.W.K. Ng, N. Zlatanov, and R. Schober (2015) Practical non-linear energy harvesting model and resource allocation for SWIPT systems. *IEEE Communications Letters,* **19**(12), 2082–2085.
6. M.R.A. Khandaker and K.-K. Wong (2015) Robust secrecy beamforming with energy-harvesting eavesdroppers. *IEEE Transactions on Wireless Communications,* **4**(1), 10–13.
7. D.W.K. Ng, E.S. Lo, and R. Schober (2014) Robust beamforming for secure communication in systems with wireless information and power transfer. *IEEE Transactions on Wireless Communications,* **13**(8), 4599–4615.
8. Z. Xiang and M. Tao (2012) Robust beamforming for wireless information and power transmission. *IEEE Wireless Communications Letters,* **1**, 372–375.
9. M.R.A. Khandaker and K.-K. Wong (2014) SWIPT in MISO multicasting systems. *IEEE Wireless Communications Letters,* **3**(3), 277–280.
10. Q. Shi, L. Liu, W. Xu, and R. Zhang (2014) Joint transmit beamforming and receive power splitting for MISO SWIPT systems. *IEEE Transactions on Wireless Communications,* **13**(6), 3269–3280.
11. H. Chen, Y. Li, Y. Jiang, Y. Ma, and B. Vucetic (2015) Distributed power splitting for SWIPT in relay interference channels using game theory. *IEEE Transactions on Wireless Communications,* **14**(1), 410–420.
12. Z. Ding, I. Krikidis, B. Sharif, and H.V. Poor (2014) Wireless information and power transfer in cooperative networks with spatially random relays. *IEEE Transactions on Wireless Communications,* **13**(8), 4440–4453.
13. C. Zhong, H.A. Suraweera, G. Zheng, I. Krikidis, and Z. Zhang (2014) Wireless information and power transfer with full duplex relaying. *IEEE Transactions on Communications,* **62**(10), 3447–3461.
14. Yoon, I.J. and Ling, H. (2011) Investigation of near-field wireless power transfer under multiple transmitters. *IEEE Antennas and Wireless Propagation Letters,* **10**, 662–665.
15. J.J. Casanova, Z.N. Low, and J. Lin (2009) A loosely coupled planar wireless power system for multiple receivers. *IEEE Transactions on Industrial Electronics,* **56**(8), 3060–3068.
16. S.A. Mirbozorgi, H. Bahrami, M. Sawan, and B. Gosselin (2014) A smart multicoil inductively coupled array for wireless power transmission. *IEEE Transactions on Industrial Electronics,* **61**(11), 6061–6070.
17. N. Hoang, J.I. Agbinya, and J. Devlin (2015) FPGA-based implementation of multiple modes in near field inductive communication using frequency splitting and MIMO configuration. *IEEE Transactions on Circuits and Systems I: Regular Papers,* **62**(1), 302–310.
18. S. Kisseleff, I.F. Akyildiz, and W.H. Gerstacker (2015) Beamforming for magnetic induction based wireless power transfer systems with multiple receivers, in *Proceedings of IEEE Globecom* 2015.
19. A. Karalis, J.D. Joannopoulos, and M. Soljacic (2008) Efficient wireless non-radiative mid-range energy transfer. *Annals of Physics,* **323**, 34–48.
20. J.I. Agbinya and M. Mashipour (2012) Power equations and capacity performance of magnetic induction communication systems. *Wireless Personal Communications Journal,* **64**(4), 831–845.
21. H. Jiang and Y. Wang (2008) Capacity performance of an inductively coupled near field communication system, in *Proceedings of IEEE International Symposium of Antenna and Propagation Society.*

22. S. Kisseleff, I.F. Akyildiz, and W.H. Gerstacker (2017) Magnetic induction based simultaneous wireless information and power transfer for single information and multiple power receivers. *IEEE Transactions on Communications*, **65**(3), 1396–1410.
23. Y. Zhang, Z. Zhao, and K. Chen (2013) Frequency splitting analysis of magnetically-coupled resonant wireless power transfer, in *Proceedings of IEEE ECCE* 2013.
24. Z. Sun, I.F. Akyildiz, S. Kisseleff, and W. Gerstacker (2013) Increasing the capacity of magnetic induction communications in RF-challenged environments. *IEEE Transactions on Communications*, **61**(9), 3943–3952.
25. S. Kisseleff, I.F. Akyildiz, and W.H. Gerstacker (2014) Throughput of the magnetic induction based wireless underground sensor networks: Key optimization techniques. *IEEE Transactions on Communications*, **62**(12), 4426–4439.
26. D. Bernstein (2005) *Matrix Mathematics*, Princeton, NJ: Princeton University Press.
27. H. Akagi, E.H. Watanabe, and M. Aredes (2007) *Instantaneous Power Theory and Applications to Power Conditioning*, Hoboken, NJ: Wiley-IEEE Press.
28. N. Tal, Y. Morag, and Y. Levron (2015) Design of magnetic transmitters with efficient reactive power utilization for inductive communication and wireless power transfer, in *Proceedings of IEEE International Conference on Microwaves, Communications, Antennas and Electronic Systems (COMCAS).*
29. R.R.A. Syms and L. Solymar (2011) Noise in metamaterials. *Journal of Applied Physics*, **109**(124909), 1–5.
30. S. Haykin (2001) *Communication Systems*, New York, NY: John Wiley & Sons, 4th edn.
31. M.V. Clark, L.J. Greenstein, W.K. Kennedy, and M. Shafi (1994) Optimum linear diversity receivers for mobile communications. *IEEE Transactions on Vehicular Technology*, **43**, 47–56.
32. X. Zhou, R. Zhang, and C.K. Ho (2013) Wireless information and power transfer: Architecture design and rate-energy tradeoff. *IEEE Transactions on Communications*, **61**(11), 4754–4767.
33. Y.-W. Liang, R. Schober, and W. Gerstacker (2007) FIR Beamforming for Frequency-Selective Channels with Linear Equalization. *IEEE Communications Letters*, **11**(7), 633–624.
34. S. Boyd and L. Vandenberghe (2004) *Convex Optimization*, Cambridge, UK: Cambridge University Press.
35. D. Tse and P. Viswanath (2005) *Fundamentals of Wireless Communication*, Cambridge, UK: Cambridge University Press.
36. A. Bunse-Gerstner, R. Byers, and V. Mehrmann (1992) A chart of numerical methods for structured eigenvalue problems. *SIAM Journal on Matrix Analysis & Applications*, **13**(2), 419–453.
37. N. Shinohara (2014) *Wireless Power Transfer via Radiowaves*, Hoboken, NJ: Wiley-ISTE.
38. R. Bansal (2004) Near-field magnetic communication, *IEEE Antennas and Propagation Magazine*, **46**, 114.
39. S. Kisseleff, I.F. Akyildiz, and W. Gerstacker (2014) Transmitter-side channel estimation in magnetic induction based communication systems, in *Proceedings of IEEE BlackSeaCom* 2014.

Chapter 3

Magnetic Communication for Human Health Monitoring

Rui Ma
Southern University of Science and Technology

Yu Lu and Fei Hu
The University of Alabama

Contents

3.1 Introduction

Biomedical sensors or devices are being used on the human body more and more as technology develops. Most of these devices such as Electroencephalography(EEG), Electrocardiography(ECG), glucose sensor for continuous glucose monitoring, or artificial heart, need long-term communication and power transfer with the external devices. Traditional radio wireless communication methods like WiFi and Bluetooth are the most popular techniques used for these sensors, but far-field communication methods still consume much power and are not good enough for long-term applications.

Researchers have developed several near-field radio communication methods for saving power. Resonant magnetic communication is one of the most promising techniques. It uses two conductive coils and builds a resonant inductive coupling between them. Just like the loop antenna, the induction field (near field, $r < \lambda/2\pi$) of magnetic coil is dominated by the magnetic field with little electrical field. In this area, the magnetic field is very strong, such that a strong coupling can be established between the two coils. Setting the circuits of the two coils to operate at the same frequency can greatly improve the power transition. Energy transfer between the resonant coupled coils has very little loss, which makes the near-field communication available at much smaller power consumption than the traditional far-field communication.

This method has been used in wireless power charging [1] and underground communication [2]. The communication path loss is tested on the human body [3], and the channel gain between the human body and external device is presented in [4]. However, few works have been reported on the fully functional digital communication system using magnetic communication method on the human body. This chapter proposes a digital communication system using USRP devices and magnetic coils as transceivers to demonstrate its feasibility on the human body. In addition, the proposed system tries to recognize the transmission medium or environment by recognizing the received waveform.

3.2 Relative Work

Magnetic resonant inductive coupling are currently most used in wireless power transfer. Shi et al. [1] proposed a MultiSpot system for the charging of six devices at distances of up to 50 cm at 1 MHz. This frequency is good for energy transfer but has limited bandwidth for communication. Increasing the frequency will largely decrease the transmission distance.

Tan et al. [2] proposed a magnetic resonant coupling system for wireless underground sensor network at 8 MHz. The system used USRP as the communication terminal, and used 3D coils to guarantee good coupling under placement disturbance. The transmission medium included sand and water. Packet error rate is tested at different distances.

Ogasawara et al. [4] presented a system of human body communication using magnetic coupling at 20 MHz. One coupling coil is deployed on the human body, and the other is fixed on the ground or wall. The channel loss was measured, but the system didn't utilize the resonant circuit. Park et al. [3] used magnetic resonant coupling for the human body communication at 21 MHz. The coupling coils are placed on the wrist or head. Network analyzer was used to measure the path loss, and the path losses of different postures are compared.

Pu et al. [5] used USRP-N210 to recognize human gestures. Human gestures will cause Doppler shifts of the wireless signal. The system extracted Doppler shifts from the WiFi signals and use them as the feature of gestures.

Few works have built a digital system for the human body communication based on magnetic resonant coupling. All the related works mentioned earlier either built the system for underground environment or the systems are only tested for path loss. This chapter has built a digital communication system for human body communication based on magnetic resonant coupling.

3.3 System Setup

The system uses two USRP N210 as the communication transceiver and uses two magnetic coils with Inductor Capacitor(LC) circuits as the antennas as shown in Figure 3.1. The coils have 9 cm diameter, 14 American Wire Gauge(AWG) (1.6 mm), and 4 turns as shown in Figure 3.2. The formula used to calculate the inductance of radio coils is as follows [6]:

$$L = \frac{r^2 n^2}{9r + 10a} \tag{3.1}$$

The units are in inches and microhenries. So the inductance of the coil is 2.18 μH. The capacitor used in the experiment is 39 pF, so the resonant frequency of the LC circuit is 17 MHz, as calculated by the following formula.

$$f_0 = \frac{1}{2\pi\sqrt{LC}} \tag{3.2}$$

The near field of this radiation is within $\lambda/2\pi$, so distance within 2.8 m is the near field covered by a strong magnetic field.

On the USRP platform, BasicTX and BasicRX daughter boards are used. They have a frequency range of 1–250 MHz, and can only transceive real-mode signals (all the

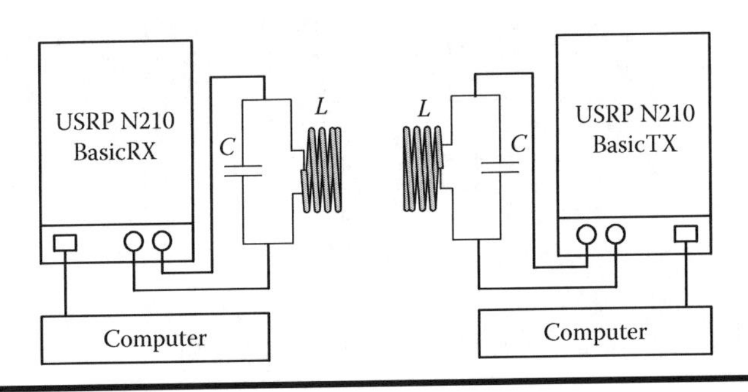

Figure 3.1 The magnetic communication system setup.

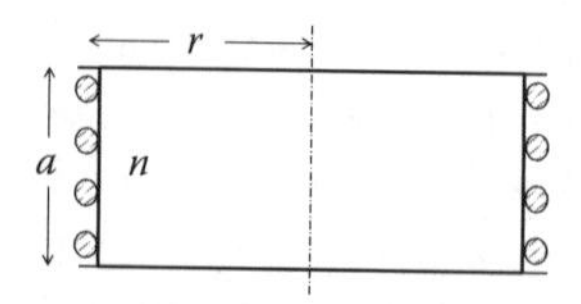

Figure 3.2 The structure of the magnetic coil.

USRP low-frequency daughter boards only work on real-mode data). The most widely used modulation scheme Quadrature Phase Shift Keying(QPSK) needs In-phase and Quadrature(I/Q) data, so we have to use BPSK, which only requires real-mode data.

3.4 Experiment and Discussion

The prototype of the proposed magnetic communication system is shown in Figure 3.3. First, we tested the system by sending strings like "123456789i," and the system can maintain a Bit Error Rate(BER) smaller than 5% at a distance of 30 in. Then, we changed the frequency, angle, distance, and medium to see their influence on the transmitted signals. The purpose of changing these conditions is to find the pattern of signal under different conditions, so that the system can recognize the environment by checking the received signals. Direct reading of the carrier wave requires too high sampling rate for the Analog-to-Digital Converter(ADC), so we inspect the digital real-mode signal of USRP instead. The constant real-mode number "1" is transmitted between the two devices. The received data is distorted, and the amplitude of distortion can reflect the signal strength (Figure 3.4).

3.4.1 Change of Frequency

The resonant frequency of the system is set to 17 MHz. Different frequencies are applied to the system, and the carrier of both sender and receiver is measured by the oscilloscope as shown in Figure 3.5. We can see at the receiver side that the signal strength is decreased at lower and higher frequencies as shown in Figure 3.6.

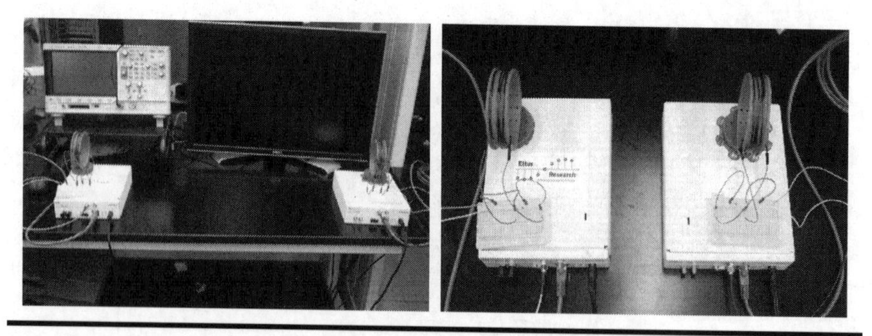

Figure 3.3 The magnetic communication system.

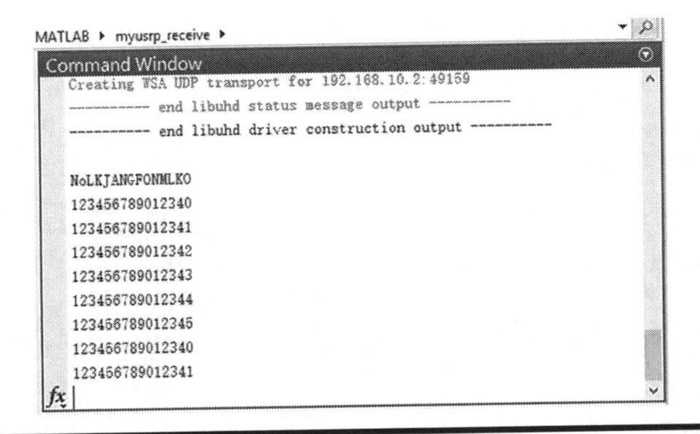

Figure 3.4 The received strings on receiver side.

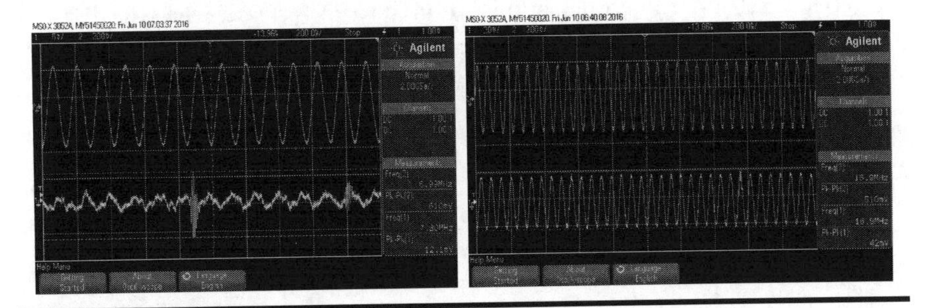

Figure 3.5 Samples of carrier wave measured by oscilloscope.

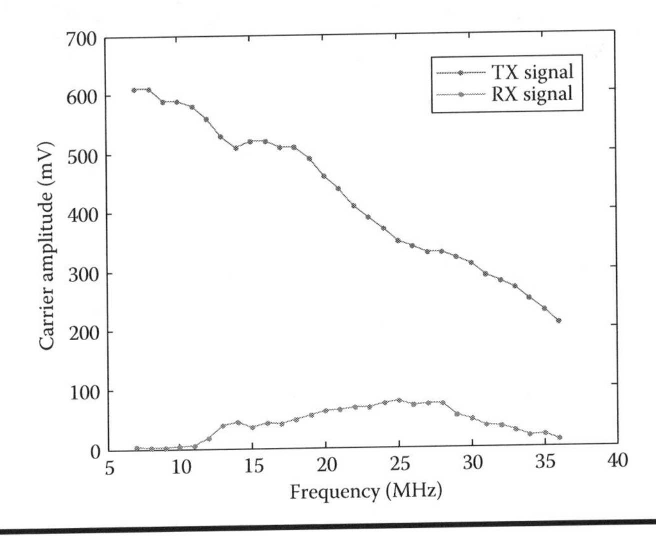

Figure 3.6 The carrier amplitude with different frequencies observed using oscilloscope.

The maximal signal strength at the receiver side should be at the true resonant frequency.

3.4.2 Change of Angle

From this section, all the measured signals are the digital real-mode data of USRP. The relative angle of the two coupling coils is changed from 0 to 90°. The received real-mode data of USRP is shown in Figure 3.7. Theoretically, the minimum amplitude should be at 90°, but the experiment shows it is at 60° as shown in Figure 3.8.

3.4.3 Change of Distance

The received real-mode data of USRP with different distance is shown in Figure 3.9. It can be seen that the amplitude of the real-mode data decreases when the distance increases as shown in Figure 3.10.

3.4.4 Change of Medium

A metal board, a box of water, and human arms are placed between the coils as different transmission medium. It can be seen from Figure 3.11 that the metal

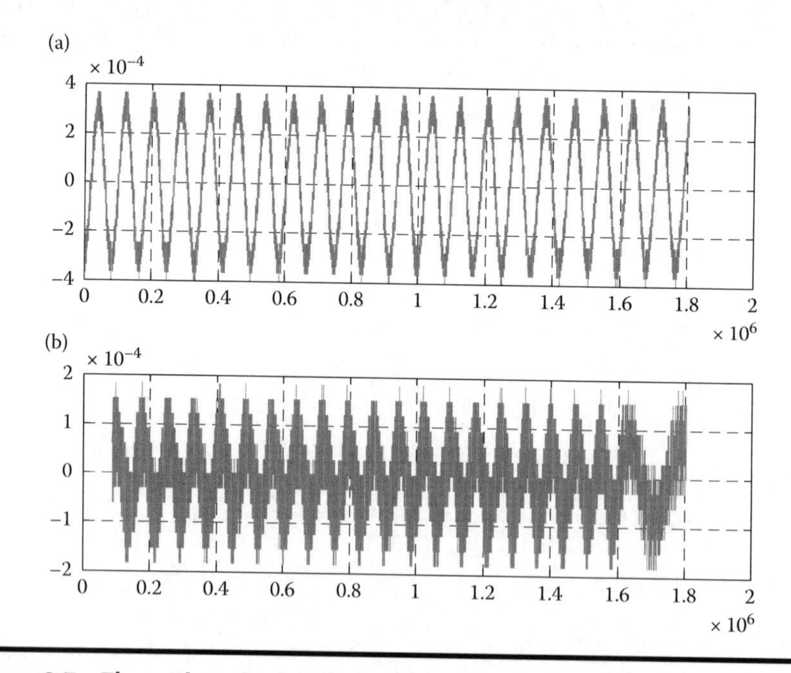

Figure 3.7 The real-mode signal of USRP with the coupling coils are relative angles: (a) 0° and (b) 60°.

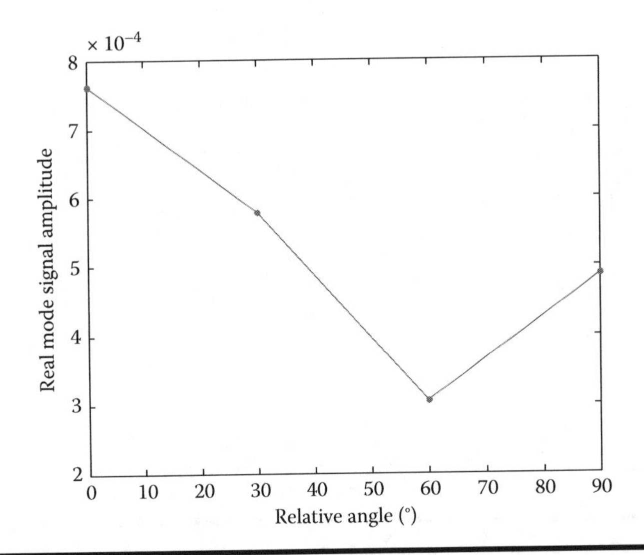

Figure 3.8 The amplitude of real-mode signal of USRP when the coupling coils are in different relative angles.

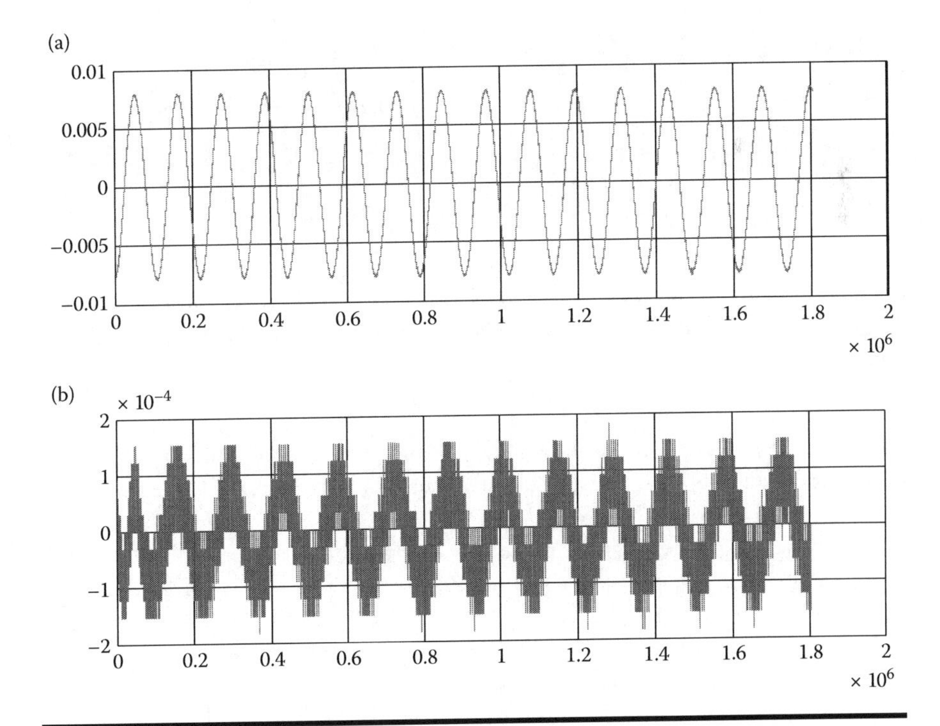

Figure 3.9 The real-mode signal of USRP with distance at (a) 0° and (b) 60°.

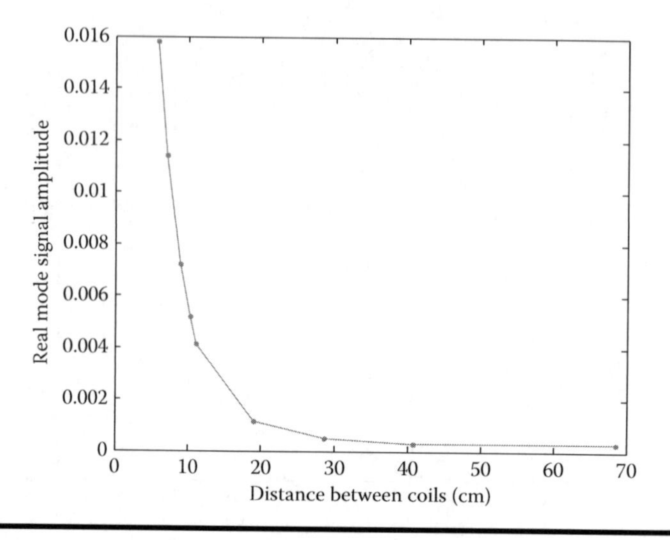

Figure 3.10 The real-mode signal of USRP with different distances.

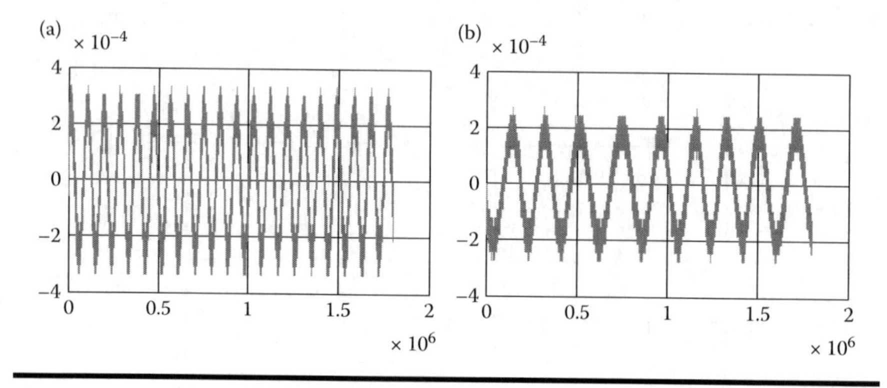

Figure 3.11 The real-mode signal of USRP with different medium: (a) air and (b) metal board in between.

board changes the pattern of the received real-mode data. The distortion frequency decreases. In Figure 3.12, the amplitude of distortion is decreased when a box of water is placed between the two coils.

The two coils are then placed on one human arm, from wrist to shoulder as shown in Figure 3.13a. It can be seen in Figure 3.14 that the path loss in air is about four times larger than in the human arm. Next, the two coils are placed at a relative angle of 120°, as shown in Figure 3.13b. The signal is first tested in the air, and then it is tested with two coils worn on two wrists. And this time the signal may go through the whole human body. The real-mode data is compared in Figure 3.15. It can be seen that the path loss of air is six times larger than the human body in this setup.

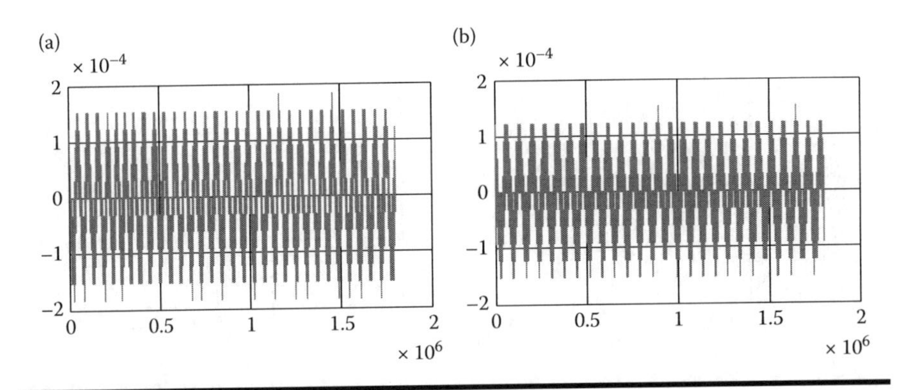

Figure 3.12 The real-mode signal of USRP with different medium: (a) air and (b) water.

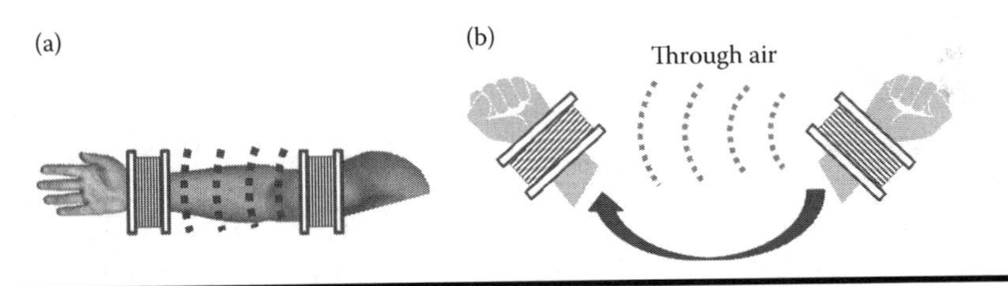

Figure 3.13 The magnetic signal has a smaller path loss in human body than in the air: (a) through one arm and (b) through air.

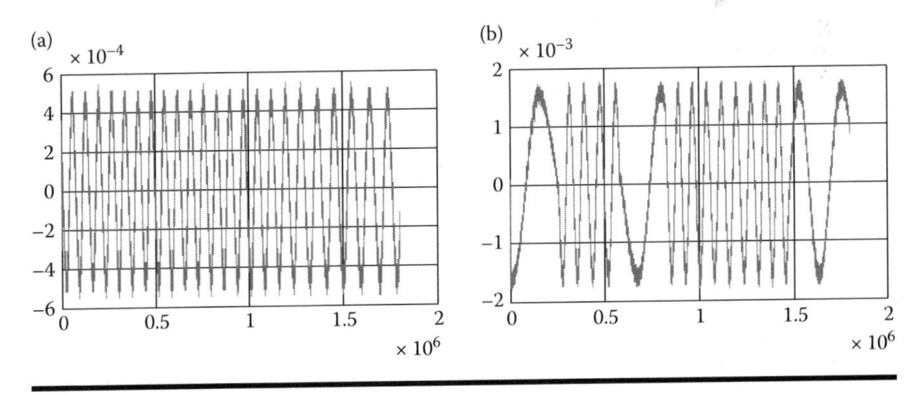

Figure 3.14 The real-mode signal of USRP with different medium: (a) air and (b) from wrist to shoulder.

From the experiments, it can be seen that different conditions like different angles and medium will change the pattern of the received real-mode signal. So it is possible to use this feature to recognize the environment of the communication system.

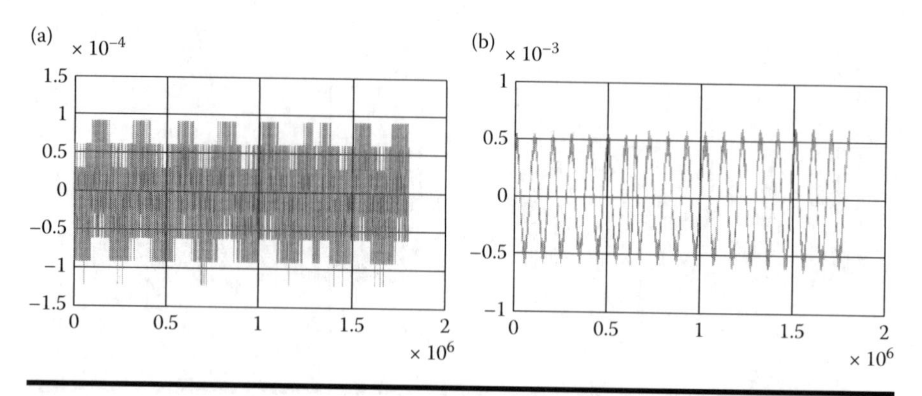

Figure 3.15 The real-mode signal of USRP with different medium: (a) air and (b) from right to left wrists.

3.5 Conclusion

This chapter has built a prototype of magnetic digital communication system and implemented multiple experiments under different conditions. The results have demonstrated the digital communication ability of the system, and they also show the possibility of using the system to recognize its surrounding environment. This feature could be used to identify the owner of the medical sensor, so that the sensor signal will only be sent to the owner's device.

References

1. L. Shi, Z. Kabelac, D. Katabi, and D. Perreault, "Wireless power hotspot that charges all of your devices," In *Proceedings of the 21st Annual International Conference on Mobile Computing and Networking (MobiCom '15)*. ACM, New York, 2015.
2. X. Tan, Z. Sun and I. F. Akyildiz, "Wireless underground sensor networks: MI-based communication systems for underground applications," *IEEE Antennas and Propagation Magazine*, vol. 57, no. 4, pp. 74–87, Aug. 2015.
3. J. Park and P. P. Mercier, "Magnetic human body communication," in *2015 37th Annual International Conference of the IEEE Engineering in Medicine and Biology Society (EMBC)*, Milan, pp. 1841–1844, 2015.
4. T. Ogasawara, A. I. Sasaki, K. Fujii and H. Morimura, "Human body communication based on magnetic coupling," *IEEE Transactions on Antennas and Propagation*, vol. 62, no. 2, pp. 804–813, Feb. 2014.
5. Q. Pu, S. Gupta, S. Gollakota, S. Patel, "Whole-home gesture recognition using wireless signals," *Proceedings of the 19th Annual International Conference on Mobile Computing and Networking (Mobicom'13)*, pp. 27–38, ACM, 2013.
6. H. A. Wheeler, "Simple inductance formulas for radio coils," *Proceedings of the Institute of Radio Engineers*, vol. 16, no. 10, pp. 1398–1400, Oct. 1928.

Chapter 4

Communication Using Magnetic Induction

Alexander D. de Sousa, Luiz F. M. Vieira,
and Marcos A. M. Vieira
Universidade Federal de Minas Gerais

Contents

4.1 Introduction

Issues related to energy represent an important research field in the actual scenario of computer science and electrical engineering. In wireless sensor networks, the lifetime of the nodes is usually limited to the autonomy of their batteries. Furthermore, the increasing use of Internet communications in the mobile context has made the charging of devices one of the most persistent concerns in the lives of users. To prolong the lifetime of devices, save resources, and even decrease the size of devices by allowing smaller batteries, the wireless power transfer (WPT) techniques are increasingly being studied.

The SWIPT—Simultaneous Wireless Information and Power Transfer—systems are those that take advantage of the same physical layer for power transfer and sending messages. These messages may carry data about the WPT itself or even about a secondary application running in the background.

There are many reasons to share the same hardware to send data and energy. Devices such as battery chargers sometimes need to send configurations to the receivers or vice versa. If they can use their power transmitter to send the data, they will not need a specialized component to do that. Furthermore, the device will not need to spend power with two separated components.

The most popular WPT techniques are divided into two main categories. The radiative approach makes use of electromagnetic radiation, such as microwave, radio frequency or laser beams, to send power through typically long distances. Depending on the used technique, the transfer range can reach up to some kilometers. These are called *far-field techniques.* Despite these advantages, statistical studies show that this approach can be harmful to the environment and human health [1]. Because of this, there were established severe restrictions to the maximal power to be transmitted via electromagnetic radiation [2]. This kind of technology also has a tradeoff between efficiency and alignment dependence. The transmission via omnidirectional waves typically has low efficiency while the laser beam based transmission requires line of sight.

The magnetic induction (MI) techniques are the most popular in domestic and industrial environments. It consists of electrically exciting a transmitting coil with alternating current and then picking up the induced current in a receiving coil. This type of technology is marked by having high efficiency up to a few centimeters or meters. After some point, the efficiency drops fast, and because of that, these technologies are known as *near-field techniques.*

The use of simulators [3] while designing MI SWIPT and MI WPT applications is highly indicated, because these are often quite hardware dependent and, therefore, any design error may generate unecessary financial cost. This chapter addresses the analytical modeling behind the MI SWIPT low-level simulation and the simulation process itself. First, a pure and simple WPT simulation technique is explained, and then, the modeling is generalized to a more complex approach, to implement the SWIPT.

4.2 The Main Concepts about MI SWIPT Systems

Figure 4.1 illustrates a typical circuit design for the two possible setups in WPT context. The *Active Circuits* are those what have their own voltage source and act as power transmitters. The *Passive* ones do not have a voltage source attached to their circuit, so they act as power receivers. An MI WPT system is usually formed by the interaction of two or more devices with at least one device of each type.

The induced current in the receiving circuit is related to the derivative of the current in each active circuit and vice versa, so typically a sinusoidal voltage source

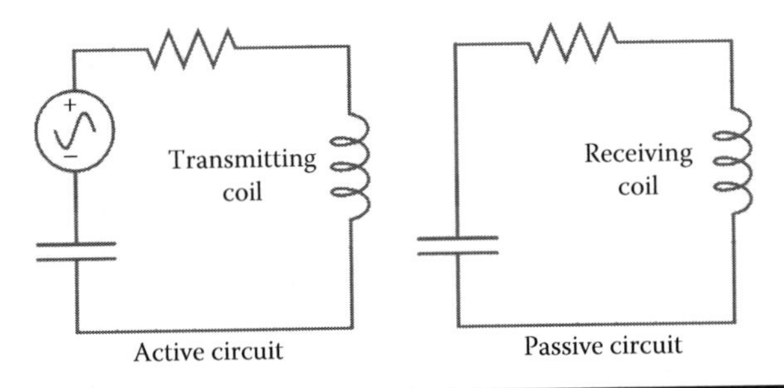

Figure 4.1 Simplified scheme of the passive and active circuits.

is used. The shape of the coils can vary for the most different applicable contexts, but the three most popular ones in the literature are the solenoid, toroidal, and planar spiral ones.

Both energy transmitter and energy receiver devices may need to send any type of data during the transmission process. Many recharging algorithms such as the Qi [4] require the receiver to send a message from time to time in order to ask the transmitter to keep the energy transfer. Configurations such as power needs, impedance information for *Impedance Matching algorithms,* and others may have to be sent from the energy transmitter or the energy receiver. This information is usually modulated with the amplitude shift keying (ASK) method [5]. The ASK modulation represents the two-bit states as two different amplitudes in a carry signal, as illustrated in Figure 4.2.

As will be explained in detail in the following sections, the current in each energy transmitter has a component related to each other WPT device in the

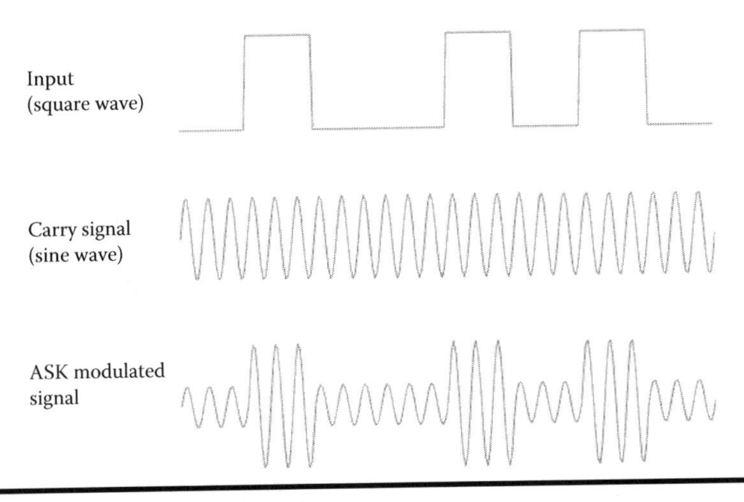

Figure 4.2 The ASK modulation.

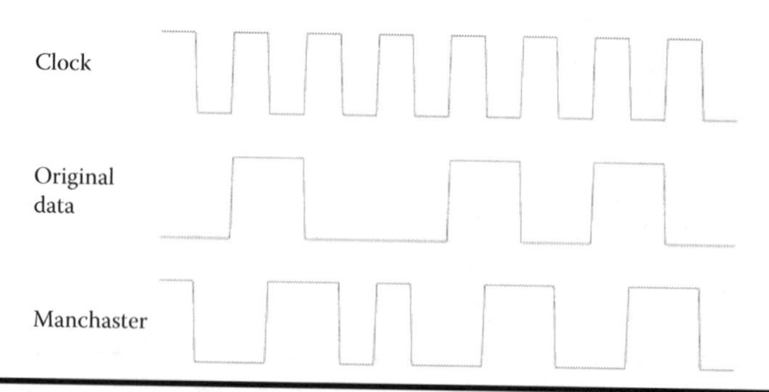

Figure 4.3 The manchester encoding.

environment. So, it is possible to transmit data passively via *Backscatter* modulation. This modulation is not only used in the *MI SWIPT* context but is also appliyed in *RFID*. The main idea behind it is to change their own current in a controlled way to cause a reflection in the other circuits of the system. The other circuits will measure the current fluctuations and then decode the message. The passive circuit can change its current by alternating between two ohmic resistances, capacitances or inductances. As the capacitance and inductance of the circuit influence directly the phase of the signal, the resistance alternation method can be preferable to allow for ASK modulation.

As the messages are encoded in the power transfer signal, the receiver of the message cannot keep constant its definitions of voltage level for each bit over time. The induced voltage can drastically vary with small changes of relative position and alignment of the coils, so the decoder settings must be adaptive. One solution is to keep the threshold value as the average between the maximum and minimum value of the last period, given a constant parameterized period. To ensure that the signal will not be kept constant for a long period and then generate an inconsistent threshold value, an encoding method like the *Manchester* must be used. The Manchester encoding applies the *exclusive or* operation between a clock signal with the double of the frequency of the data signal and the data itself at each moment, as illustrated in Figure 4.3.

It is usual to keep the voltage level high when there is no message in the channel since the signal is simultaneously being applied to collect energy. Because of that, in the period between two messages, the threshold definitions usually assume wrong values, as the device is receiving the same bit for a long time. Thus, each message must have a preamble with a high variating wave, such as a clock signal.

4.3 Analytical Modeling

The process of simulating an MI system is basically a circuit analysis. Each component of the system can be modeled as an RLC (Resistor-Inductor-Capacitor) circuit, with the coil being the inductor, as shown in Figure 4.1. So, for low frequencies, usually up to some megahertz, the calculations can be approximated by using the *Kirchhoff Laws*. Thus, the source voltage of an alone active circuit must be equals to the sum of the voltage drop across the resistor, the capacitor, and the inductor, as shown in equation (4.1). Here $V_i(t)$ is a function that describes the source voltage of circuit i over time, $I_i(t)$ a function that describes the current and R_i, L_i, and C_i are, respectively, the resistance, inductance, and capacitance of circuit i.

$$V_i(t) = R_i I_i(t) + L_i \frac{\partial I_i(t)}{\partial t} + \frac{1}{C_i} \int_0^t I_i(\tau) d\tau \quad (4.1)$$

The relationship between each system can be modeled by adding just one more inductor to the circuit, which represents the mutual inductance between the coils. This addition abstracts all issues related to the permeability of the environment, position and alignment of the two coils. The mutual inductance M_{ij} between the coils i and j can be expressed by Neumann's formula, shown in equation (4.2). The μ_0 factor corresponds to the magnetic permeability of the medium, ds is the infinitesimal slice of one coil, and $D_{ij}|$ is the absolute distance between ds_i and ds_j.

$$M_{ij} = \frac{\mu_0}{4\pi} \oint_{P_i} \oint_{P_j} \frac{ds_i \cdot ds_j}{|D_{ij}|} \quad (4.2)$$

Each RLC system interacts with all other RLC systems. Using the already mentioned abstraction of the virtual inductor and assuming $M_{ii} = L_i$, equation (4.1) can be generalized by equation (4.3), with n being the number of circuits in the system.

$$V_i(t) = R_i I_i(t) + \frac{1}{C_i} \int_0^t I_i(\tau) d\tau + \sum_{j=1}^{n} M_{ij} \frac{\partial I_j(t)}{\partial t} \quad (4.3)$$

4.3.1 Pure WPT

Assuming that all circuits resonate at the same angular frequency ω and the source voltage of all circuits are in phase and are sinusoidal, the complexity of the simulation becomes relatively simple, both in terms of mathematical analysis and in asymptotic complexity. The resonance property is reached when the source frequency is equals to the natural oscillating frequency of the circuit, defined by $\frac{1}{\sqrt{LC}}$. Under this property, the capacitive and inductive reactances cancel each other, producing a purely resistive impedance.

This simplification is convenient to represent the voltage and current variables in phasor notation, which means that the sinusoidal voltage in the time-domain form $V_i(t) = V_{0_i} \sin(\omega t)$ becomes $V_i = V_{0_i}$ and the sinusoidal current

$I_i(t) = I_{0_i} \sin(\omega t + \phi)$ becomes $I_i = I_{0_i} + \alpha\phi$. Here V_{0_i} and I_{0_i} are, respectively, the amplitudes of the voltage and the current, ϕ the phase variation between voltage and current, and $\alpha = \sqrt{-1}$.

The voltage V_i across each RLC system has a component $\hat{V}_{ij}$ generated as a function of the current induced by each other system in the environment. According to elementary circuit analysis, the component $\hat{V}_{ij}$ over the circuit i can be expressed by equation (4.4), where M_{ij} is the mutual inductance between the coils i and j and I_j is the current over the circuit j.

$$\hat{V}_{ij} = -\alpha\omega M_{ij} I_j \tag{4.4}$$

Using equation (4.3) in phasor notation and the resonant property, we can express the voltage over the active circuit i as $V_i = R_i I_i + \sum_{j \neq i} \hat{V}_{ij}$. Applying equation (4.4), we find equation (4.5). If the same is done to the passive circuit i, we find equation (4.6).

$$V_i = R_i I_i - \sum_{j \neq i} (\alpha\omega M_{ij} I_j) \tag{4.5}$$

$$0 = R_i I_i - \sum_{j \neq i} (\alpha\omega M_{ij} I_j) \tag{4.6}$$

The set of equations (4.5) or (4.6) for each circuit i became a system of linear equations. Let V_i be the source voltage of the active circuit i, I_{A_i} be the current on the active circuit i, I_{P_i} be the current on the passive circuit i, n_A be the number of active circuits, n_P be the number of passive circuits, and $n = n_A + n_P$. Putting the system on matrix notation, we get something like equation (4.7), where V, I_A, and I_P are column matrices. 0_{n_p} represents a column vector with n_p zeros.

$$\begin{bmatrix} V \\ 0_{np} \end{bmatrix} = Z \cdot \begin{bmatrix} I_A \\ I_P \end{bmatrix} \tag{4.7}$$

The Z matrix holds all impedance information and can be defined as on equation (4.8).

$$Z = \begin{bmatrix} R_1 & -\alpha\omega M_{12} & -\alpha\omega M_{13} & \cdots & -\alpha\omega M_{1n} \\ -\alpha\omega M_{21} & R_2 & -\alpha\omega M_{23} & \cdots & -\alpha\omega M_{2n} \\ \vdots & \vdots & \vdots & \ddots & \vdots \\ -\alpha\omega M_{n1} & -\alpha\omega M_{n2} & -\alpha\omega M_{n3} & \cdots & R_n \end{bmatrix} \tag{4.8}$$

Multiplying both sides of equation (4.7) by Z^{-1}, we get a formula to calculate the current of each circuit. As the power over the circuit i can be evaluated by

$P_i = R_i \cdot ||I_i||^2$, we reach a formula to evaluate the sum of the power over the active circuits [equation (4.9)] and one analog for the passive circuits [equation (4.10)]. Here R_A corresponds to the diagonal matrix that holds the resistances of the active circuits in order and R_P the one with the resistances of the passive circuits. The superscript "*" represents the *Hermitian Transpose* of the matrix, that is the transposed matrix in which each value corresponds to the complex conjugate of the corresponding value in the original matrix.

$$P_A = I_A^* \cdot R_A \cdot I_A \tag{4.9}$$

$$P_P = I_P^* \cdot R_P \cdot I_P \tag{4.10}$$

The power transfer efficiency η of all circuits is given by the ratio between the power effectively used by the devices attached to the receivers and the total spent power. In other words, $\eta = \dfrac{P_P}{P_P + P_A}$.

4.3.2 Wireless Information via MI in Active Circuits

The simplest way to transmit data through a MI channel from active circuits is using amplitude modulation. The function $V_i(t)$ that describes the source voltage of the circuit i in the time domain is formed by a sinusoidal signal with at least two well-defined amplitude values—one for the *low* binary state and one for the *high* binary state. So, the voltage function of the transmitter can be represented as in equation (4.11), where ω_i is the angular frequency of this wave, ϕ_i the initial phase, V_{i0} the value of the low-voltage level, V_{i1} the value of the high-voltage level, and $p(t)$ is the square-wave representation of the binary string that must be transmitted.

$$V_i(t) = \sin(\omega_i t + \phi_i) \cdot \left[(V_{i1} - V_{i0}) p(t) + V_{i0} \right] \tag{4.11}$$

The $p(t)$ function can be expressed as a combination of *Heaviside* functions. The *Heaviside* function $U(t-a)$ is defined as 0 for $t < a$, 0.5 for $t=a$ and 1 otherwise, as illustrated on Figure 4.4. The expression $U(t-a)-U(t-b)$ for $a, b \in \mathrm{IR}$ and $b > a$ results in a pulse starting at a and ending at b. So, the pulse corresponding to the jth bit b_j, starting transmission at time t_0 and at a baud rate of B, will be given by $b_j\left(U\left(t-\left(\tau_j - \frac{1}{2B}\right)\right) - U\left(t-\left(\tau_j + \frac{1}{2B}\right)\right)\right)$, where τ_j is the central moment of the pulse, expressed by $\tau_j = t_0 + \frac{1}{B}(j-1) + \frac{1}{2B}$.

Defining the base voltage V_{b_i} as $V_{b_i} = V_{i0} \cdot \sin(\omega_i t + \phi_i)$ and the pulse voltage $V_{P_{ij}}$ as $V_{P_{ij}} = V_{u_{ij}}^0 - V_{u_{ij}}^1$, where $V_{u_{ij}}^0$ and $V_{u_{ij}}^1$ are expressed by equations (4.12) and (4.13), the voltage function can be expressed as in equation (4.14). The decomposition of the voltage function as a sum of simpler elements is mainly important due to

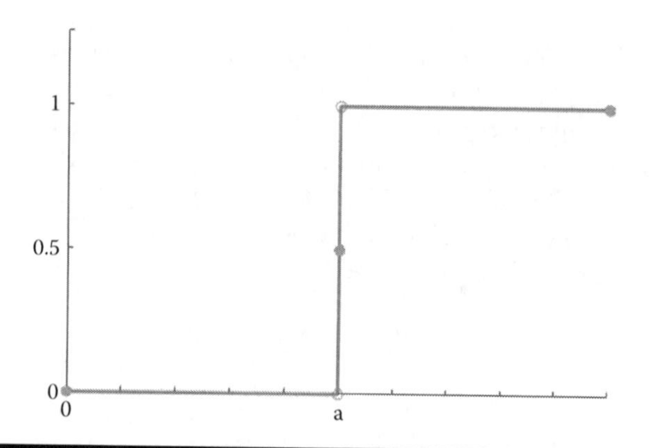

Figure 4.4 Heaviside function *U*(*t* − *a*).

precision issues related to the process of numerically determining the value of the current in each circuit, as will be discussed in detail at the end of this section.

$$V_{u_{ij}}^{0} = b_j\left(V_{i_1} - V_{i_0}\right)\sin\left(\omega_i t + \phi_i\right)U\left(t - \left(\tau_j - \frac{1}{2B}\right)\right) \tag{4.12}$$

$$V_{u_{ij}}^{1} = b_j\left(V_{i_1} - V_{i_0}\right)\sin\left(\omega_i t + \phi_i\right)U\left(t - \left(\tau_j + \frac{1}{2B}\right)\right) \tag{4.13}$$

$$V_i(t) = V_{b_i}(t) + \sum_{j|b_j=1} V_{p_{ij}}(t) = V_{b_i}(t) + \sum_{j|b_j=1} V_{u_{ij}}^{0}(t) - \sum_{j|b_j=1} V_{u_{ij}}^{1}(t) \tag{4.14}$$

Although simpler than the transmission with frequency modulation, the transmission with amplitude modulation does not admit simplifications like resonance or the phasor representation. Thus, its analysis requires that we return to equation (4.3). Without the phasor notation, the process of calculating the current across each circuit becomes a resolution of a system of second-order differential equations.

Applying the Laplace transform to both sides of equation (4.3), we get $\mathcal{L}\left[V_i(t)\right](s) = R_i\mathcal{L}\left[I_i(t)\right](s) + \frac{1}{C_i}\mathcal{L}\left[\int_0^t I_i(t)\,dt\right](s) + \sum_{j=1}^{n} M_{ij}\mathcal{L}\left[\frac{\partial I_j(t)}{\partial t}\right](s)$. Admitting $I_i(0) = 0\ \forall i$, we can reduce the expression to the general equation of each *i* circuit in the frequency domain, as shown in equation (4.15).

$$\mathcal{L}\left[V_i(t)\right](s) = R_i\mathcal{L}\left[I_i(t)\right](s) + \frac{1}{C_i s}\mathcal{L}\left[I_i(t)\right](s) + \sum_{j=1}^{n} M_{ij} s\mathcal{L}\left[I_j(t)\right](s) \tag{4.15}$$

To calculate the current of each circuit, we must first solve the system of linear equations formed by the general equations of all circuits. Using matrix notation, we get something like equation (4.16).

$$\begin{bmatrix} \mathcal{L}[V_1(t)](s) \\ \vdots \\ \mathcal{L}[V_n(t)](s) \end{bmatrix} = Z \cdot \begin{bmatrix} \mathcal{L}[I_1(t)](s) \\ \vdots \\ \mathcal{L}[I_n(t)](s) \end{bmatrix} \tag{4.16}$$

Here Z is an impedance matrix analog to the one used in the pure WPT simulation, but without considering phasors and resonance. This matrix is shown in equation (4.17) in detail.

$$Z(s) = \begin{bmatrix} R_1 + \frac{1}{C_1 s} + sM_{11} & sM_{12} & sM_{13} & \cdots & sM_{1n} \\ sM_{21} & R_2 + \frac{1}{C_2 s} + sM_{22} & \cdots & & sM_{2n} \\ \vdots & \vdots & \vdots & \ddots & \vdots \\ sM_{n1} & sM_{n2} & sM_{n3} & \cdots & R_n + \frac{1}{C_n s} + sM_{nn} \end{bmatrix} \tag{4.17}$$

To easily invert the matrix of equation (4.17) and solve the system, the decomposition of equation (4.18) can be used, where z_i is a matrix with the same dimensions of Z in witch $z_{i_{ab}} = R_a + \frac{1}{C_a s}$ for $a=b=i$ and $z_{i_{ab}} = 0$ otherwise. Since sM is easily invertible and z_i has rank 1 for all i, the technique proposed in [6] can be easily applied.

$$Z(s) = sM + \sum_{i=1}^{n} z_i \tag{4.18}$$

Multiplying both sides of equation (4.16) by $Z^{-1}(s)$ and using the definition of equation (4.14), we get equation (4.19).

$$\begin{bmatrix} \mathcal{L}[I_1(t)](s) \\ \vdots \\ \mathcal{L}[I_n(t)](s) \end{bmatrix} = Z^{-1}(s) \begin{bmatrix} \mathcal{L}[V_{b_1}(t)](s) + \sum \mathcal{L}[V^0_{u_{1j}}(t)](s) - \sum \mathcal{L}[V^1_{u_{1j}}(t)](s) \\ \vdots \\ \mathcal{L}[V_{b_n}(t)](s) + \sum \mathcal{L}[V^0_{u_{nj}}(t)](s) - \sum \mathcal{L}[V^1_{u_{nj}}(t)](s) \end{bmatrix} \tag{4.19}$$

This formula is quite hard to derive analytically in order to get a closed formula for representing the current in each circuit. In particular, the difficulty rises increasingly more when the involved matrices get larger and lies mainly in finding the inverse Laplace transform of each element, resulting from the multiplication of Z by a vector with the Laplace transform of the voltages.

The best solution when modeling a scalable system is to use a numerical algorithm to solve each inverse Laplace transform. There are many options, such as *Talbot method* [7]. The inconvenient of using them is that most of those methods have serious limitations concerning the class of functions that can be inverted or the achievable accuracy [8].

However, the decomposition of each source voltage function, as done in equation (4.19), allows the Laplace transform of each current to be expressible as an extensive sum of relatively simple expressions. Using the property of linearity of the inverse Laplace transform, the formula of equation (4.20) can be derived from equation (4.19). This formula allows the solver to compute a set of independent simple inverse Laplace transforms instead of calculating as a whole, which may even be used to parallelize the calculations.

$$
\begin{aligned}
I_i(t) = {} & \mathcal{L}^{-1}\left[Z_{i1}^{-1}\mathcal{L}\left[V_{b_1}\right]\right] + \sum \mathcal{L}^{-1}\left[Z_{i1}^{-1}\mathcal{L}\left[V_{u_{1j}}^{0}\right]\right] - \sum \mathcal{L}^{-1}\left[Z_{i1}^{-1}\mathcal{L}\left[V_{u_{1j}}^{1}\right]\right] \\
& + \mathcal{L}^{-1}\left[Z_{i2}^{-1}\mathcal{L}\left[V_{b_2}\right]\right] + \sum \mathcal{L}^{-1}\left[Z_{i2}^{-1}\mathcal{L}\left[V_{u_{2j}}^{0}\right]\right] - \sum \mathcal{L}^{-1}\left[Z_{i2}^{-1}\mathcal{L}\left[V_{u_{2j}}^{1}\right]\right] \\
& \vdots \\
& + \mathcal{L}^{-1}\left[Z_{in}^{-1}\mathcal{L}\left[V_{b_n}\right]\right] + \sum \mathcal{L}^{-1}\left[Z_{in}^{-1}\mathcal{L}\left[V_{u_{nj}}^{0}\right]\right] - \sum \mathcal{L}^{-1}\left[Z_{in}^{-1}\mathcal{L}\left[V_{u_{nj}}^{1}\right]\right]
\end{aligned}
\tag{4.20}
$$

The definitions of $\mathcal{L}\left[V_{b_i}(t)\right](s)$, $\mathcal{L}\left[V_{u_{ij}}^{0}(t)\right]$, and $\mathcal{L}\left[V_{u_{ij}}^{1}(t)\right](s)$ can be achieved after some algebraic manipulations from simple identities as $\mathcal{L}\left[U(t-a)f(t)\right](s) = e^{-as}\mathcal{L}\left[f(t+a)\right](s)$ and $\mathcal{L}\left[\sin(at+b)\right](s) = \frac{s\cdot\sin(b)+a\cdot\cos(b)}{s^2+a^2}$, as in equations (4.21)–(4.23).

$$
\mathcal{L}\left[V_{b_i}(t)\right](s) = V_{i0}\cdot\frac{s\cdot\sin(\phi_i)+\omega_i\cdot\cos(\phi_i)}{s^2+\omega_i^2} \tag{4.21}
$$

$$
\mathcal{L}\left[V_{u_{ij}}^{0}(t)\right](s) = \left(V_{i1}-V_{i0}\right)e^{-(\tau_j-1/2B)s}\cdot\left(\frac{s\sin\left(\omega_i\left(\tau_j-\frac{1}{2B}\right)+\phi_i\right)+\omega_i\sin\left(\omega_i\left(\tau_j-\frac{1}{2B}\right)+\phi_i\right)}{s^2+\omega_i^2}\right) \tag{4.22}
$$

$$\mathcal{L}\left[V_{u_{ij}}^{0}(t)\right](s)=\left(V_{i1}-V_{i0}\right)e^{-(\tau_j+1/2B)s}\cdot$$
$$\left(\frac{\sin\left(\omega_i\left(\tau_j+\frac{1}{2B}\right)+\phi_i\right)+\omega_i\sin\left(\omega_i\left(\tau_j+\frac{1}{2B}\right)+\phi_i\right)}{s^2+\omega_i^2}\right) \tag{4.23}$$

4.3.3 Wireless Information via MI in Passive Circuits

As described in Section 4.2, the transmission of data from passive circuits uses the *backscatter* modulation, where the data is modulated via ASK and sent to the active circuit by controllably changing the current over the passive circuit itself.

As discussed earlier in this section, each circuit i has a voltage drop component $\hat{V}_{ij}$ when analyzed using the Kirchhoff principles. This component is proportional to the derivative of the current over the circuit j. If the circuit j, therefore, encodes the data by changing its own current, the circuit i can decode it by measuring the small pulses caused by the voltage drop component in its own current.

Any circuit can directly change its internal current by changing its impedance. The impedance consists of the resistance and the capacitive and inductive reactances. The data can be modulated by changing any of those three parameters over time. Our analysis will consider only the modulation through resistance variation from this point to ahead, for reasons of simplification. The system will also be limited to one active circuit and one passive circuit because that is the most common situation of passive transmission. This kind of communication is only applicable to strongly coupled coils, due to its fast attenuation over distance.

The function $R(t)$ can be defined analogously to $V(t)$ in Section 4.5. Being $U(t-a)$ the *Heaviside* function that inflects at $t=a$, the jth bit in the square wave can be defined as $b_j\left(U\left(t-\left(\tau_j-\frac{1}{2B}\right)\right)-U\left(t-\left(\tau_j+\frac{1}{2B}\right)\right)\right)$, where B is the baud rate of the transmission, b_j is the value of the bit, and τ_j is the central moment of the pulse, expressed by $\tau_j=t_0+\frac{1}{B}(j-1)+\frac{1}{2B}$. So, the $R(t)$ as a whole can be expressed as in equation (4.24). Here R_0 is the resistance assumed by the circuit when idle or sending low bits and R_0 is the resistance when sending high bits.

$$R(t)=R_0+(R_1-R_0)\sum_{j|b_j=1}\left(U\left(t-\left(\tau_j-\frac{1}{2B}\right)\right)-U\left(t-\left(\tau_j+\frac{1}{2B}\right)\right)\right) \tag{4.24}$$

As the voltage drop due resistance will keep being $R(t)I_i(t)$, the system of second-order differential equations of equation (4.25) can be derived from equation (4.3).

Here, both circuits are assumed to have the same capacitance C, initial resistance R_0, and self-inductance L.

$$
\begin{aligned}
V(t) &= R_0 I_1(t) + \frac{1}{C}\int_0^t I_1(\tau)\,d\tau + L\frac{\partial I_1(t)}{\partial t} + M\frac{\partial I_2(t)}{\partial t} \\
0 &= R(t) I_2(t) + \frac{1}{C}\int_0^t I_2(\tau)\,d\tau + L\frac{\partial I_2(t)}{\partial t} + M\frac{\partial I_1(t)}{\partial t}
\end{aligned}
\tag{4.25}
$$

The analytical resolution of this system of ordinary differential equations is not always achievable, so the simplest solution is to use a numerical method. The *Runge-Kutta method* [9], for example, is an iterative algorithm applicable to a great variety of families of ODE's and much more efficient than most of the other methods, such as Euler's.

To be easily solvable by the *Runge-Kutta method*, the system must be constituted by a set of equations with a first-order differential in each left-hand side and a non-differential function in each right-hand side. Inserting the artificial variables Q_1 and Q_2 and doing some algebraic manipulations, equation (4.25) becomes equation (4.26). Here the parameters of the functions were omitted to facilitate the visualization of the expressions, and the notation f' was used to denote the first derivative of some function $f(t)$.

$$
\begin{aligned}
Q_1' &= I_1 \\
Q_2' &= I_2 \\
I_1' &= \frac{RM}{L^2 - M^2} I_2 + \frac{M}{C\left(L^2 - M^2\right)} Q_2 + \frac{L}{L^2 - M^2} V - \frac{LR_0}{L^2 - M^2} I_1 - \frac{L}{C\left(L^2 - M^2\right)} Q_1 \\
I_2' &= \frac{R_0 M}{L^2 - M^2} I_1 + \frac{M}{C\left(L^2 - M^2\right)} Q_1 + \frac{LR}{L^2 - M^2} I_2 - \frac{L}{C\left(L^2 - M^2\right)} Q_2 - \frac{M}{L^2 - M^2} V
\end{aligned}
\tag{4.26}
$$

4.4 Simulation

This section deals with the results of WPT and MI SWIPT simulations for a typical system and serves as an illustration for the algorithms discussed in the previous sections. The algorithms were implemented in MATLAB (MATrix LABoratory) environment and tested comparatively when possible.

Figure 4.5 shows the results of the application of the Neumann's formula in a pair of identical solenoid coils with 12.25 cm of radius, eight turns, and 0.3 cm of

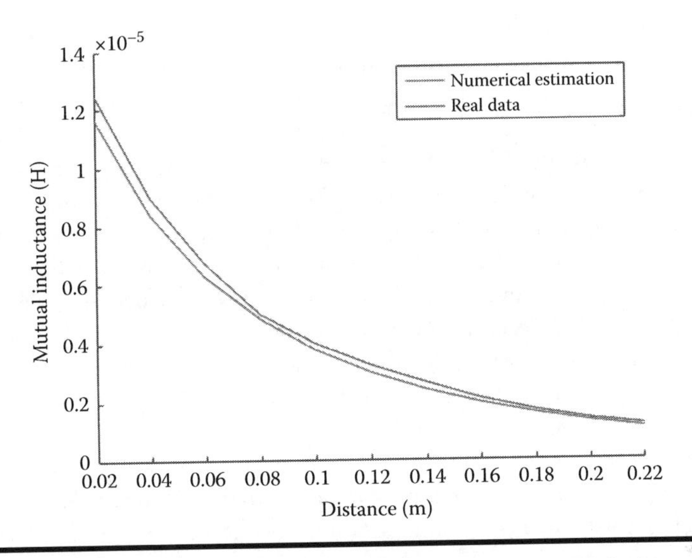

Figure 4.5 Comparison between real measures of mutual inductance and estimates obtained with the formula of Neumann.

pitch between the spires. The coils were kept coaxial while the spacing between them was increasing. The results were compared with real data and had between 5% and 7.5% of error.

Figure 4.6 shows the comparative results between the *Phasor* and the *Laplace Transform* formulas to compute the power over the same two coils described for Figure 4.5. The circuits were kept resonating at 100 kHz and with an inner resistance

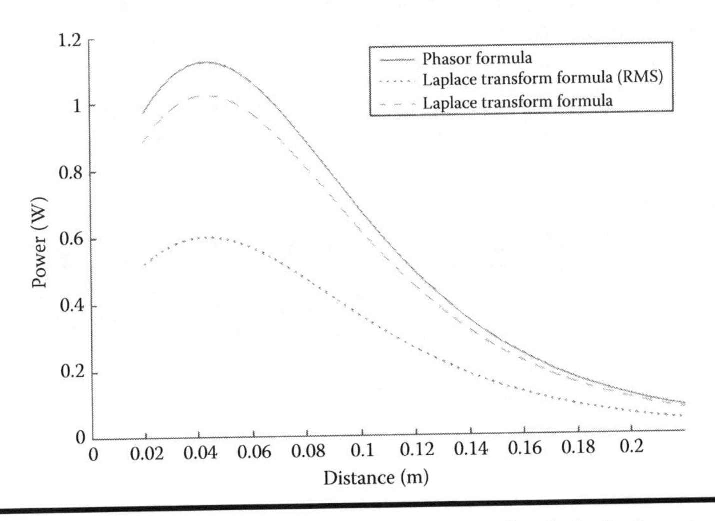

Figure 4.6 Comparison between the outcomes obtained via laplace transform and phasor-based formulas.

of 50 Ω. The active circuit was subjected to a voltage of 15 V. Considering only the base voltage in the Laplace transform-based formula, it only describes WPT and must, therefore, generate results consistent with the *Phasor-based formula*. Since the *Laplace transform-based formula* returns a set of points of the current in each circuit over time and the *Phasor-based* returns a scalar value, two metrics were considered for the comparison of the results. The RMS (Root Mean Square) current was calculated by the mean of the squares of all values of current and represents the most accurate method of calculating the power dissipation when compared with the direct current model. The absolute current was calculated by the maximal absolute value between each two sequential roots of the function that defines the current. In particular, the absolute current is close to the real value of the amplitude of the sinusoidal wave. The power was calculated by multiplying the square of the average current by the inner resistance.

The next set of images illustrates the process of simulating data transmission from the active circuit to the passive one at the same conditions described in Figure 4.6, but with a constant mutual inductance of 10 μH. Figure 4.7 shows the current over the passive circuit at the moment of data transmission. The active circuit sent a clock signal starting at 40 μs after the initialization of the system. Using ASK modulation, the controller of the active circuit excited its coil with 15 V when idle or sending the low bits and with 10 V when sending the high bits. Figure 4.8 shows the details of a single pulse carried by the sinusoidal signal.

Figure 4.9 shows the absolute amplitude of the signal of Figure 4.7. The decoding of the signal considered an adaptive amplitude threshold. Given a window of time, the threshold was calculated by the mean between the maximal and minimal amplitude values of the period. The threshold was then applied to the signal of Figure 4.9, obtaining the square wave of Figure 4.10.

The decoding process of the signal of Figure 4.9 still considered an adaptive baud rate estimate, rather than a fixed rate. For this, the same time window considered for threshold maintenance was used. The baud rate was approximated by the inverse of the longest period inside the window in which there was no change of

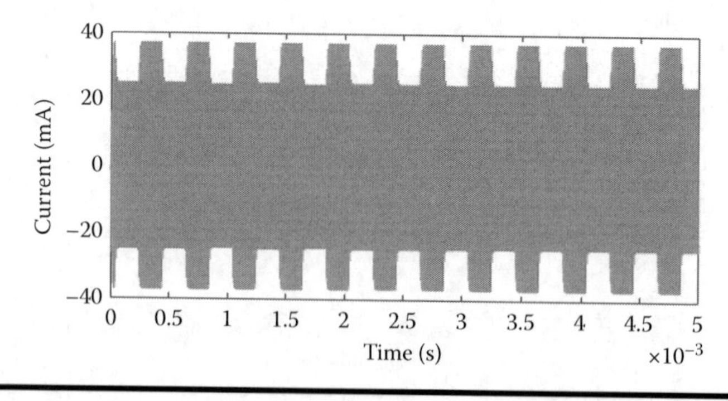

Figure 4.7 Current over the passive circuit while receiving a clock signal from the active one.

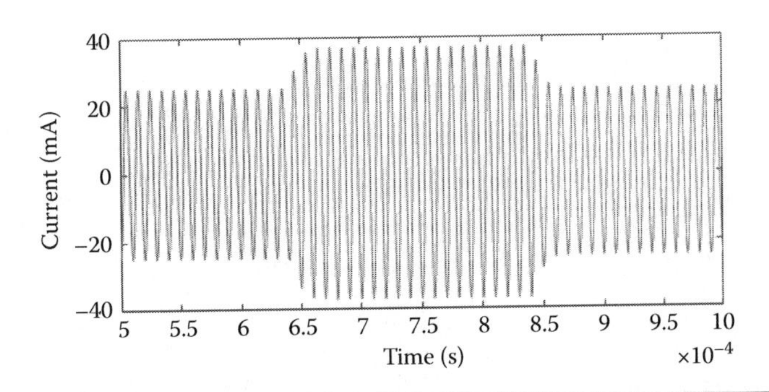

Figure 4.8 A single current pulse received by the passive circuit in detail.

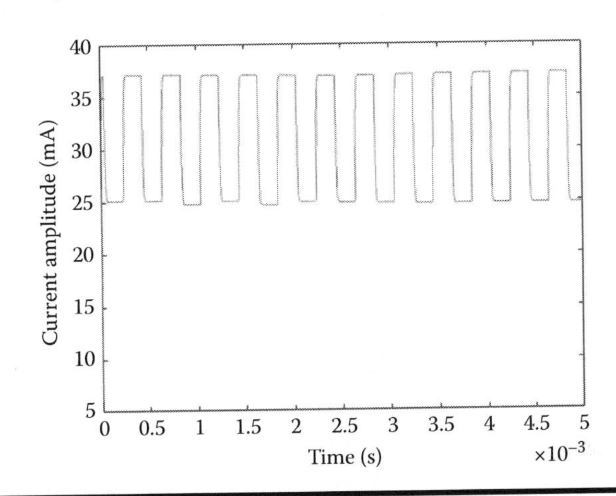

Figure 4.9 Absolute amplitude of the sinusoidal wave of Figure 4.7.

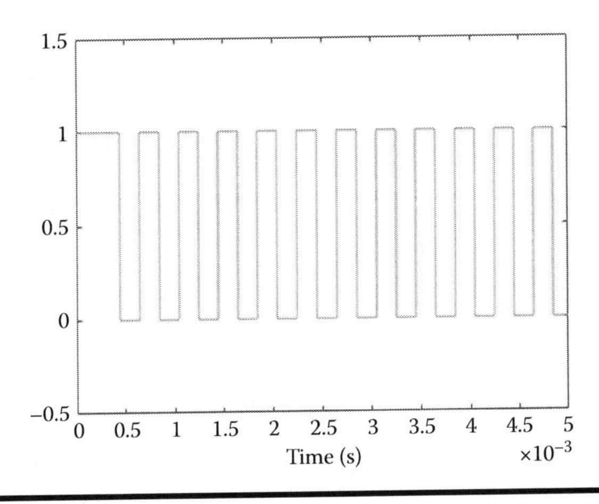

Figure 4.10 Square-wave signal obtained from the signal of Figure 4.9.

state in the square wave. In order to avoid that signals with low variability degenerate the baud rate estimate, it is calculated by the weighted mean between the last estimate and the current sample value. The progression of the baud rate estimate is shown in the chart of Figure 4.11.

Figure 4.12 compares the Laplace transform-based formula with the *Runge-Kutta solver-based algorithm*. In this simulation, capacitance, inductance, and resistance conditions were maintained the same as those described for the last charts. The

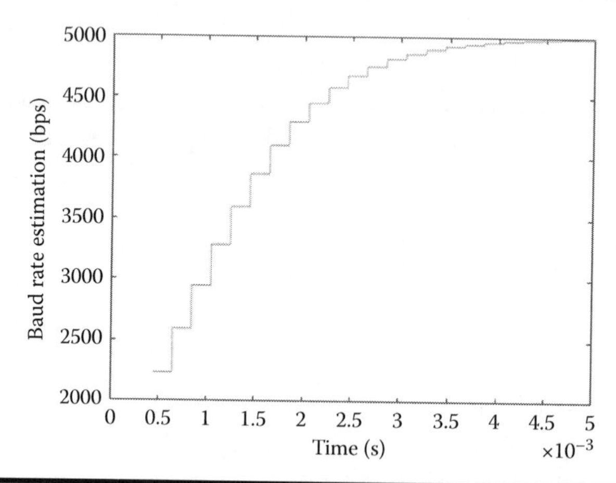

Figure 4.11 Progression of the baud rate estimate over time. The value stabilizes at 5 kbps, which matches the real rate used by the transmitter.

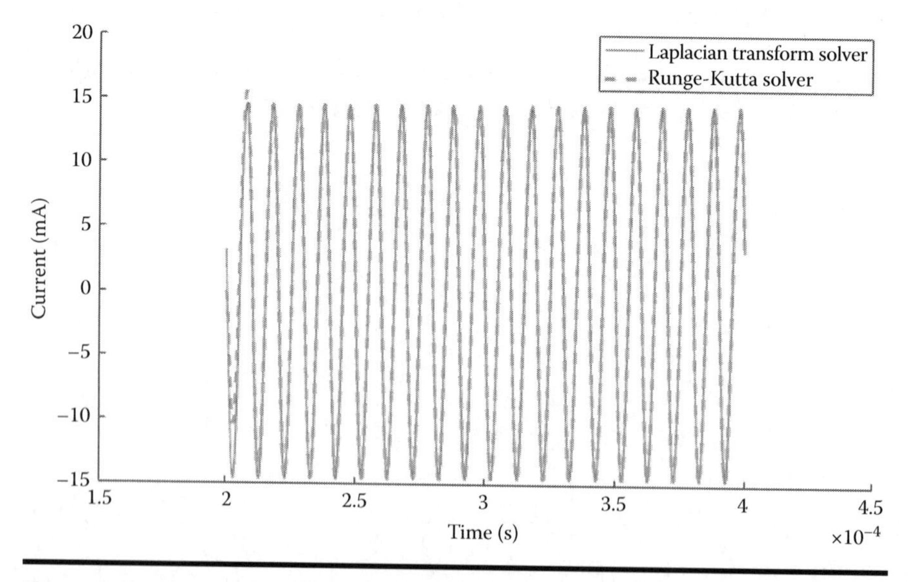

Figure 4.12 Current readings over the passive circuit while receiving power via WPT.

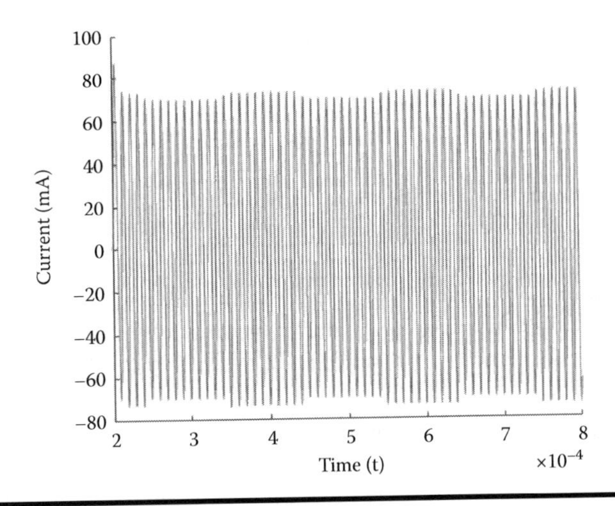

Figure 4.13 Current readings over the active circuit while receiving a clock signal via passive SWIPT.

active circuit was excited with 15 V, generating a pure WPT signal. The chart shows the readings of current on the passive circuit over time, in which the similarity of the results provided by the two methods is evident.

Finally, the last two images deal with the transmission of a clock signal passively. For this simulation, the same conditions as before were considered, except for the coils' self-inductance and the mutual inductance between them, which were changed to 1 mH and 100 μH, respectively. Figure 4.13 shows the current readings over the active circuit—the receiver of the message—and Figure 4.14 shows the

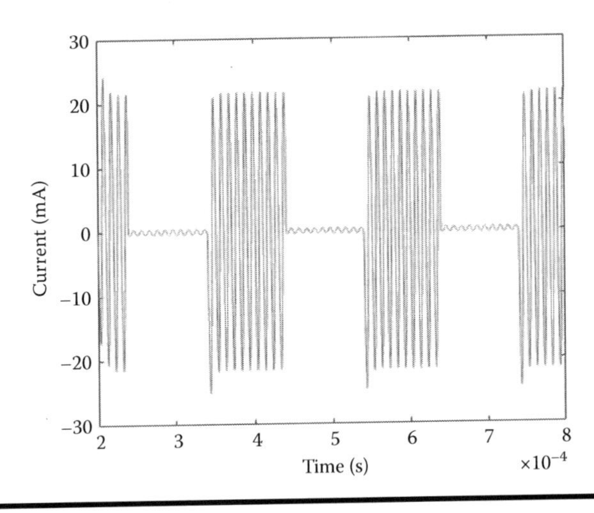

Figure 4.14 Current readings over the passive circuit while sending a clock signal via passive SWIPT.

measurements over the passive circuit—the transmitter of the message. The *Runge-Kutta solver-based algorithm* was employed, and the data were encoded as variations of the inner resistance in the passive circuit. It is evident that the passive transmission is much more strongly coupled coil-dependent than the active one, mainly due the square attenuation that the signal suffers—first from the active circuit to the passive circuit as a WPT signal and then in the way back as a SWIPT signal.

4.5 Conclusion

This chapter aimed to describe the operation of the wireless energy transmission via MI, with emphasis on the simultaneous transmission of modulated data via amplitude. Three algorithms were described, implemented, and compared to highlight their accuracy and physical meaning.

The *Phasor-based* algorithm stands out for its simplicity of implementation but is limited to the transfer of energy in resonant systems. The *Laplace transform-based* allows data transmission in an active way and is highly parallelizable but does not have direct support for passive communication. The *Runge-Kutta solver-based*, in turn, is complete but more complex than the others, also allowing passive communication.

References

1. Haipeng Dai, Yunhuai Liu, Guihai Chen, Xiaobing Wu, and Tian He. Safe charging for wireless power transfer. *INFOCOM, 2014 Proceedings IEEE*, pages 1105–1113. IEEE, 2014.
2. Sotiris Nikoletseas, Yuanyuan Yang, and Apostolos Georgiadis. *Wireless Power Transfer Algorithms, Technologies and Applications in Ad Hoc Communication Networks*. Springer, 2016.
3. Alexander D. Sousa, Luiz FM Vieira, and Marcos AM Vieira. Modeling, analysis and simulation of wireless power transfer. *Proceedings of the 6th ACM Symposium on Development and Analysis of Intelligent Vehicular Networks and Applications*, 2017.
4. Wireless Power Consortium. The Qi wireless power transfer system: Power class 0 specification, 2016. [online] https://www.wirelesspowerconsortium.com/downloads/download-wireless-power-specification.html.
5. Fuqin Xiong. Amplitude shift keying. *Encyclopedia of RF and Microwave Engineering*, 2005.
6. Kenneth S Miller. On the inverse of the sum of matrices. *Mathematics Magazine*, 54(2):67–72, 1981.
7. Alan Talbot. The accurate numerical inversion of Laplace transforms. *IMA Journal of Applied Mathematics*, 23(1):97–120, 1979.
8. Juraj Valsa and Lubomír Brančik. Approximate formulae for numerical inversion of Laplace transforms. *International Journal of Numerical Modelling: Electronic Networks, Devices and Fields*, 11(3):153–166, 1998.
9. John R Dormand and Peter J Prince. A family of embedded Runge-Kutta formulae. *Journal of Computational and Applied Mathematics*, 6(1):19–26, 1980.

Chapter 5

Capacity Maximization of Magnetic Communication

Kisong Lee
Chungbuk National University

Dong-Ho Cho
Korea Advanced Institute of Science and Technology

Contents

Magnetic communication (MC) is a promising alternative technology for transferring information under dense media, including underground, water, and soil, which establishes a wireless communication link through magnetic induction between transmitter (Tx) coil antenna and receiver (Rx) coil antenna. In contrast with traditional wireless communication using electromagnetic (EM) waves, the propagation loss of magnetic fields owing to absorption is relatively small in MC because the magnetic permeability of dense media is similar to that of air. Therefore, MC enables a reliable communication for short transmission distance in dense media. However, as the transmission distance (d) increases, the magnetic fields experience a significant attenuation in inverse proportion to d. For example, when predominantly magnetic fields with wavelength λ are generated by a coil antenna, the power of magnetic fields falls seriously in proportion to $1/d^6$ while that of electric fields drops in proportion to $1/d^4$ in the near field where $d \leq \frac{\lambda}{2\pi}$. In addition, the power of EM waves decreases in proportion to $1/d^2$ in the far field where $d > \frac{\lambda}{2\pi}$ [1]. Such a severe attenuation of signal strength for magnetic fields restricts the usage of MC to only short-range communications. For a wide use of MC, it is important to ensure reliable communication and range extension by maximizing the capacity of MC in strongly and loosely coupled regions.

In this chapter*, we reveal major reasons behind degrading the capacity of MC and introduce methods for solving the problems in each region. In the strongly coupled region, we investigate the effect of frequency splitting on the capacity, and suggest a frequency tracking scheme and an impedance matching scheme for improving the capacity of MC. In the loosely coupled region, we evaluate the relationship between the quality factor of coil antennas and 3 dB bandwidth, and the effect of those factors on the capacity in MC based on two-coil systems. Moreover, we extend this analysis to MC relay systems, where relay coils are deployed between Tx and Rx coils to increase communication distance. We also show that there is an optimal quality factor to maximize the capacity for both MC systems. Finally, we introduce the multiple-antenna based MC technologies, including magnetic multiple-input and multiple-output (MagMIMO) and MultiSpot, and their possibility, which is expected to improve a degree of freedom for the position of Rxs with enhancement of the capacity for MC.

5.1 System Model

As shown in Figure 5.1, we consider simple MC systems with a Tx resonator and a Rx resonator. The Tx and Rx are aligned on a single axis, and d is the distance between the two resonators. The Tx is connected to an external alternating voltage, V_s, and a source resistor, R_s. The Rx is connected to a load resistor, R_L. The self-inductances of the Tx and Rx can be denoted as L_0 and L_1, respectively. In addition, C_0 and C_1 are the capacitances that make two resonators resonate at the following same frequency.

$$\omega_o = 2\pi f_o = \frac{1}{\sqrt{L_0 C_0}} = \frac{1}{\sqrt{L_1 C_1}}. \tag{5.1}$$

* This chapter is written based on [2].

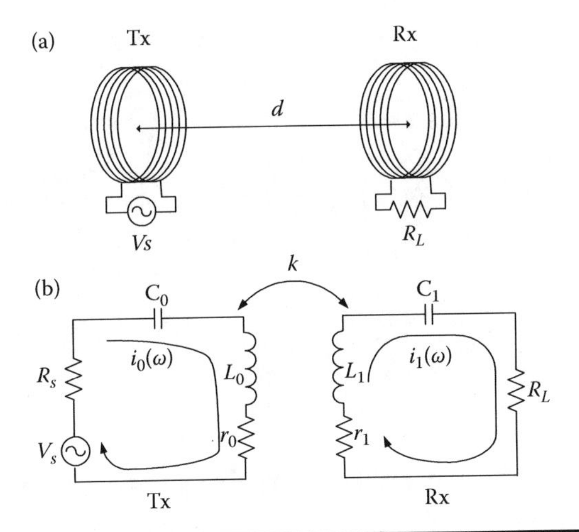

Figure 5.1 Model for MC systems. (a) MC systems. (b) Equivalent circuit model.

In MC systems, the communication link is formed by mutual induction between two resonators. For example, the voltage source V_s generates a sinusoidal current $i_0(\omega)$ in the Tx. Then, the magnetic fields created by $i_0(\omega)$ penetrate into Rx and induce an alternating current $i_1(\omega)$. This magnetic induction between two resonators can be represented by mutual inductance M [3].

$$M = \frac{\pi \mu_0 N_0 N_1 a_0^2 a_1^2 \left|\cos\theta_{01}\right|}{2(a_0^2 + d^2)^{3/2}}, \tag{5.2}$$

where μ_0 is the permeability of free space, N_0 and N_1 are the number of turns for Tx and Rx resonators, respectively, and, θ_{01} is the angle between the axes of Tx and Rx resonators, which pass through the centers of the resonators. Finally, the two resonators are magnetically linked with the following coupling coefficient k, which makes MC possible.

$$k = \frac{M}{\sqrt{L_0 L_1}}, \tag{5.3}$$

where M is a mutual inductance. Then, we can build the following equations from the Kirchhoff's voltage law (KVL).

$$\begin{aligned} V_S &= \left(R_S + r_0 + j\omega L_0 + \frac{1}{j\omega C_0} \right) i_0(\omega) - j\omega k \sqrt{L_0 L_1}\, i_1(\omega), \\ 0 &= -j\omega k \sqrt{L_0 L_1}\, i_0(\omega) + \left(R_L + r_1 + j\omega L_1 + \frac{1}{j\omega C_1} \right) i_1(\omega). \end{aligned} \tag{5.4}$$

From (5.4), the currents generated in Tx and Rx can be expressed as follows.

$$i_0(\omega)=\frac{V_s}{\left(R_S+r_0+j\omega L_0+\frac{1}{j\omega C_0}\right)+\left(\frac{\omega^2k^2L_0L_1}{R_L+r_1+j\omega L_1+\frac{1}{j\omega C_1}}\right)},$$
$$i_1(\omega)=\frac{j\omega k\sqrt{L_0L_1}}{R_L+r_1+j\omega L_1+\frac{1}{j\omega C_1}}\cdot i_0(\omega). \tag{5.5}$$

5.2 Capacity Maximization in Strongly Coupled Region

If the following condition, $\frac{k^2\omega_o^2L_0L_1}{r_0r_1}>1$, is satisfied, it is called that the two resonators are strongly coupled [4]. In a strongly coupled region, the effect of coupling by Rx is shown in Tx, so the term $\left(Z_{in}=\frac{\omega^2k^2L_0L_1}{R_L+r_1+j\omega L_1+\frac{1}{j\omega C_1}}\right)$ in the denominator of (5.5) should be considered. In this region, if k increases beyond a threshold (k_s), strong coupling between two resonators causes frequency splitting [5,6]. The frequency splitting is the phenomenon that the peak of transferred power is observed at two different frequencies, rather than the resonant frequency, when two resonators are strongly coupled. As a result, the received power at the resonant frequency is seriously deteriorated as k increases. This is a major cause to decrease the capacity of MC in a strongly coupled region; therefore, we will deal with the methods to overcome this problem.

5.2.1 Frequency Tracking

When frequency splitting occurs, the degradation of capacity can be compensated if the Tx uses an optimal frequency where the maximum received power occurs instead of the original resonant frequency. Therefore, we introduce a frequency tracking scheme, which finds the optimal frequency and utilizes that as a center frequency for data transmission.

To manipulate (5.5), we define the following function for Tx, $f_0(\omega)=j\omega L_0+\frac{1}{j\omega C_0}$. Then, the first derivation of $f(\omega)$ can be obtained as $f_0'(\omega)=j\left(L_0+\frac{1}{\omega^2C_0}\right)$. Using the first-order Taylor series expansion, $f_0(\omega)$ near the resonant frequency ω_0 can be approximated as

$$\begin{aligned} f_0(\omega) &= f_0(\omega_o) + f_0'(\omega_o)(\omega - \omega_o) \\ &= 0 + j\left(L_0 + \frac{1}{\omega_0^2 C_0}\right)(\omega - \omega_o) \\ &= j2\left(\frac{\omega - \omega_o}{\omega_o}\right)\left(\frac{\omega_o L_0}{R_S + r_0}\right)(R_s + r_0) \\ &= j2(\Delta\omega)(Q_0)(R_S + r_0), \end{aligned} \tag{5.6}$$

where the quality factor, Q_0, indicates the strength of mutual coupling near the resonant frequency. For example, as the quality factor is large, the coupling between resonators becomes strong at the same distance. From the similar approach, we can obtain $f_1(\omega) = j2(\Delta\omega)(Q_1)(R_L+r_1)$ for Rx. Then, we can translate (5.5) into the following equation.

$$\begin{aligned} i_0(\omega) &= \frac{V_S}{(R_S + r_0)(1 + j2\Delta\omega Q_0) + \left(\frac{\omega^2 k^2 L_0 L_1}{(R_L + r_1)(1 + j2\Delta\omega Q_1)}\right)}, \\ i_1(\omega) &= \frac{j\omega k\sqrt{L_0 L_1}}{(R_L + r_1)(1 + j2\Delta\omega Q_1)} \cdot i_0(\omega). \end{aligned} \tag{5.7}$$

The consumed power at the load resistor of the Rx can be expressed as $P_L(\omega) = |i_1(\omega)|^2 R_L$. Then, using (5.7), we can represent $P_L(\omega)$ as follows.

$$P_L(\omega) = \frac{P_S Q_0 Q_1 \eta_0 \eta_1 k^2}{\left(\sqrt{(1 + (2\Delta\omega)^2 Q_0^2)(1 + (2\Delta\omega)^2 Q_1^2)} + Q_0 Q_1 k^2\right)^2}, \tag{5.8}$$

where $P_S = \frac{V_S^2}{R_S}$, $\eta_0 = \frac{R_S}{R_S + r_0}$, and $\eta_1 = \frac{R_L}{R_L + r_1}$. Here, P_S is the available transmission power, and η_0 and η_1 are the circuit efficiencies of the Tx and Rx, respectively. In addition, $\Delta\omega$ becomes zero at resonant frequency ω_o, so (5.8) can be simplified as

$$P_L(\omega_o) = \frac{P_S Q_0 Q_1 \eta_0 \eta_1 k^2}{(1 + Q_0 Q_1 k^2)^2}. \tag{5.9}$$

From (5.9), we can also find the equivalent S_{21} parameter [7,8].

$$\begin{aligned} S_{21}(\omega_o) &= 2\frac{V_L}{V_S}\sqrt{\frac{R_S}{R_L}} \\ &= 2\sqrt{\frac{Q_0 Q_1 \eta_0 \eta_1 k^2}{(1 + Q_0 Q_1 k^2)^2}}. \end{aligned} \tag{5.10}$$

Here, S_{21} indicates the ratio of the output voltage at the load resistor, V_L, and the input voltage at the source resistor, V_S.

We also denote the coupling coefficient just before frequency splitting occurs as a splitting coupling point, which is represented by k_s. Taking the derivative of (5.10) with respect to k, we can find k_s from the following conditions: $\frac{\partial S_{21}(\omega_o)}{\partial k} = 0$ and $\frac{\partial^2 S_{21}(\omega_o)}{\partial k^2} < 0$. Then, k_s can be obtained as follows.

$$k_s = \frac{1}{\sqrt{Q_0 Q_1}}. \tag{5.11}$$

When $k > k_s$, the peaks of S_{21} are observed at the following split frequencies [9].

$$\omega_{\pm} = \frac{\omega_o}{\sqrt{1+k}}, \frac{\omega_o}{\sqrt{1-k}} \tag{5.12}$$

Figure 5.2 shows the phenomenon of frequency splitting. Here, we used the parameters described in Table 5.1 to obtain the results. When $k < k_s$, $S_{21}|$ at the resonant frequency increases as k increases. However, the peaks of $S_{21}|$ appear at split frequencies, i.e., approximately 9.5 and 10.5 MHz, when $k > k_s$, while $S_{21}|$ at the resonant frequency, i.e., 10 MHz, decreases with increasing k. This means that both $S_{21}|$ and P_L can be degraded severely at the resonant frequency as the Tx is close to the Rx in a strongly coupled region where $k > k_s$.

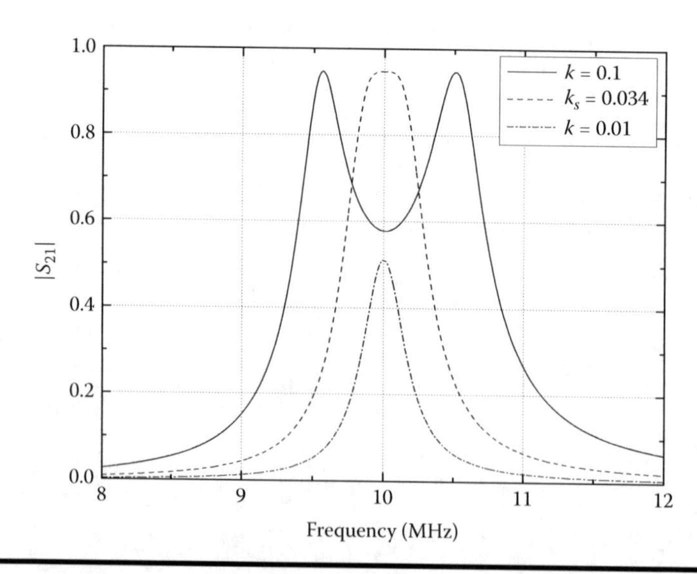

Figure 5.2 Frequency splitting.

Table 5.1 Parameters for Evaluation

Parameters	*Values*
L_0, L_1 (uH)	4.9
C_0, C_1 (pF)	51.3
R_S, R_L (Ω)	10
r_0, r_1 (Ω)	0.55
P_S (W)	1
f_o (MHz)	10
N_0 (dBm)	–103

When $k > k_s$, we can use a frequency tracking scheme to maximize the capacity of MC [10]. Figure 5.3 shows the flow chart for describing the procedures of the frequency tracking scheme. At first, the Tx transmits a signal using ω_o as a center frequency, then the Rx measures k from the received signal power and sends the measured k to the Tx. The Tx determines whether the frequency splitting happens or not by comparing k with k_s. If $k < k_s$, the Tx utilizes the original resonant frequency, ω_o. Otherwise, comparing $|S_{21}|$ at $\omega_+ = \frac{\omega_o}{\sqrt{1+k}}$ and $|S_{21}|$ at $\omega_- = \frac{\omega_o}{\sqrt{1-k}}$, the Tx finds the optimal frequency ω_{max} between ω_+ and ω_-, where the maximum value of $|S_{21}|$ appears. After the Tx sends the information of ω_{max} to the Rx and receives the acknowledgement from the Rx, it uses ω_{max} as a center frequency for communication.

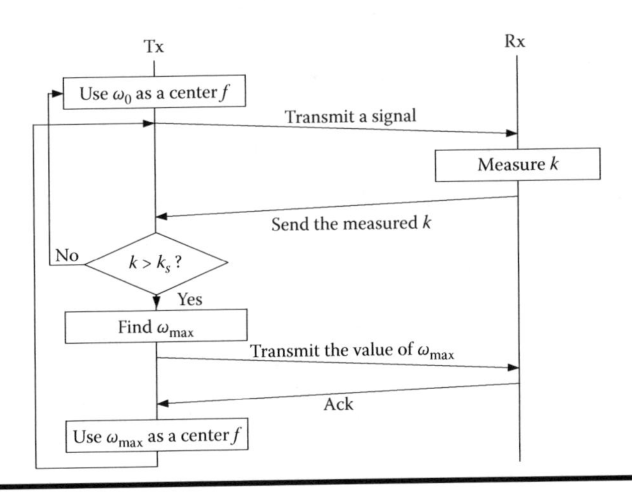

Figure 5.3 Frequency tracking scheme.

5.2.2 Impedance Matching

At each distance, if we find an optimal load resistance and adjusts R_L to its optimal value, we can achieve the maximum capacity as well as prevent the occurrence of frequency splitting. We call this process as an impedance matching.

To find the optimal load resistance, we translate $S_{21}(\omega_o)$ in (5.10) into

$$S_{21}(\omega_o) = 2\sqrt{\frac{k^2\omega_o^2 L_0 L_1 R_S R_L}{((R_S + r_0)(R_L + r_1) + k^2\omega_o^2 L_0 L_1)^2}}. \tag{5.13}$$

Taking the derivative of (5.13) with respect to R_L, we can find the optimal load resistance at a given k ($R_{L,\text{opt}}$) from the following conditions: $\frac{\partial S_{21}(\omega_o)}{\partial R_L} = 0$ and $\frac{\partial^2 S_{21}(\omega_o)}{\partial R_L^2} < 0$. Then, $R_{L,\text{opt}}$ can be obtained as follows.

$$R_{L,\text{opt}} = r_1 + \frac{k^2\omega_o^2 L_0 L_1}{R_S + r_0}. \tag{5.14}$$

If the Rx measures the coupling coefficient from the signal power transmitted by the Tx and sets the value of load resistance to $R_{L,\text{opt}}$ based on a measured k, the capacity of MC can be significantly improved even when $k > k_s$.

Figure 5.4 shows $|S_{21}|$ against the coupling coefficient k, which provides the performance comparison for three schemes, such as frequency tracking, impedance matching, and conventional scheme. In the conventional scheme, the original

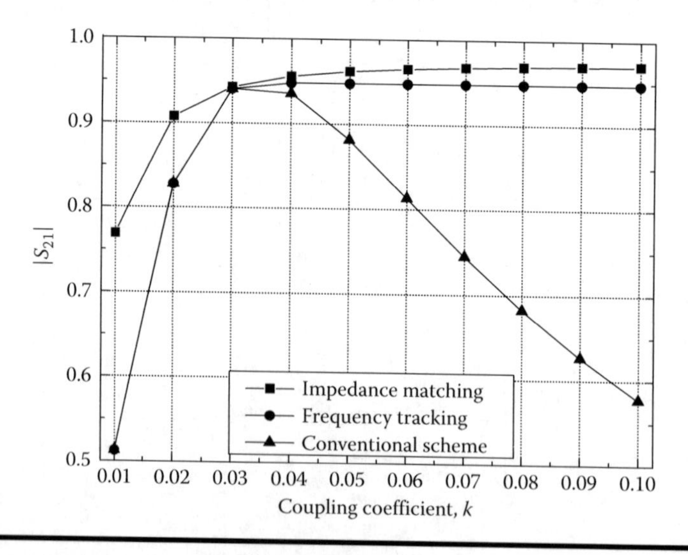

Figure 5.4 $|S21|$ **versus coupling coefficient** k.

resonant frequency is used for transmission, even though frequency splitting occurs. As a result, $|S_{21}|$ is degraded severely when $k \geq 0.03$. In contrast with the conventional scheme, the frequency tracking scheme can achieve large $|S_{21}|$ by tracking and utilizing the optimal frequency $\omega_{\max}$ when $k \geq 0.03$. In short, as the MC systems use $\omega_{\max}$ as a center frequency for data transmission instead of the original resonant frequency, its capacity can be improved significantly in a strongly coupled region. In addition, the impedance matching scheme can achieve the maximum $|S_{21}|$ for all k regions, which provides the upper bound of performance for MC.

5.3 Capacity Maximization in Loosely Coupled Region

In a loosely coupled region where the following condition is satisfied, $\frac{k^2\omega_o^2 L_0 L_1}{r_0 r_1} \leq 1$, the frequency splitting does not appear since $k \ll k_s$. However, as the distance between resonators increases, the capacity of MC decreases because the coupling strength becomes weak. In addition, the effect of coupling by Rx, that is seen in the Tx, can be neglected, because two resonators are sufficiently separated. In the loosely coupled region, it is important to increase the possible range of communication, so we will consider MC relay systems as well as simple MC systems with only Tx and Rx.

5.3.1 MC Systems

When $k \ll k_s$, we can simplify $i_0(\omega)$ and $i_1(\omega)$ near the resonant frequency by neglecting Z_{in} seen in the Tx as follows.

$$\begin{aligned} i_0(\omega) &= \frac{V_S}{(R_S + r_0)(1 + j2\Delta\omega Q_0)}, \\ i_1(\omega) &= \frac{j\omega k\sqrt{L_0 L_1}}{(R_L + r_1)(1 + j2\Delta\omega Q_1)} \cdot i_0(\omega). \end{aligned} \tag{5.15}$$

In the context of MC systems, $i_1(\omega)$ can be represented as $|i_1(\omega)|^2 = |h|^2 \cdot |i_0(\omega)|^2$, where $|h|^2$ is the channel gain of magnetic inductive link that $i_0(\omega)$ experiences during propagation. Thus, $|h|^2$ can be expressed as follows.

$$\begin{aligned} |h|^2 &= \left| \frac{j\omega k\sqrt{L_0 L_1}}{(R_L + r_1)(1 + j2\Delta\omega Q_1)} \right|^2 \\ &= \frac{R_S + r_0}{R_L + r_1} \cdot \frac{k^2 Q_0 Q_1}{(1 + (2\Delta\omega)^2 Q_1^2)}. \end{aligned} \tag{5.16}$$

Here, coupling coefficient, k, can be approximated as a function of the distance between the resonators, d [11].

$$k(d) = \frac{a_0^2 a_1^2}{\sqrt{a_0 a_1}\left(\sqrt{d^2 + a_0^2}\right)^3}, \tag{5.17}$$

where a_0 and a_1 are the radii of the Tx and Rx, respectively. Then, the channel gain $|h|^2$ can be represented by

$$|h|^2 = \frac{R_S + r_0}{R_L + r_1} \cdot \frac{a_0^3 a_1^3 Q_0 Q_1}{(d^2 + a_0^2)^3 \left(1 + (2\Delta\omega)^2 Q_1^2\right)}. \tag{5.18}$$

From (5.18), we can know that $|h|^2$ is proportional to $1/d^6$, which indicates that the path loss exponent is 6 [12]. In addition, $\Delta\omega$ goes to 0 as ω approaches to ω_o, as a result, $|h|^2$ has a large value near the resonant frequency.

Then, the received power $P_L(\omega)$ can be obtained as follows.

$$\begin{aligned} P_L(\omega) &= |i_0(\omega)|^2 |h|^2 R_L \\ &= \frac{P_S Q_0 Q_1 \eta_0 \eta_1 k^2}{\left(1 + (2\Delta\omega)^2 Q_0^2\right)\left(1 + (2\Delta\omega)^2 Q_1^2\right)}. \end{aligned} \tag{5.19}$$

When $\omega = \omega_o$, the received power is also simplified as

$$P_L(\omega_o) = P_S Q_0 Q_1 \eta_0 \eta_1 k^2. \tag{5.20}$$

Here, we can find that $P_L(\omega_o)$ is proportional to $Q_0 Q_1$ as well as k^2 [13]. Large quality factor and coupling coefficient lead to a strong coupling near the resonant frequency, as a result, the received power increases.

From the observation that the received power can be maximized when $Q = Q_0 = Q_1$ [14], we consider identical Tx and Rx. Also, the 3-dB bandwidth is used as the communication bandwidth B for MC systems. Here, the 3-dB bandwidth is defined as a range of frequencies where the spectral power density of signal is larger than the half of its maximum value. According to the definition of 3-dB bandwidth, we can formulate the following relation, $P_L\left(\omega_o - \frac{B}{2}\right) : P_L(\omega_o) = \frac{1}{2} : 1$; finally, B can be derived as follows [11].

$$B = \sqrt{\sqrt{2} - 1} \cdot \frac{f_o}{Q}. \tag{5.21}$$

From (5.21), we can see that B is inversely proportional to Q.

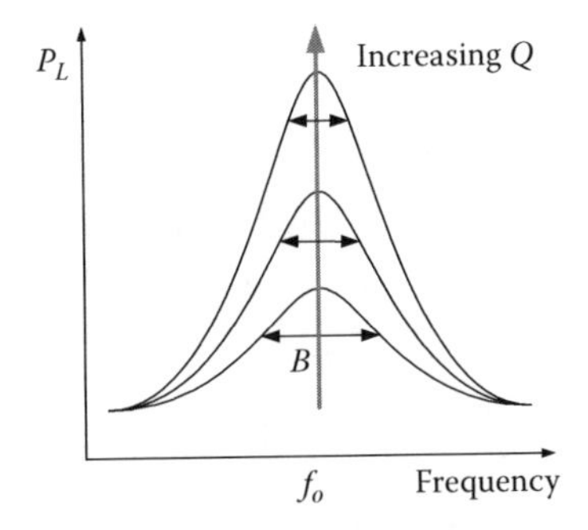

Figure 5.5 Relationship between P_L versus B with increasing Q.

From the Shannon capacity, the capacity of MC can be expressed as follows.

$$C = B\log_2\left(1+\frac{P_L}{BN_0}\right), \tag{5.22}$$

where N_0 is a noise spectral density. In (5.22), P_L and B, which are the functions of Q, are dominant factors to determine capacity, C. Therefore, the capacity of MC can be determined by the value of Q. As Q increases, most signal power is concentrated at near the resonant frequency while the bandwidth becomes narrow. As a result, P_L at f_o increases but B decreases. Figure 5.5 shows this tradeoff relationship between P_L and B with increasing Q. From this observation, we can know intuitively that there is an optimal value of Q for maximizing C. In a loosely coupled region, the optimal value of quality factor Q_{opt} can be found numerically, and the maximum capacity can be achieved by designing the quality factor of resonators to Q_{opt} at a given distance.

5.3.2 MC Relay Systems

In a loosely coupled region, it is difficult to ensure large capacity at long distances, because the power of magnetic fields decreases in proportion to $1/d^6$. Thus, relay resonators are deployed between the Tx and Rx to amplify a transmitted signal. As shown in Figure 5.6, we consider MC relay systems, which consists of Tx, Rx, and n relays. The total distance between the Tx and Rx is D, and the n relays are arranged at equal distance d between the Tx and Rx. For the sake of simplicity, we consider identical Tx, relay, and Rx with same inductance and capacitance, so that they will resonate at the same resonant frequency. In MC relay systems, the alternating current in the Tx induces another alternating current in the adjacent relay through magnetic induction. The induced alternating current in this relay also creates another alternating current in the second relay, and so on. In this way,

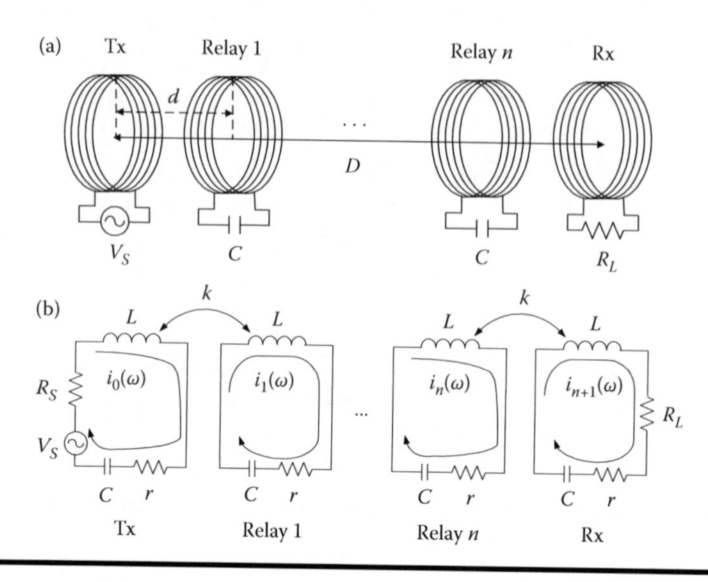

Figure 5.6 Model for MC relay systems. (a) MC systems. (b) Equivalent circuit model.

the signal transmitted by the Tx is amplified and relayed by intermediate relays, and finally delivered to the Rx.

From KVL, we can build the following equations.

$$
\begin{aligned}
V_S &= \left(R_S + r + j\omega L + \frac{1}{j\omega C} \right) i_0 - j\omega k L i_1 - \cdots - j\omega k L i_n - j\omega k L i_{n+1}, \\
0 &= -j\omega k L i_0 + \left(r + j\omega L + \frac{1}{j\omega C} \right) i_1 - \cdots - j\omega k L i_n - j\omega k L i_{n+1}, \\
&\vdots \\
0 &= -j\omega k L i_0 - j\omega k L i_1 - \cdots + \left(r + j\omega L + \frac{1}{j\omega C} \right) i_n - j\omega k L i_{n+1}, \\
0 &= -j\omega k L i_0 - j\omega k L i_1 - \cdots - j\omega k L i_n + \left(R_L + r + j\omega L + \frac{1}{j\omega C} \right) i_{n+1}.
\end{aligned}
\tag{5.23}
$$

Here, we assume that adjacent resonators are coupled loosely, because relays are generally used to provide reliable communication at long distances. As a result, similar to (5.15), we also ignore the effect of coupling by the Rx (Z_{in}), that is seen in the Tx, as well as a cross-coupling between nonadjacent resonators. Then, the currents in the Tx, relays, and Rx can be obtained as

$$
\begin{aligned}
i_0(\omega) &= \frac{V_S}{(R_S + r)(1 + j2\Delta\omega Q)}, \\
i_1(\omega) &= \frac{j\omega kL}{r(1 + j2\Delta\omega Q_r)} \cdot i_0(\omega), \\
i_2(\omega) &= \left\{\frac{j\omega kL}{r(1 + j2\Delta\omega Q_r)}\right\}^2 \cdot i_0(\omega), \\
&\vdots \\
i_n(\omega) &= \left\{\frac{j\omega kL \mid}{r(1 + j2\Delta\omega Q_r)}\right\}^n \cdot i_0(\omega), \\
i_{n+1}(\omega) &= \frac{j\omega kL}{(R_L + r)(1 + j2\Delta\omega Q)} \cdot \left\{\frac{j\omega kL}{r(1 + j2\Delta\omega Q_r)}\right\}^n \cdot i_0(\omega).
\end{aligned} \tag{5.24}
$$

Here, k is a coupling coefficient between adjacent resonators, the quality factor of Tx and Rx is defined as $Q = \frac{\omega L}{R + r}$, and the quality factor of relays is defined as $Q_r = \frac{\omega L}{r}$. Then, the received power is represented by

$$
\begin{aligned}
P_L(\omega) &= | i_{n+1}(\omega) |^2 R_L \\
&= \frac{P_S \eta^2 Q^2 k^2}{(1 + (2\Delta\omega)^2 Q^2)^2} \cdot \left\{\frac{Q_r^2 k^2}{1 + (2\Delta\omega)^2 Q_r^2}\right\}^n,
\end{aligned} \tag{5.25}
$$

For simplicity, if we assume $Q = Q_r$, the received power can be simplified as

$$
P_L(\omega) = \frac{P_S \eta^2 (Q^2 k^2)^{n+1}}{(1 + (2\Delta\omega)^2 Q^2)^{n+2}}. \tag{5.26}
$$

In addition, at resonant frequency ω_o, the received power can be expressed simply as follows.

$$
P_L(\omega_o) = P_S \eta^2 (Q^2 k^2)^{n+1}. \tag{5.27}
$$

Using a similar approach to MC systems, the optimal bandwidth of MC relay systems can be obtained from the relation, $P_L\left(\omega_o - \frac{B}{2}\right) : P_L(\omega_o) = \frac{1}{2} : 1$, as follows.

$$B=\sqrt{\sqrt[n+2]{2}-1}\cdot\frac{f_o}{Q}. \tag{5.28}$$

Here, P_L increases proportionally to Q while B is inversely proportional to Q. This indicates that the tradeoff relationship between P_L versus B with increasing Q is also met in MC relay systems. From a numerical method, we can find the optimal value of quality factor Q_{opt} at a given k. When relays are deployed to increase the range of communication, it is possible to improve the capacity of MC relay systems by adjusting the quality factor of resonators to Q_{opt}.

Figure 5.7 shows capacity C against quality factor Q when D = 3 m. Large Q increases the received power at the Rx, as a result, C also increases. However, when Q increases beyond a threshold, the bandwidth becomes extremely narrow although the received power increases. In consequence, C drops gradually with increasing Q. This observation explains that there are optimal quality factors that maximize the capacities of MC and MC relay systems. The deployment of relays between the Tx and Rx can prevent the serious attenuation of the signal strength. As a result, the maximum capacity of MC relay systems, which is made at Q_{opt}, is larger than that of MC systems. For the same reason, the maximum capacity of MC relay systems also increases as the number of relays, n, increases. In addition, we can know that the maximum capacity can be achieved at smaller Q_{opt} when n is large in D = 3 m. At the same D, as n increases, adjacent relays are coupled strongly. So, the signal transmitted by the Tx can be transferred to the Rx reliably even with

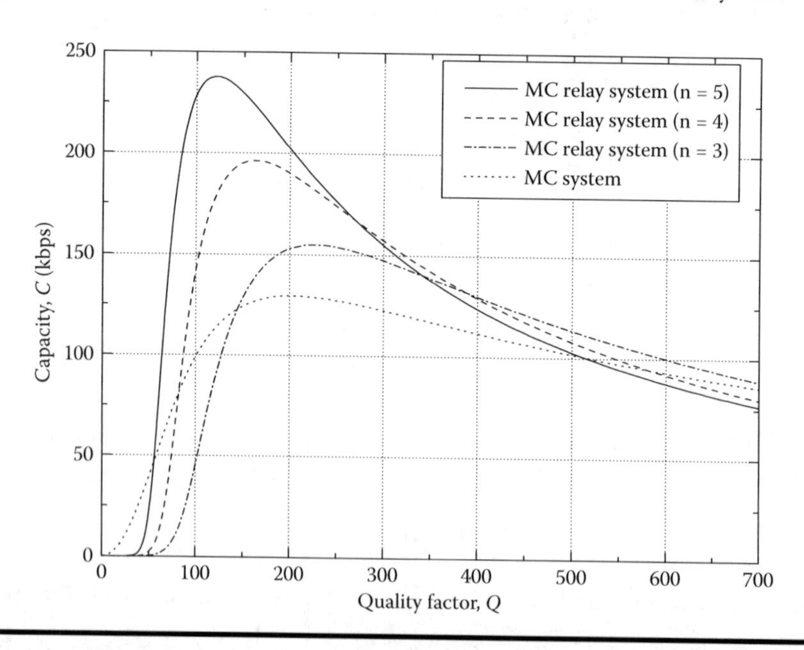

Figure 5.7 Capacity C versus quality factor Q when D = 3 m.

small Q. This indicates that small Q with large B is good for maximizing capacity as n increases in the same total distance.

5.4 Possibility of Multiple-Antenna Based MC Technology

In wireless communications, MIMO technology was proposed as a method for improving the capacity of a communication link using multiple transmit and receive antennas to exploit multipath propagation. Recently, MIT researchers suggested a novel wireless charging technology called MagMIMO to detect and cast a cone of energy toward a device located anywhere [15]. In MC, we can adapt this MagMIMO technology to give a degree of freedom for the position of Rx as well as enhance the capacity of MC.

Figure 5.8 shows the conceptual diagram of MC MagMIMO systems and its equivalent circuit model. We consider an $N \times 1$ system where N antennas are aligned vertically in the Tx and only one antenna is implemented in the Rx. The mutual inductance among the antennas in the Tx, $M_{t_{ij}}$, is independent of the Rx and does not change over time. Therefore, $M_{t_{ij}}$ can be calculated when the antennas are set up in the Tx, in the absence of Rx. As a result, $M_{t_{ij}}$ can be treated as a constant and simply neglected for the following analysis [15]. Then, the relation between the Tx and Rx can be built from KVL, as following.

$$\begin{aligned} V_S &= (Z_1 + Z_2 + \cdots + Z_N)i_T - j\omega(M_{1r} + M_{2r} + \cdots + M_{Nr})i_R, \\ 0 &= -j\omega(M_{1r} + M_{2r} + \cdots + M_{Nr})i_T + Z_r i_R. \end{aligned} \tag{5.29}$$

Here, i_T and i_R are currents in the Tx and Rx, respectively, and M_{nr} is the mutual inductance between the antenna n in the Tx and the antenna in the Rx. In addition, Z_n is the input impedance of the antenna n in the Tx while Z_r is the input impedance of the antenna in the Rx.

Then, i_R can be obtained from (5.29) as follows.

$$\begin{aligned} i_R &= \frac{j\omega(M_{1r} + M_{2r} + \cdots + M_{Nr})}{Z_r} \cdot i_T \\ &= \sum_{n=1}^{N} m_{nr} \cdot i_T, \end{aligned} \tag{5.30}$$

where $m_{nr} = \frac{j\omega M_{nr}}{Z_r}$. From (5.30), we can know that the MC MagMIMO can achieve a spatial diversity. For example, the other magnetic links compensate the degradation of received signal even though one magnetic link experiences deep fading due to the misalignment of the Tx and Rx. Therefore, the quality and

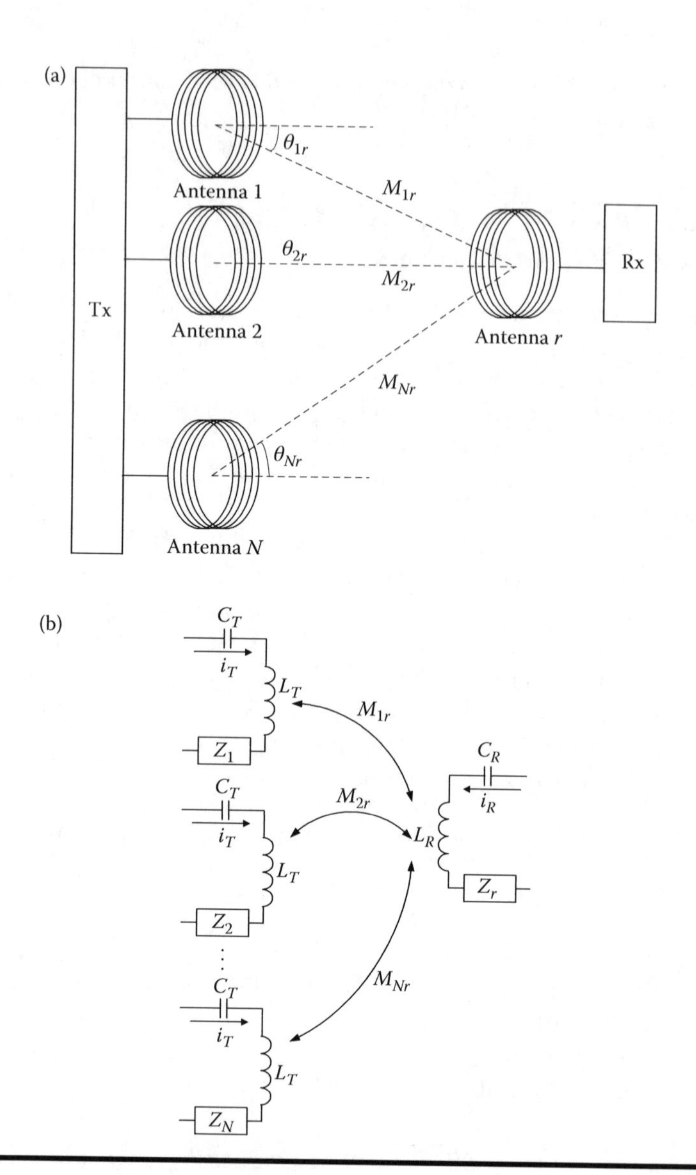

Figure 5.8 Conceptual diagram of MC MagMIMO systems.

reliability of a wireless magnetic link can be improved regardless of the location of Rx. A beamforming is an important research issue in the MC MagMIMO. For example, based on the estimated value of magnetic links, the Tx can find the beamforming vectors of antennas and steer beam to concentrate the energy of a transmitted signal, depending on the location of the Rx. Then, the Rx can receive a directional magnetic signal intensively, as a result, the capacity of MC MagMIMO systems can be enhanced.

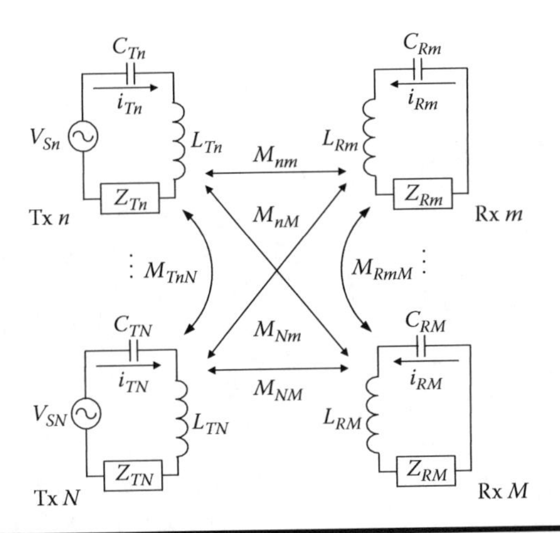

Figure 5.9 MC MultiSpot systems.

In addition to MC MagMIMO, a MultiSpot technology can be used in MC for communicating with multiple devices at the same time [16]. As shown in Figure 5.9, there are N antennas in the Tx and M antennas in the Rx for the MC MultiSpot systems. Then, the KVL equations for Tx n and Rx m can be expressed as follows.

$$V_{Sn} = Z_{Tn} i_{Tn} + \underbrace{\sum_{i \neq n} j\omega M_{Tni} i_{Ti}}_{\text{from the other Txs}} - \underbrace{\sum_{m=1} j\omega M_{nm} i_{Rm}}_{\text{from the Rxs}}$$
$$0 = \underbrace{\sum_{n=1}^{N} j\omega M_{nm} i_{Tn}}_{\text{from the Txs}} + Z_{Rm} i_{Rm} + \underbrace{\sum_{k \neq m} j\omega M_{Rmk} i_{Rk}}_{\text{from the other Rxs}}. \tag{5.31}$$

Finally, the KVL equations for overall MC MultiSpot can be obtained in the following matrix forms.

$$\begin{aligned} \vec{V}_S &= \left(\vec{Z}_T + \omega^2 \vec{M}^T \vec{Z}_R^{-1} \vec{M}\right) \vec{i}_T \\ \vec{i}_R &= j\omega \vec{Z}_R^{-1} \vec{M} \vec{i}_T. \end{aligned} \tag{5.32}$$

Here, $\vec{V}_S$, $\vec{i}_T$, $\vec{i}_R$, $\vec{Z}_T$, $\vec{Z}_R$, and $\vec{M}$ are defined as follows.

$$\vec{V}_S = \left[V_{S1}, V_{S2}, \ldots, V_{SN}\right]_{N\times 1}^T$$

$$\vec{i}_T = \left[i_{T1}, i_{T2}, \ldots, i_{TN}\right]_{N\times 1}^T$$

$$\vec{i}_R = \left[i_{R1}, i_{R2}, \ldots, i_{RM}\right]_{M\times 1}^T$$

$$\vec{Z}_T = \begin{bmatrix} Z_{T1} & j\omega M_{T12} & \cdots & j\omega M_{T1N} \\ j\omega M_{T21} & Z_{T2} & \cdots & j\omega M_{T2N} \\ \vdots & \vdots & \ddots & \vdots \\ j\omega M_{TN1} & j\omega M_{TN2} & \cdots & Z_{TN} \end{bmatrix}_{N\times N}$$

$$\vec{Z}_R = \begin{bmatrix} Z_{R1} & j\omega M_{R12} & \cdots & j\omega M_{R1M} \\ j\omega M_{R21} & Z_{R2} & \cdots & j\omega M_{R2M} \\ \vdots & \vdots & \ddots & \vdots \\ j\omega M_{RM1} & j\omega M_{RM2} & \cdots & Z_{RM} \end{bmatrix}_{M\times M} \quad (5.33)$$

$$\vec{M} = \begin{bmatrix} M_{11} & M_{21} & \cdots & M_{N1} \\ \vdots & \vdots & \ddots & \vdots \\ M_{1M} & M_{2M} & \cdots & M_{NM} \end{bmatrix}_{M\times N}$$

Since the MC MultiSpot can support MC links for multiple devices simultaneously, it has a potential to increase the capacity of MC dramatically. Therefore, it is worth attempting to study the following issues as further works.

- Signal processing techniques for multiple-antenna based MC: It is required to analyze the achievable capacity of $N \times M$ MC MultiSpot and propose efficient techniques for processing signals from multiple antennas, i.e., data combining or interference alignment schemes in accordance with magnetic properties.
- Antenna array optimization: It is needed to investigate the optimal structure of antenna array for MC MultiSpot in consideration of magnetic couplings among all antennas including Tx-to-Tx, Tx-to-Rx, and Rx-to-Rx. This can achieve not only further improvement on the capacity of MC but also the miniaturization of Tx and Rx systems.

References

1. H. W. Ott, *Noise Reduction Techniques in Electronic Systems*, 2nd ed. New York: Wiley, 1988.
2. K. Lee and D.-H. Cho, "Maximizing the capacity of magnetic induction communication for embedded sensor networks in strongly and loosely coupled regions," *IEEE Trans. Magn.*, vol. 49, no. 9, pp. 5055–5062, Sep. 2013.
3. Z. Sun and I. Akyildiz, "Underground wireless communication using magnetic induction," *Proc. 2009 IEEE International Conference on Communications*, pp. 1–5, June 2009.

4. A. Kurs, A. Karalis, R. Moffatt, J. D. Joannopoulos, P. Fisher, and M. Soljacic, "Wireless power transfer via strongly coupled magnetic resonances," *Sci. Express*, vol. 317, no. 5834, pp. 83–86, July 2007.
5. Y. D. Tak, J. M. Park, and S. W. Nam, "Mode-based analysis of resonant characteristics for near-field coupled small antennas," *IEEE Antennas Wireless Propag. Lett.*, vol. 8, pp. 1238–1241, Nov. 2009.
6. A. P. Sample, D. T. Meyer, and J. R. Smith, "Analysis, experimental results, and range adaptation of magnetically coupled resonators for wireless power transfer," *IEEE Trans. Ind. Electron.*, vol. 58, no. 2, pp. 544–554, Feb. 2011.
7. R. Mongia, *RF and Microwave Coupled-Line Circuits*. Norwood, MA: Artech House, 2007.
8. J. Chen, *Feedback Networks: Theory and Circuit Application*. Singapore: World Scientific, 2007.
9. D. Ahn and S. Hong, "A study on magnetic field repeater in wireless power transfer," *IEEE Trans. Ind. Electron.*, vol. 60, no. 1, pp. 360–371, Jan. 2013.
10. N. Y. Kim, K. Y. Kim, J. Choi, and C. W. Kim, "Adaptive frequency with power-level tracking system for efficient magnetic resonance wireless power transfer," *Electron. Lett.*, vol. 48. No. 8, pp. 452–454, Apr. 2012.
11. H. Jiang and Y. Wang, "Capacity performance of an inductively coupled near field communication system," *Proc. IEEE Antennas and Propagation Society International Symposium (AP-S 2008)*, pp. 1–4, July 2008.
12. Z. Sun and I. F. Akyildiz, "Magnetic induction communications for wireless underground sensor networks," *IEEE Trans. on Antenna and Propag.*, vol. 58, no. 7, pp. 2426–2435, July 2010.
13. C.-J. Chen, T.-H. Chu, C.-L. Lin, and Z.-C. Jou, "A study of loosely coupled coils for wireless power transfer," *IEEE Trans. Circuits Syst. II, Exp. Briefs*, vol. 57, no. 7, pp. 536–540, July 2010.
14. J. I. Agbinya and M. Masihpour, "Power equations and capacity performance of magnetic induction body area network nodes," *Proc. Broadband and Biomedical Communications (IB2Com '10)*, pp. 1–6, Dec. 2010.
15. J. Jadidian and D. Katabi, "Magnetic MIMO: How to charge your phone in your pocket," *Proc. The 20th Annual International Conference on Mobile Computing and Networking*, pp. 495–506, Sep. 2014.
16. L. Shi, Z. Kabelac, D. Katabi, and D. Perreault, "Wireless power hotspot that charges all of your devices," *Proc. The 21th Annual International Conference on Mobile Computing and Networking*, pp. 2–13, Sep. 2015.

Chapter 6

Routing Challenges and Associated Protocols in Acoustic Communication

Muhammad Khalid and Yue Cao
Northumbria University

Muhammad Arshad and Waqar Khalid
Institute of Management Sciences

Naveed Ahmad
University of Peshawar

Contents

6.1 Introduction

Underwater wireless sensor network (UWSN) is a newly emerging wireless sensor technology that is used to provide the most promising mechanism and methods that are used for discovering aqueous environment. It is used in various key applications in underwater environment. It works efficiently in many situations like commercial, military, emergency monitoring, data collection, and environmental monitoring purposes. In this kind of networks, small sensors nodes are deployed in sea water. These nodes are equipped with a central processing unit, antenna, and battery. Batteries in these sensor nodes are nonrechargeable and nonreplaceable. These sensors collect the required data and send it to sinks which are installed offshore [1]. Autonomous underwater and unmanned vehicles are equipped with sensors that are specially designed for underwater communication [2] and are mostly used in areas where humans are unable to explore underwater resources directly. Information about natural resources that lie underwater is obtained by unmanned vehicles and forwarded to sinks [3,4]. Radio waves cannot be used in underwater communication; therefore, acoustic communication is needed [5]. Communication through acoustic links is costly when compared with radio links. Acoustic links have high end-to-end delay and low bandwidth. Once data packet is received at the sink, it is forwarded through radio waves to other sinks and base stations [6]. Underwater networks have limited resources in comparison to terrestrial wireless sensor networks. Protocols suites that are used in other networks cannot be directly applied to underwater networks [7]. Till date, many protocols have been proposed for underwater sensor networks. These are mainly divided into two types: localization-based and localization-free protocols [8], where the term localization means knowledge of nodes and sink in network. The routing protocols which need prior geographic information of other nodes and sinks are localization-based routing (LBR) protocols, while those routing protocols that do not need any earlier geographic information for routing can be categorized as localization-free routing (LFR) protocols [7,9]. The rest of the chapter is organized as follows. Section 6.2 discussed the architecture of terrestrial wireless sensor network. In Section 6.3, the architecture of UWSN is explained. Section 6.4 has defined the related work, while LBR and LFR protocol are discussed in Sections 6.5 and 6.6, respectively, and finally conclusion is drawn in Section 7.

6.2 Basic Architecture of Acoustic Communication

UWSN is a wireless technology that has gained worldwide attention these days. It provides the most promising mechanism used for discovering aqueous environment very efficiently for many scenarios like military [19], emergency, and commercial purposes. Autonomous underwater and unmanned vehicles are equipped with sensors that are specially designed for underwater communication, which are mostly used in areas where exploration for natural resources which lie underwater is needed [20]. These unmanned vehicles gather data of resources that lie underwater and send back to offshore sinks, which is forwarded to other stations for further processing. Radio waves cannot be used in underwater communication; therefore, acoustic communication is used. Once data packet reaches sink, then it is forwarded through radio waves to other sinks and stations [3].

Underwater wireless sensor environment is much different from that of terrestrial. Acoustic waves are used in underwater communication while terrestrial network uses radio waves [4]. Normally, the problems that occur during communication in underwater communication are due to dense salty water, and electromagnetic as well as optical signal does not work in UWSN [6]. Due to high attenuation and absorption effect, signals cannot travel long distances [21]. Hence to overcome these problems, acoustic communication is used. It can overcome these problems and provide a better transfer rate in underwater environment [6]. Due to limitations of acoustic communication, the communication speed slows down to 1500 m/s, that is, speed of sound to speed of light. Due to lower speed, there is usually long propagation delay and higher end-to-end time [4]. In acoustic communication, bandwidth is very limited, which is less than 100 kHz [1]. In underwater scenarios, sensor nodes are usually considered static but it is also considered that they may move from 1 to 3 m/s because of flow of water [1]. Sensor nodes used in underwater network are battery operated, and it is almost impossible to replace their batteries. In underwater applications, a multihop or multipath network is required and data is forwarded by passing all nodes towards sink. Once data is received at any of the sinks, then data is forwarded to the concerned node through radio transmission [1]. Figure 6.1 represents a network architecture of UWSN.

While using those routing protocols that require higher bandwidth [1] that usually have higher delay at the node's end, it is known that acoustic communication does not support higher bandwidth, using routing protocols that are used in terrestrial network will not perform good due to their higher delay and high energy consumption [2]. Using underwater network, topology does not remain the same as node moves due to flow of water [3]. In localization-based protocol,

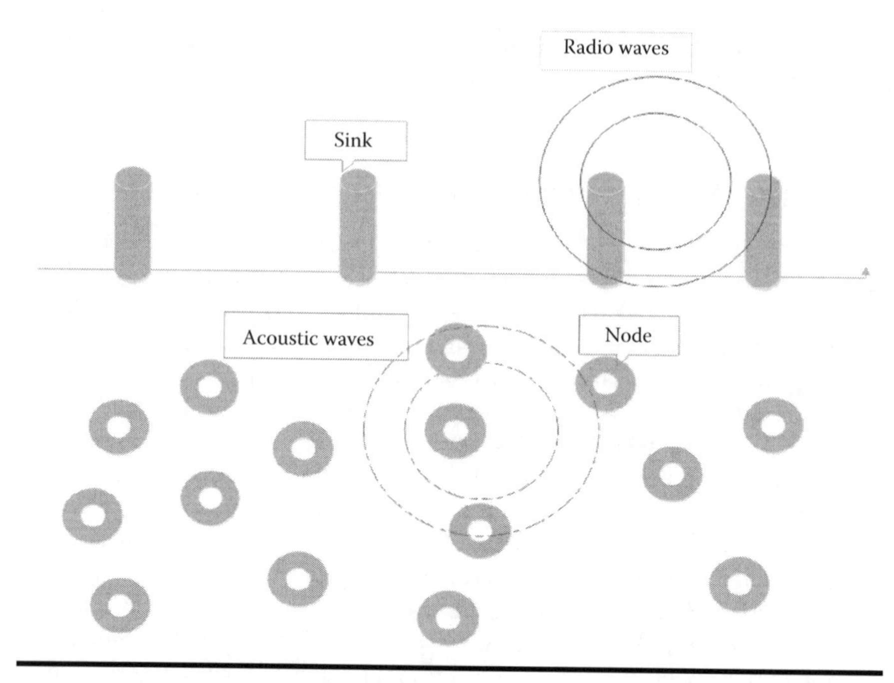

Figure 6.1 A UWSN network architecture.

geographical network information is necessary, so it possesses more control messages than localization-free protocol, in which no prior network information is necessary [4]. Figure 6.2 shows the state-of-the-art LBR and LFR protocols in UWSN. Oceans are vast and cover around 140 million square miles, which is more than 70 percent of Earth's total surface. Not only has it been considered a major source of nourishment but also with span of time it is taking a good role in transportation stuffs, defense as well as adventurous purposes, and presence of natural resources [20]. As underwater resources play a vital role in human life, very less of this area is explored [2]. Less than 10 percent of the whole ocean volume is investigated, while a large amount of area is still not explored. The increase in roles of the oceans in the lives of humans [3] and the largely unexplored areas have a lot of importance [4]. On one hand, the traditional approaches for underwater monitoring has got several disadvantages, while on the other side, human presence is not considered to be feasible for underwater environments [19].

6.2.1 Node Architecture

A general architecture of underwater wireless sensor node is composed of five main elements, which are energy management unit, data sensing unit, depth measuring unit, communication unit, and a central processing unit [18]. As depicted in Figure 6.2, processing unit is responsible for all kinds of data processing where

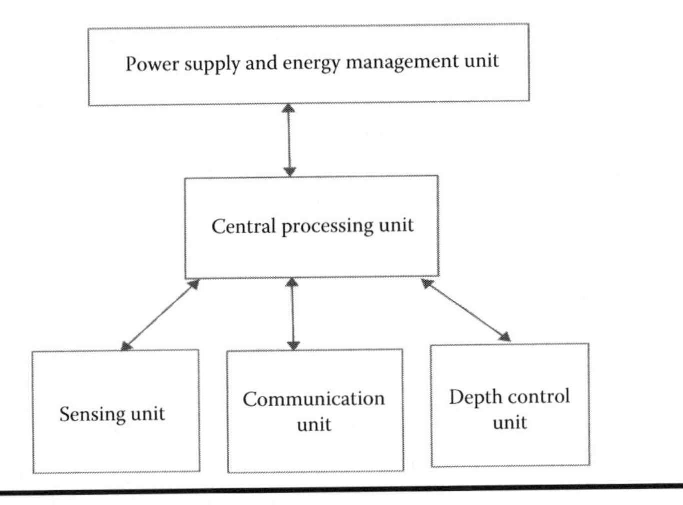

Figure 6.2 Architecture of a typical underwater sensor node.

energy management unit has the responsibility to manage the remaining energy of the node and consumption of energy in run time [4]. Data sensing unit is used to sense data. It always remains active even when the node is in sleep mode [7]. Communication unit is responsible for all kinds of data communication, whereas depth measuring unit is used for measuring depth of node when it is deployed in sea [14].

6.2.2 Constraints in Acoustic Communication

UWSN carries multiple differences in comparison with terrestrial area network, where nodes are stable or move in a specified direction, while in underwater networks, they usually displace their positions with the flow of water. Acoustic communication is used for underwater transmission, which minimizes the bandwidth for data transferring. A few constraints are discussed as follows.

i. Limited Bandwidth. Acoustic channels offer limited amount of bandwidth, as radio transmission cannot be used for underwater communication [3]. Acoustic communication requires more energy to send a small amount of data due to its lower bandwidth.
ii. Propagation Delay. Due to the use of acoustic communication, propagation speed becomes five times slower than that of radio frequency, that is, 1500 m/s [4], which obviously results in high propagation delays in the network.
iii. Limited Energy. Nodes that are used in underwater communication are larger in size [3]; hence, they require larger amount of energy for communication. Furthermore, acoustic channels also required more energy for communication than terrestrial network. Batteries in UWSN cannot be recharged or

replaced; therefore, use of energy-efficient communication is always a need to provide a network with higher lifetime.

iv. Limited Memory. In UWSN, nodes are smaller in size and therefore have a limited amount of storage and processing capacity [6].

v. Variable Topology. UWSN does not have a specific or static topology as flow of water makes it difficult for the node to remain static in one place; therefore, the node moves randomly.

6.3 Related Work

6.3.1 The Architecture of Acoustic Communication

In [3], 2D and 3D underwater communication and different layers of communication in underwater networks are discussed. Multiple open researches have been provided in this survey article. However, discussion about routing protocol in underwater networks and their comparison have not been discussed. In [3], the differences between terrestrial and underwater network were presented. Like UWSN, low bandwidth, propagation delay, high bit error rate, floating of node, and limited energy have been discussed. Multiple unique characteristics of UWSN, their benefits, and flaws were also discussed. Similarly, no proper discussion about routing protocols has been carried out. Multiple schemes of routing in underwater communication have been discussed in this survey. They also discussed about multiple routing protocols. Detailed diagrams were presented to get a good understanding of the different routing protocols, but still no comparison was carried out. In [1], the term localization has been discussed. Localization is a phenomenon in which the location of the node is already known to other nodes and sinks, which make it easier for the sink to locate and communicate it. Multiple schemes like area-based scheme, area localization scheme, and hop count based scheme have been discussed [5]. In energy-efficient dynamic address based (EE-DAB) routing [16], every node is assigned Node-ID, S-HopID, and C-HopID. Node-ID shows the physical address of node; S-HopID consists of two digits which show how many hops one or two sinks are away. Left hop is considered as the highest priority and is selected as the primary route. The C-HopID also consists of two digits which show how many hops the receiving nodes are away from courier nodes. Acoustic communication uses more energy than that of radio communication. As wireless sensor nodes are battery operated and higher energy consumption leads to a serious problem, energy efficiency has become a major problem in UWSNs.

The proposed delay-tolerant protocol is also called delay-tolerant data dolphin scheme. This proposed scheme is designed for delay-tolerant systems and applications. In these protocols, all the sensing nodes stay static and data sensed by static nodes are passed on to data dolphin that acts as courier nodes. So in this

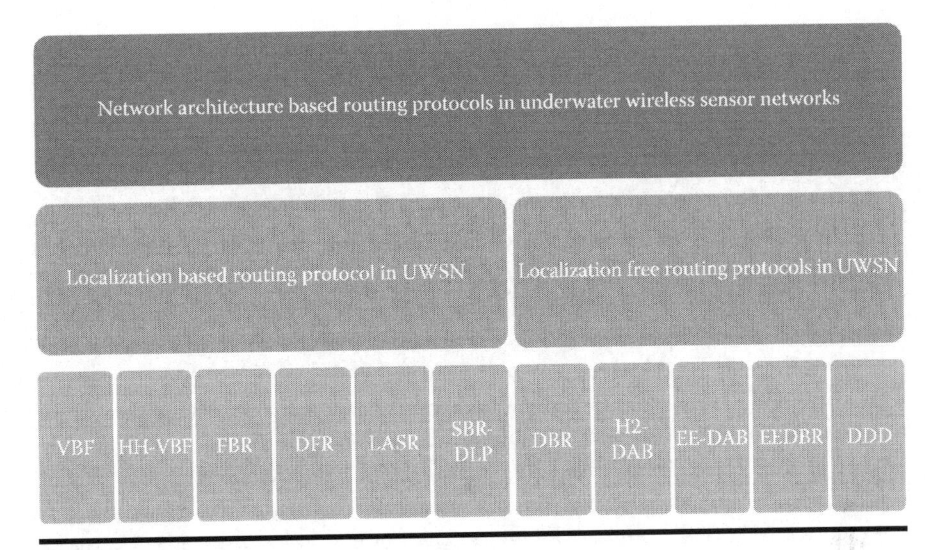

Figure 6.3 Routing protocols in UWSN.

methodology, high energy-consumed hop-by-hop communication is avoided. Data dolphins that act as courier nodes are provided with continuous energy. In the architecture, all static nodes are deployed in the sea bed. These static sensors go into sleep mode if there is no data to sense, and they periodically wake up when they sense some data. After sensing some kind of desired data, they simply forward this data to courier nodes which are also called data dolphins. These data dolphins take this data and deliver it to the base station or sink. The number of dolphin nodes depends on the kind of network and its application and the number of nodes deployed in the network. In the proposed virtual sink architecture, sinks are connected with each other through radio communication. In this scheme, every sink broadcasts a hello packet that is also known as hop count update packet. After receiving hello packet by nodes, a hop count value is assigned to every sensor. These hop counts are used for selection of forwarding nodes while sending data packet from one node to another. However, the proposed scheme has few limitations, which include redundant transmission, that is, transmission of the same packet many times.

This chapter has divided routing protocols into LBR and LFR protocols [1], as depicted in Figure 6.3.

6.4 Localization Based Routing Protocols

Routing protocols that need prior network information before sending any data over the network are called LBR protocols. These protocols usually need geographical information of all nodes in the network as well as information about sink location. These protocols are considered to be less energy efficient; most of the energy is wasted in collecting their geographical information. These records are updated dynamically after a fixed interval of time, as the node's position may change due to water flow. Routing protocols basically need the assumption of sensor nodes in underwater sensor networks [16]. In LBR protocols, a node needs the information of all the network nodes as well as of sink, such that in this scenario prior network information is needed for a node [1,2]. In [9], focused beam routing (FBR) protocol requires geographical information of itself and of destination. It uses request in order to send (RTS)/Clear To Send (CTS) mechanism to forward data. Sender protocol transmits the RTS and receiver of the packet sends back CTS. In vector-based forwarding (VBF) [7], a source node develops a vector-based routing pipe starting from sender node towards sink. Various times, it is hard to find an available node in the routing pipe for data forwarding. SBR-DLP [12], also known as sector-based routing with destination location prediction, is an LBR algorithm where node is not required to have the information of its neighbor nodes. It only needs to carry its own information and preplanned movement of sink, although it decreases the flexibility of the network and will only move around in a scheduled manner. Table 6.1 provides a detailed overview of LBR protocols in UWSN.

6.4.1 Vector-Based Forwarding

VBF [7] is a routing scheme that requires maintenance and frequent recovery of routing paths. This is a position-based routing protocol in which less amount of nodes is actually involved in the routing process. Therefore, a small number of nodes play their role in the operation of data forwarding, as the important phenomenon in routing is data forwarding, where a small number of sensor nodes take part in this data-forwarding operation. In this scheme, a sensor already knows its location and the location of the destination. It is also considered that a node already knows all the nodes that are involved in the routing process or forwarding of a node, which include the source node, forwarding intermediate nodes, and the final node or the destination. The idea of this protocol is based on a virtual routing pipe and all forwarding data is sent through this pipe. As routing pipe phenomenon is involved in this scheme, most of the time the nodes used during routing process are the nodes that lie in the area of the pipe.

6.4.2 Hop-by-Hop Vector Based Forwarding

Hop-by-hop VBF (HH-VBF) [8] is an advanced version of VBF. In this scheme, the main focus is on robustness and problems faced by its earlier version [1]. The same concept as was used in VBF is also used here. Concept of virtual pipe is deployed here. However, instead of a single virtual routing pipe that is used by VBF, a single routing pipe is used for every forwarder, which means a single pipe

Table 6.1 Localization-Based Routing Protocols

Architecture	*Technique*	*Performance Metrics*				*Knowledge Required*
		Packet Delivery Ratio	*Multiple Sinks*	*Energy Efficiency*	*Packet Overhead*	
VBF [7]	Based on localization, geographic routing scheme	Low	N	Fair	High	Whole network
HH-VBF [8]	hop by hop, geographic routing algorithm	Fair	Y	Low	Medium	Whole network
FBR [9]	Route is being established dynamically in this distributed algorithm	Fair	N	High	High	Own and sink location
DFR [10]	Directional flooding routing approach	Fair	N	Low	Medium	Own, sink, and one-hop neighbor
LASR [11]	Link quality metrics and location awareness technique	Fair	Y	Fair	Medium	Own and sink location
SBR-DLP[12]	Sector-based routing with destination location prediction	Low	N	Fair	High	Movement of sink and own location

for every forwarding hop, as we observed that only a few nodes are involved in VBF, while in HH-VBF, multiple routing pipes are created, which ultimately result in lower end-to-end delay and higher energy efficiency. Using this mechanism, every node can make a decision about the direction of pipe that is based on the node's current location.

6.4.3 Directional Flooding-Based Routing

Directional flooding-based routing (DFR) [10] is an LBR protocol. In DFR, the flooding phenomenon is used, where packet is sent through a flooding mechanism to the final destination. In this protocol, it is assumed that every node must know the location of itself and one-hop-away node and final destination. Only a limited number of nodes take part in the routing process. In this scheme, the flooding zone is decided by FS and FD, where S is the source node and D is the destination node while F is final node also considered as sink.

6.4.4 Location-Aware Source Routing

Location-aware source routing (LASR) [11] is an advanced version of DSR. Link quality metrics and location awareness technique are used by LASR routing scheme. Earlier protocol only depended upon the shortest path metrics and in the end it led to bad performance.

6.4.5 Focused Beam Routing

FBR [9] is an LFR protocol in which sender node knows only its own location information and the location information of the final destination. No further geographical information of other nodes is necessary, which results in less control messages and high throughput. The mechanism that FBR has adopted for data forwarding is that the next hop is selected keeping in view the final destination. First of all, an RTS packet is multicast in its neighbors, which contains the location of sender and final destination. This multicast operation is performed at a low power level. If sender does not receive any response, then the level is increased. Figure 6.4 explained data forwarding method which is used in FBR, where node A has a data packet that is required to be sent to the destination node which is D. To complete this operation, node A has to multicast an RTS packet to its neighboring nodes which lie in its range, as this RTS packet contains the location of node A and that of final destination D. Initially, this multicast action will be performed at the lowest power level, which can be increased if neither of the nodes is found as next hop in the transmission range. For this purpose, they define finite power levels, which are P1–PN. In FBR, if no node lies in the sender's range, then it has to rebroadcast RTS, which results in consumption of high energy. The working of FBR is depicted in Figure 6.4.

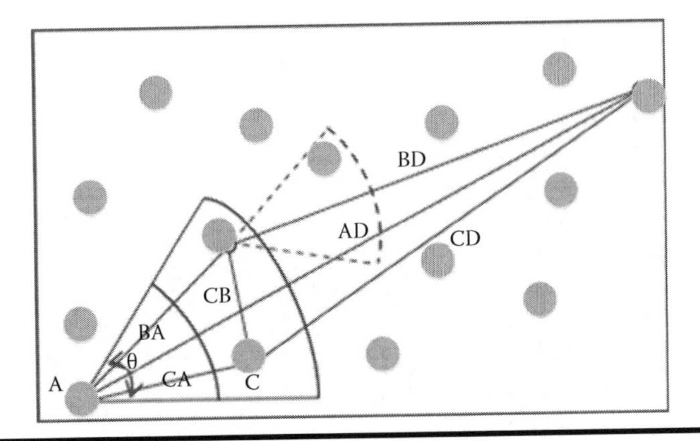

Figure 6.4 Working of FBR protocol.

6.4.6 *Sector-Based Routing with Destination Location Prediction*

In UWSN, many LBR algorithms have been introduced, and it is considered that network with already known geographical location of other nodes improves energy efficiency. It helps in minimizing control messages and network overhead. SBR-DLP [12] is an LBR protocol. In this protocol, not only other nodes but also destination nodes are considered to be mobile. In SBR-DLP, sensor does not need to carry information about neighbors. In this algorithm, it is considered that every node must know its own location information and preplanned movement of destination nodes. Hop-by-hop mechanism is used to forward data to destination nodes. In Figure 6.5, a node S has a data packet that is needed to be sent to the destination D. In order to do so, next hop is found by broadcasting a Chk Ngb packet which has its current location as well as its Node-ID. The neighbor node will receive Chk Ngb, whether it is near to destination node D. The nodes that meet these conditions will reply to the node S by sending a Chk Ngb Reply packet. Forwarding node selection in SBR-DLP is depicted in Figure 6.5.

6.5 Localization Free Routing Protocols

This category includes those routing protocols that do not require any earlier geographical information of the network. These protocols perform their operation without having location information of other nodes. In these kinds of routing protocols, a sensor node does not require any prior network information of other network nodes. Most of the localization protocols work on flooding phenomenon and are considered to have fast packet delivery ratio and low end-to-end delay [10,11]. In depth-based routing (DBR) protocol [13], prenetwork information is

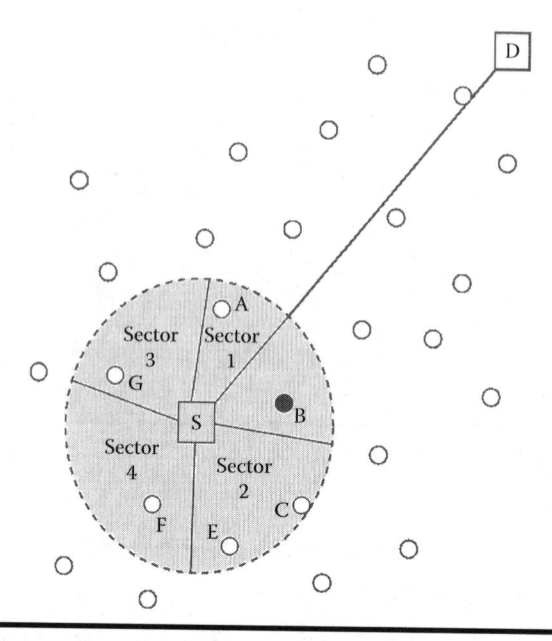

Figure 6.5 Forwarder node selection in SBR-DLP.

not needed. It just takes the depth of sensor nodes into account and forwards a packet. It actually compares the depth of sending the node with that of the receiving node, so if the depth of the sender node is higher than that of the receiver node, then it will forward the data; otherwise, it will ignore that node. Similarly in [14], energy-efficient DBR (EE-DBR) takes into account the depth information as well as residual energy of the node at the time of sending data. A detailed summary of LFR protocol has been provided in Table 6.2.

6.5.1 Depth-Based Routing

Many routing protocols in UWSN need geographic location of the nodes to communicate. Localization itself requires much energy and calculations. DBR protocol [13] does not need any earlier information. DBR needs depth information of each node. When a node with the highest depth senses some movement, it starts sending data to higher nodes, such that it compares its depth with neighbor nodes if it sends packets to only those nodes whose depth is lower than the sender node. The same process continues till the packet is received by the sink. This protocol is mainly concerned about the depth of the node. Sinks are provided with continuous power.

Figure 6.6 defines the next node selection in DBR protocol, where the three nodes N1–N3 are in the communication range of sender S. In the first step, the depth of the receiver node is checked. N1 and N2 are found eligible for data forwarding, as their depth is less than the sender node S. DBR does not take into

Table 6.2 Localization Free Routing Protocols

Architecture	*Technique*	*Performance Metrics*				*Knowledge Required*
		Packet Delivery Ratio	*Multiple Sinks*	*Energy Efficiency*	*Packet Overhead*	
DBR [13]	Only depth information is needed for comparison in routing	High	Y	Low	Low	No network information required
EE-DBR [14]	Compare depth as well as routing while performing routing operations	Low	Y	Fair	Medium	One-hop neighbor
H2-DAB [15]	Assigns dynamic addresses to all nodes in the network	High	N	Fair	Low	One hop neighbor
EE-DAB [16]	Assigns dynamic addresses and those addresses are compared during routing	Fair	Y	Fair	High	One-hop neighbor and sink
Delay-tolerant Data Dolphin (DDD) [17]	Sensing nodes stay static and data sense by them are forwarded to courier nodes	Low	Y	High	Low	Presence of dolphin nodes

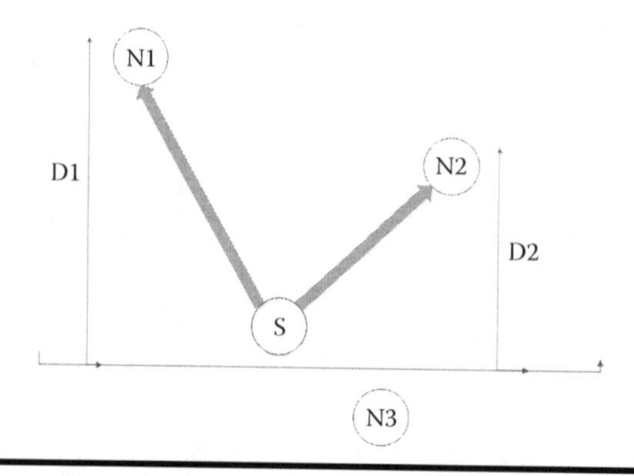

Figure 6.6 Node selection in DBR.

account any other parameters other than depth, which leads to a few drawbacks. Lifetime of the network where DBR is used will be less as it will always send data to the same higher node as no check has been observed. This will lead to the death of the node. Path selection in DBR has no proper mechanism, as no proper strategy is used for efficient or short path selection.

6.5.2 Energy-Efficient Depth Based Routing

In EE-DBR [14] protocol, when a node forwards its data, it takes into account the depth of the receiver node and its residual energy. When a node forwards data, it first compares the depth of the receiver node with itself; if the depth of receiver node is smaller than the sender, it checks the residual energy of receiver node. The node with higher residual energy and less depth among the neighbors is selected as the next hop for communication. Every node has information on depth and residual energy about their neighbors, so the node with the most suitable parameter is selected for communication. EE-DBR has not defined any mechanism for multipath communication. A node may forward data to node, which is far away from sender, and will result in higher energy consumption. Similarly, no parameter has been taken into account to define the shortest and efficient path towards sink.

6.5.3 Hop-by-Hop Dynamic Addressing Based Routing

In Hop-by-hop-dynamic addressing based (H2-DAB) routing [15], dynamic addresses are assigned to nodes and destination ID is set to "0" for all nodes. No prenetwork information is required in this protocol. As a first step of the network setup, a Hop-ID is assigned to each node. Every node in the network will have two types of addresses: Node-ID and Hop-ID. Node-ID is the physical address of the node while Node-ID changes with the change in location. Hop-IDs are assigned from top to bottom. Nodes having lower depth are assigned lower Hop-ID, like node which is nearest will have Hop-ID of 1. Similarly, nodes having higher depth are assigned higher Hop-IDs. H2-DAB supports multisink architecture, where multiple sinks are installed on shore. Those sinks are connected with each other through radio communication. Data packet received at any sink is considered as received.

However, this approach might create problems where a node cannot find any node in range, which has lower Hop-ID from sender node. In case of failure at finding a suitable node in the first attempt, the sender will retransmit data packet and then wait again for a specified amount of time. If results were still the same, then the sender node will forward data to a node having nearly or equal Hop-ID as sender node. This process results in energy wastage.

6.5.4 Energy-Efficient Dynamic Addressing Based Routing

In EE-DAB routing [16] scheme, every node is assigned Node-ID, S-Hop-ID, and C-Hop-ID. Node-ID shows the physical address of the node; S-Hop-ID consists of two digits which show how many hops one or two sinks are away. Left hop is considered as the highest priority and is selected as the primary route. The C-Hop-ID also consists of two digits which show how many hops the receiving nodes are away from courier nodes. Figure 6.7 describes how to make the selection of nodes for sending data packets. As source node N23 is having a data packet, with their

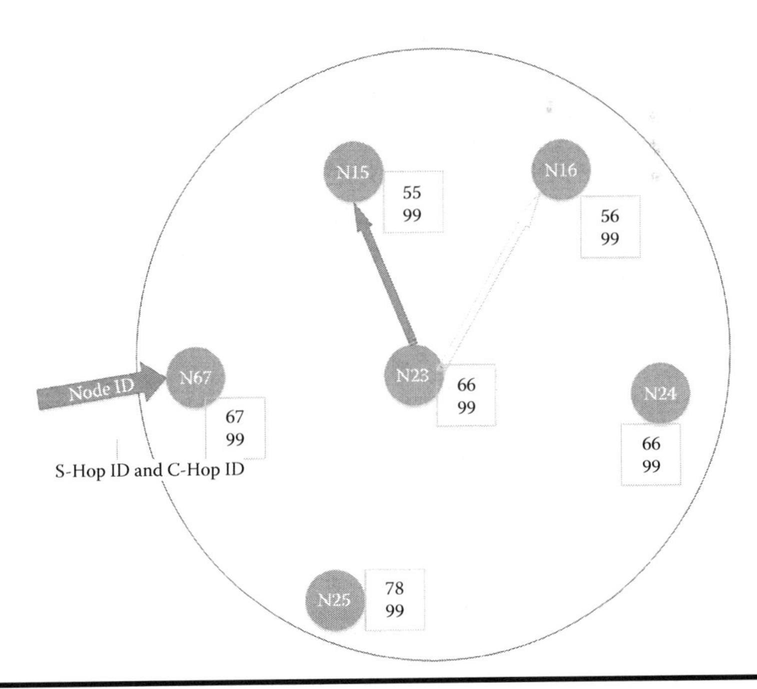

Figure 6.7 EE-DAB next hop selection.

own Hop-IDs 66 and 99, as C-Hop-IDs for all the nodes is 99 because of non-availability of courier node in the area. In result, simple query message will be sent asking neighbor nodes about their Hop-ID. In its reply, an inquiry reply packet is sent back to the sender node, which contains only three fields, that is, Node-ID, S-Hop-ID, and C-Hop-ID of replying nodes, where nodes N15, N16, N22, N24, and N25 lie in the communication range and will reply with their Node-ID and Hop-ID. After receiving, N23 sorts out these inquiry replies and gets the minimum Hop-ID. As the diagram shows, nodes N15 and N16 are declared as candidates for the next hop, because both of them have smaller Hop-ID when compared with the Hop-ID of the source node, but N15 qualifies for this competition because of its backup link, which is also smaller than N16. The source node will forward the data packet with N15 Node-ID as a next hop. In other cases, if two nodes respond with the same minimum Hop-ID, then the node that replied earlier will be selected as the next hop for further communication.

6.5.5 Mobile Delay Tolerant Routing

As acoustic communication uses more energy than that of radio communication, wireless sensor nodes are battery operated, and higher energy consumption leads to a serious problem, thus energy efficiency has become a major problem in UWSNs. In [17], a delay-tolerant protocol is proposed, which is called delay-tolerant data dolphin scheme. This proposed scheme is designed for delay-tolerant systems and applications. In these protocols, all the sensing nodes stay static, and data sensed by static nodes are passed on to data dolphin which acts as courier nodes. So in this methodology, high energy-consumed hop-by-hop communication is avoided. Data dolphins that act as courier nodes are provided with continuous energy. In the architecture, all static nodes are deployed in the sea bed. These static sensors go into sleep mode if there is no data to sense and they periodically wake up when they sense some data. After sensing some kind of desired data, they simply forward this data to courier nodes which are also called data dolphins. These data dolphins take this data and deliver it to the base station or sink. The number of dolphin nodes depends on the kind of network and its application and the number of nodes deployed in the network.

6.6 Conclusion

In this chapter, state-of-the-art routing protocols in UWSN have been presented. Almost all routing protocols in UWSN are presented in tabular form. UWSN environment is very different when compared with terrestrial wireless sensors network. Acoustic channels consume a large amount of energy with very less amount of data transferred. Furthermore, the flow of water makes it quite difficult for the sensor nodes to forward data in a stable scenario. Routing in UWSN is considered to be a very important part with respect to energy efficiency. Among all defined protocols given earlier, one cannot be selected the best because every protocol has some pros and cons. As a newly emerging technology, a lot of work needs to be done with respect to energy efficiency, end-to-end delay, propagation delay, and path loss. Energy-efficient routing schemes play a vital role in extending the lifetime of network and efficient path selection for data forwarding. Keeping in view the limitations in UWSN, energy-efficient schemes are encouraged. Underwater sensors are used in multiple application scenarios, and a separate mechanism is adopted when proposing a new routing scheme. In recent years, routing in UWSN has attracted a larg number of researchers in this area. Still this area carries certain challenges like topology management, energy efficiency, data retransmission, and path loss, which needs researcher's attention.

References

1. M. Khalid, Z. Ullah, N. Ahmad, M. Arshad, B. Jan, Y. Cao, & A. Adnan, "A survey of routing issues and associated protocols in underwater wireless sensor networks," *Journal of Sensors*, 2017(7539751), 2017.
2. S. M. Ghoreyshi, A. Shahrabi, & T. Boutaleb, "Void-handling techniques for routing protocols in underwater sensor networks: Survey and challenges," *IEEE Communications Surveys & Tutorials*, 19(2), 800–827, 2017.
3. N. Li, J.F. Martínez, J.M. Meneses Chaus, & M. Eckert, "A survey on underwater acoustic sensor network routing protocols," *Sensors*, 16(3), 414, 2016.
4. N. Ilyas, M. Akbar, R. Ullah, M. Khalid, A. Arif, A. Hafeez, & N. Javaid, "SEDG: Scalable and efficient data gathering routing protocol for underwater WSNs," *Procedia Computer Science*, 52, 584–591, 2015.
5. M. Khalid, Z. Ullah, N. Ahmad, H. Khan, H.S. Cruickshank, & O.U. Khan, "A comparative simulation based analysis of location based routing protocols in underwater wireless sensor networks," in *Recent Trends in Telecommunications Research (RTTR)*, Workshop on (pp. 1–5). 2017 IEEE.
6. M. Khalid, Z. Ullah, N. Ahmad, A. Adnan, W. Khalid, & A. Ashfaq, "Comparison of localization free routing protocols in underwater wireless sensor networks," *International Journal of Advanced Computer Science and Applications*, 8(3), 408–414, 2017.
7. P. Xie, J.H. Cui, & L. Lao, "VBF: Vector-based forwarding protocol for underwater sensor networks," *Networking*, 3976, 1216–1221, 2006.
8. N. Nicolaou, A. See, P. Xie, J.H. Cui, & D. Maggiorini, "Improving the robustness of location-based routing for underwater sensor networks," in *Oceans 2007-Europe* (pp. 1–6). IEEE.
9. J.M. Jornet, M. Stojanovic, & M. Zorzi, "Focused beam routing protocol for underwater acoustic networks," in *Proceedings of the Third ACM International Workshop on Underwater Networks*, pp. 75–82, 2008.
10. D. Hwang & D. Kim, "DFR: directional flooding-based routing protocol for underwater sensor networks," in *Proceedings of the IEEE Oceans*, pp. 1–7, IEEE, Quebec, Canada, September 2008.
11. E.A. Carlson, P.-P. Beaujean, & E. An, "Location-aware routing protocol for underwater acoustic networks," in *Proceedings of Oceans 2006*, September 2006.

12. N. Chirdchoo, W.-S. Soh, & K.C. Chua, "Sector-based routing with destination location prediction for underwater mobile networks," in *Proceedings of the International Conference on Advanced Information Networking and Applications Workshops (WAINA '09)*, pp. 1148–1153, IEEE, Bradford, UK, May 2009.
13. H. Yan, Z. J. Shi, & J.-H. Cui, "DBR: Depth-based routing for underwater sensor networks," *NETWORKING 2008: Ad Hoc and Sensor Networks, Wireless Networks, Next Generation Internet*, Lecture Notes in Computer Science 4982, 72–86, 2008.
14. A. Wahid & D. Kim, "An energy efficient localization-free routing protocol for underwater wireless sensor networks," *International Journal of Distributed Sensor Networks*, 2012, Article ID 307246, 11 pages, 2012.
15. M. Ayaz & A. Abdullah, "Hop-by-hop dynamic addressing based (H2-DAB) routing protocol for underwater wireless sensor networks," in *Proceedings of the International Conference on Information and Multimedia Technology (ICIMT '09)*, pp. 436–441, December 2009.
16. M. Ayaz, A. Abdullah, I. Faye, & Y. Batira, "An efficient dynamic addressing based routing protocol for underwater wireless sensor networks," *Computer Communications*, 35, 4, 475–486, 2012.
17. E. Magistretti, J. Kong, U. Lee, M. Geria, P. Bellavista, & A. Corradi, "A mobile delay-tolerant approach to long-term energy-efficient underwater sensor networking," in *Proceedings of the IEEE Wireless Communications and Networking Conference (WCNC '07)*, pp. 2868–2871, IEEE, March 2007.
18. Y. Noh, U. Lee, S. Lee, P. Wang, L.F. Vieira, J.H. Cui, & K. Kim, "Hydrocast: Pressure routing for underwater sensor networks," *IEEE Transactions on Vehicular Technology*, 65(1), 333–347, 2016.
19. H. Luo, K. Wu, R. Ruby, F. Hong, Z. Guo, & L.M. Ni, "Simulation and experimentation platforms for underwater acoustic sensor networks: Advancements and challenges," *ACM Computing Surveys (CSUR)*, 50(2), 28, 2017.
20. R.W. Coutinho, A. Boukerche, L.F. Vieira, & A.A. Loureiro, "Geographic and opportunistic routing for underwater sensor networks," *IEEE Transactions on Computers*, 65(2), 548–561, 2016.
21. Y. Noh, U. Lee, P. Wang, B.S.C. Choi, & M. Gerla, "VAPR: Void-aware pressure routing for underwater sensor networks," *IEEE Transactions on Mobile Computing*, 12(5), 895–908, 2013.

Chapter 7

Opportunistic Routing Protocols in Underwater Acoustic Sensor Networks: Issues, Challenges, and Future Directions

Varun G. Menon

SCMS School of Engineering and Technology

Contents

7.1 Introduction

Underwater acoustic sensor networks (UASNs) consist of a group of sensor nodes deployed under the ocean to collect data for many applications in ocean exploration, underwater surveillance, pollution detection, etc. With plenty of highly valuable hidden resources, exploration and study of the ocean is getting more attention everyday. Moreover, constant underwater surveillance has become an inevitable responsibility of military in many countries, especially in today's world with rising tension among nations. Traditional approaches used for underwater exploration and monitoring suffer from numerous limitations such as higher access time, increased cost, increased delay, and limitations in data transfer. This has extensively increased the demand for using group of sensor nodes and its related devices and applications for the monitoring and exploration purposes.

As radio signals do not work well under water due to its quick attenuation, all the nodes in underwater wireless acoustic sensor networks are equipped with acoustic transceivers to communicate with each other [1,2]. UASNs differ from traditional terrestrial sensor networks in numerous ways. Table 7.1 summarizes the differences between UASNs and traditional sensor networks. The sensor nodes in traditional

Table 7.1 Difference between TSNs and USNs

Attribute	*Terrestrial Sensor Networks (TSNs)*	*Underwater Sensor Networks (USNs)*
Deployment of sensor nodes	Densely deployed	Sparsely deployed due to challenges in cost
Cost	Sensor nodes deployed in the networks are less expensive	Sensor nodes deployed underwater are expensive
Power	Less power required in communication.	Power required is higher due to greater distance and more complex signal processing at the receiver
Data transfer rate	High	Low
Delay in data transfer	Less	More
Energy consumption	Low	High
Storage capacity	Sensor nodes have limited storage capacity	Underwater sensor nodes have more storage capacity as they require more data catching due to intermittent channel requirement
Mobility	More number of static nodes	Dynamic nodes
Error probability	Less error probability in communication	High error probability in communication

sensor networks are less mobile while the sensor nodes deployed under the ocean are dynamic due to sudden changes in the ocean and ocean currents. The delay in data transmission is higher in underwater sensor networks (UWSNs), and the data transfer rate is much lower compared with traditional sensor networks [3–4].

Over these years, numerous researches have been carried in the area of communication and data transfer in traditional sensor networks. Routing of data from the source to the destination nodes has been well investigated in traditional sensor networks. A number of efficient routing protocols have been designed so far for traditional sensor networks and mobile ad hoc networks. From traditional topology-based routing protocols to the latest geographic routing protocols [5] and opportunistic routing (OR) protocols [6–10], researchers have been successful in designing efficient protocols for traditional sensor networks. But due to the unique features discussed in Table 7.1, these protocols cannot be used efficiently for UWSNs. Moreover, all these protocols have been developed based on the working of radio signals. Thus, extensive efforts have been made by researchers worldwide in designing efficient communication protocols, considering the unique characteristics of UASNs.

In recent years, a number of advanced routing protocols have been proposed for efficient data delivery and reliable routing of data packets between the nodes in UASNs [11]. OR protocols [12] are the latest and most efficient class of protocols proposed for UASNs. OR utilizes the broadcasting nature of the wireless medium to increase the number of probable forwarder nodes in the network and improves the packet delivery rate. Opportunistic protocols proposed for UASNs are divided into two categories: geographic routing protocols [12] and pressure-based routing protocols [13]. Geographic routing protocols utilizes the location information of nodes, while pressure-based routing protocols [13] uses the pressure-level information for making routing decisions in UASNs. Most of the protocols in these two categories suffer from few issues and problems leading to their limited performance in UASNs.

The objective of this chapter is to study and analyze the working of the latest opportunistic protocols proposed for UASNs and to identify their issues and drawbacks. The analysis and study would help researchers in designing more efficient and optimized protocols for UASNs in the future. The chapter is organized as follows. Section 7.2 discusses the concept of OR in detail. Section 7.3 discusses the working of the most popular OR protocols proposed for UASNs. The various advantages, disadvantages, and issues faced by these protocols are discussed in this section. Section 7.4 discusses the performance evaluation of some of the major protocols using simulations. Open issues, challenges, and future research directions existing in designing efficient protocols for UWSNs are discussed in Section 7.5 and the chapter concludes in Section 7.6.

7.2 Opportunistic Routing

The concept of OR was first proposed for wireless ad hoc networks with extremely OR protocol [14]. The protocol aimed at exploiting and taking advantage of the broadcasting nature of the wireless channel to improve performance of data delivery in the network. The OR utilized the reception of the same broadcasted packet at multiple devices in the network and selected one best forwarder device dynamically from the set of multiple receivers. The most important advantage of this class of protocols is that they do not commit to a fixed route before data transmission. The next forwarder device and the route are only determined dynamically based on current network conditions, leading to its better performance compared with all previous classes of routing protocols proposed for ad hoc networks. When a sender device wants to send a data packet to a particular destination device, it broadcasts the data packet to a list of candidate devices that are in its transmission range. Now, these candidate relay devices are prioritized based on some metrics like expected transmission count (ETX) [14] or expected transmission time (ETT) [15] calculated dynamically from the network. The candidate devices that receive the data packet run a coordination scheme to determine the best forwarder for the current data packet. Thus, the forwarder device is selected dynamically from the network based on current network characteristics. The data packet is then forwarded by the best forwarder device, and this OR strategy continues till the data packet reaches the destination.

7.2.1 Stages in Opportunistic Routing

Working of OR can be divided into four stages. Every stage has equal importance in achieving good performance in dynamic networks. The four stages are as follows.

7.2.1.1 Selection of Candidate Set

In the first phase selection of candidate set, the sender device generates a list of probable forwarder nodes referred as candidate devices from the neighboring devices that are in its transmission range. The source device may use periodic or nonperiodic message broadcasts or beacon messages to discover and maintain the list of candidate devices in the network. Every device in the network constantly updates this candidate list and keeps the candidate set dynamic for every transmission.

7.2.1.2 Data Broadcast

Once the candidate set is selected, the data packet is broadcasted by the sender to all devices in the candidate set. This is the major advantage of having multiple forwarders with OR, as more than one candidate device receives the data packet and is ready to forward the data packet.

7.2.1.3 Prioritization of the Forwarder Devices

In the next phase, OR sorts the devices in the candidate set based on a particular metric calculated dynamically from the network. A number of metrics such as ETX [14], packet advancement [5], expected any path transmission [16], ETT [15], expected any path transmission time [17], expected duty cycle [18], etc. are used for prioritization of forwarder devices. Based on the metric, the best forwarder device is selected to forward the data packet to the destination.

7.2.1.4 Data Forwarding by the Best Forwarder Device

Once the priority of devices is generated using the specific metric calculated dynamically from the network, the data packet is forwarded by the best forwarder device in the list. If the best forwarder device is unable to forward the data packet, the next best forwarder device in the priority list forwards the data packet. This strategy is used by OR till the packet reaches the destination device. Thus, forwarding of data packet is ensured as long as there is one device in the forwarder candidate set leading to excellent packet delivery rate in the network.

Figure 7.1 depicts the working of OR protocol in mobile ad hoc networks. Here the node S wants to send a data packet to the destination node D. Node S broadcasts the data packet intended for the destination into the network. This data packet is received by nodes X, Y, and Z which are in the transmission range of node S. Based on the particular OR strategy used, one node is selected as the forwarding node. Here we use packet advancement [5] as the deciding factor. So node Z which is the nearest to the destination is selected to forward the data packet. Node Z broadcasts the data packet into the network. Node X and node Y receive a copy of the data packet and thus understand that the data packet has already been forwarded by the best forwarder. So they discard the data packet. Node A and B also receive the data packet, and based on the nearness to the destination node, B is selected to forward the data packet to the destination. Node B checks for the destination in the list of neighbors and finds the destination D in it. So node B directly delivers the data packet to the destination node D.

UWSNs collect data from the environment and transfer them to the sonobuoys on the surface to send them to a center for further processing. The acoustic channels common to UWSNs have low bandwidth, high error probability, and longer

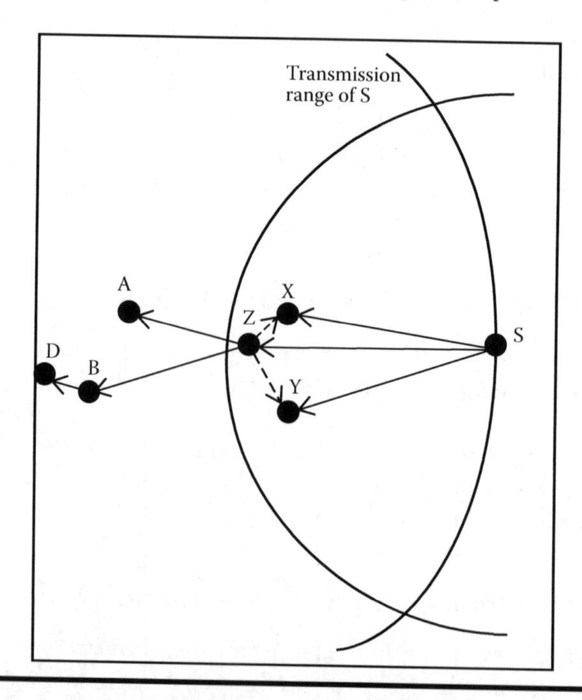

Figure 7.1 An example of working of opportunistic routing.

propagation delay compared with radio channels, and these properties of UWSNs make them good candidates for using OR concepts to deliver packets to the destination. Recently, the focus of research in UASNs has been in improving the packet delivery rate and reliability of data transfer in the network. This has led to the increased popularity of adopting the OR concept in many recent protocols designed for efficient data delivery in UASNs. Most of these protocols proposed for UASNs use the advantages of OR to increase link reliability and path stability between sensor nodes and sink nodes [12]. Using multiple potential forwarding candidate nodes, the packets have improved progress towards the destination node, which results in fewer retransmissions. Next section discusses the latest protocols proposed for UASNs using the OR concept and highlights their issues and challenges.

7.3 Opportunistic Routing Protocols in UASNs

The OR protocols proposed for UASNs are mainly classified into two major categories: location-based protocols and pressure-based protocols. Location-based category of protocols uses the position information of the deployed sensor nodes in the network for generating the probable forwarding candidate set, in prioritizing the candidate set and in making the decision on the best forwarder device. Pressure-based protocols on the other hand use the depth or pressure information of nodes for making the best forwarder decision in the network.

7.3.1 *Location-Based Opportunistic Routing Protocols*

Location-based protocols uses the position information of the deployed sensor nodes in the network for generating the probable forwarding candidate set, in prioritizing the candidate set and in making the decision on the best forwarder device. These protocols commonly need geographical information of all nodes in the network as well as information about the sink location. These protocols use most of the energy in nodes in obtaining the location information and are considered less energy efficient. This section discusses the working of the popular and latest location-based protocols proposed for UASNs.

7.3.1.1 Vector-Based Forwarding

The vector-based forwarding (VBF) protocol was introduced by Xie et al. [19] in 2006 to provide robust, flexible, and scalable routing in UASNs. Being a location-based routing protocol, VBF uses the vector-based routing approach in which the nodes close to the vector from the source node to the destination node will forward the data packet and only a tiny fraction of the nodes are involved in routing. The performance of VBF is further enhanced by the use of a localized and distributed self-adaptation algorithm that allows nodes to benefit the forwarding packets. The major enhancement is the advantage offered in reduced energy consumption by discarding the low-benefit packets. The most important advantage offered by VBF is that no state information is required to be maintained by each node.

In UWSN, the VBF protocol initially obtains the position information of each node in the network using the location algorithm. It then creates a vector pipe from the source node to destination, with a certain angle of arrival and signal strength to find the forwarder. Once the packet is being delivered, the particular intermediate node acts as the sender and finds the next forwarder using the same strategy. This strategy is used until the data packet reaches the destination. Another major advantage of VBF is that it forwards the data packets along multiple redundant paths that make the protocol reliable against packet loss. Each packet header contains the positions of the source, destination, and forwarder. Using the positions of source and destination, a virtual pipe (vector) between the two nodes is created and the packets are forwarded through it.

The working of VBF is illustrated in the right side of Figure 7.2. S1 is the source node and D1 is the sink node in our UASN. Using VBF, a vector S1D1 is drawn between source and destination nodes. The nodes located in the virtual pipe A, B, E, F, H, and J are potential candidates for forwarding the packets while sensor nodes C and D doesn't take part in the forwarding process. VBF is a receiver-based and stateless routing protocol that needs only the destination position of the node. When a sensor node receives the broadcasted data packet, it checks whether it is close enough to the line between the source and destination. If a node determines that it is close enough to the routing vector, then it includes its position in the packet header as the forwarder and thereby transmits the packet. The nodes that

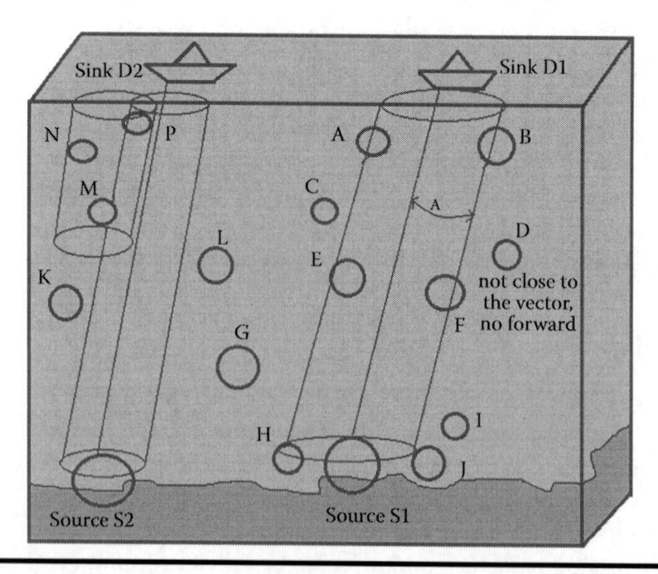

Figure 7.2 An illustration of working of VBF and HH-VBF.

have received the packet but are not in the virtual pipe will discard the packet. In the figure, sensor nodes C and D are not in the virtual pipe and they would discard the data packet if received.

The major issues with the VBF protocol is the low delivery rate offered in sparse UASNs. When the numbers of nodes are few, VBF results in low packet delivery rate because of the fixed virtual pipe between the source and sink. VBF also does not have any recovery mechanism in the event that a packet reaches a void node. The performance of VBF comes down considerably in networks with communication voids. Sometimes error in candidate coordination mechanism leads to multiple duplicate transmissions that lead to performance degradation.

7.3.1.2 Hop-by-Hop Vector-Based Forwarding (HH-VBF)

Hop-by-hop VBF (HH-VBF) is the enhanced version of VBF that creates the virtual pipe for each hop as the packets move from the source node to the destination node [20]. The unique characteristic of HH-VBF compared with VBF is that, instead of having a single virtual pipe between the source and destination node, HH-VBF makes a virtual pipe per hop between the current forwarder to the sink node. HH-VBF proposes that, by creating a virtual pipe from the current forwarder to the destination, there is a better chance of reliable data delivery. The radius of the pipeline is similar to the transmission range of the node. HH-VBF changes the direction of the forwarding pipe hop by hop in the entire lifetime and, in this way, every forwarding node can make routing decisions based on the current local topology information. The working of HH-VBF is illustrated in the left side of Figure 7.2. The source node S2 wants to send the message to the destination sink D2 in the UASN. The node S2 creates a virtual pipe to the destination D2. Once the packet reaches the intermediate sensor node M, it creates a virtual pipe to the destination. Now there are more nodes (N and P) included for forwarding the data packet to the destination. HH-VBF uses this forwarding strategy till the packet reaches the destination. HH-VBF improves the performance in the network by dynamically changing the direction of the flooding pipeline. HH-VBF improves the packet delivery ratio (PDR) compared with VBF because it increases the chance of finding a more suitable forwarder within the hop-by-hop virtual pipeline. In a sparse network, there is a higher chance of delivering the packet to the destination through nodes in each virtual pipe, which is defined by each candidate.

When a constant pipeline radius is set, compared with VBF, HH-VBF has a relatively low performance in sparse network. However, one major issue existing with both HH-VBF and VBF is the interference of the marine mammals. When the marine mammals block the virtual pipe, the transmission often gets interrupted leading to data loss. However, both VBF and HH-VBF are vulnerable to marine mammals, since the transmission occurs only in the pipe. Once the pipe is blocked by marine mammals, then the transmission can possibly be interrupted. Duplicate transmissions arising from incorrect candidate coordination is another major issue in HH-VBF. Duplicate transmissions have a great impact in performance degradation of data transfer in UASNs, resulting in extremely low quality of service (QoS). Duplicate data transmission increases packet collisions and waste network bandwidth, which is a major performance parameter in UWSNs. HH-VBF does not provide energy fairness to the sensor nodes during the routing and decision-making processes.

7.3.1.3 Geographic Partial Network Coding

Geographic partial network coding (GPNC) combines the advantage of location-based routing schemes with partial network coding [21]. GPNC uses the location information of the nodes in the network for routing and partial network coding for data transmission. Partial network coding helps to decrease the number of sending packets and also reduces the collision between packets. To forward a packet, GPNC selects a forwarding node that is nearest to the destination node using the packet advancement metric. Packet advancement is the estimated progress of the packet to the destination that is calculated based on the distance of the intermediate forwarder node to the destination. Working of GPNC is simple and efficient. When a data packet needs to be transmitted by a forwarding node, it encodes the packet using partial network coding, and with the distance decreasing to the sink node, the encoded packet can finally arrive at the sink node. GPNC uses the greedy forwarding strategy to select the best forwarder node for each data packet. As soon as the destination node receives a group of linear independently encoded data packets, it performs decoding to retrieve the packets. The major advantage of GPNC is in its reduced energy consumption and network delay. GPNC takes into consideration the amount of energy left with the intermediate nodes during the decision-making process in routing. GPNC protocol improves the performance of data transfer considerably compared with the previous protocols. However, the protocol has many limitations in handling the communication void problem in the network. The performance of this protocol is highly affected by the communication void regions in the network.

7.3.1.4 Opportunistic Void Avoidance Routing

Opportunistic void avoidance routing (OVAR) is one of the advanced protocols using the OR concept and location information of the nodes to improve the performance of data transfer even with communication holes in the network [22]. The protocol also aims to address the energy efficiency issue among the sensor nodes in the network. It tries to increase the transmission reliability and network throughput in the network by excluding all routes leading to a communication void area. OVAR is a soft-state routing protocol in which some reachability information such as hop count distance and forwarding direction are provided and kept in each node. During the packet forwarding, if an intermediate forwarding node cannot find a qualified node with positive progress towards the destination, the data packet may be dropped. This is the communication void or communication hole problem. Communication void problem has led to the low performance of a number of protocols in UASNs. Most of the geographic routing protocols using the location information of the nodes suffer from this problem in the network. To address the void problem and the lossy nature of the underwater acoustic channel, forwarding nodes locally collaborate on packet forwarding with very low overhead.

OVAR works with a hop-by-hop forwarding set selection to deliver packets to the sink. The forwarding set is determined by using a combination of packet advancement and hop distance metrics. The forwarding set prevents the hidden terminal problem, by including the nodes that are out of range of each other. The energy is managed using the number of collaborative nodes that can be adjusted according to the density of the network. Each node considers its depth as the second metric to set a relaying timer and thereby prioritizes multiple forwarding nodes. The node with the highest priority or lowest depth transmits the packet and other low-priority nodes can drop the packet by hearing the transmission. This mechanism of selecting a path with a lower hop count leads to more energy savings and a higher delivery ratio. OVAR is able to select the forwarding set in any direction from the sender, which increases its flexibility to bypass any kind of void area with the minimum deviation from the optimal path. Major advantages of OVAR are that the protocol is highly scalable and also minimizes the number of retransmissions in the network (Figure 7.3).

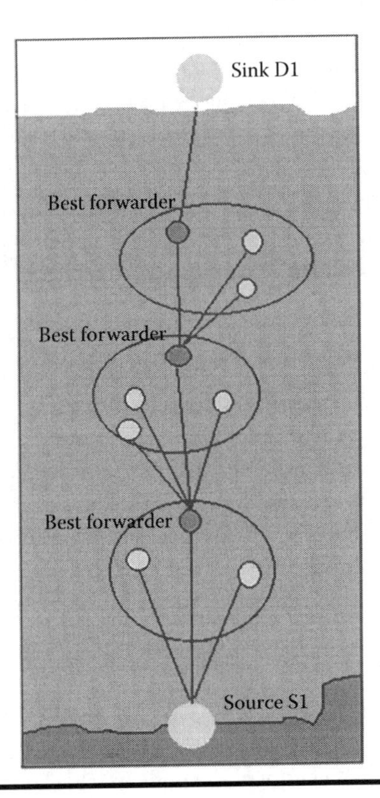

Figure 7.3 An illustration of working of OAVR.

7.3.1.5 Geographic and Opportunistic Routing with Depth Adjustment-Based Topology Control for Communication Recovery over Void Regions

Geographic and OR with Depth Adjustment-based topology control for communication Recovery over void regions (GEDAR) is an anycast, geographic, and OR protocol that routes data packets from sensor nodes to multiple sonobuoys (sinks) at the sea's surface [23]. In normal operation mode, GEDAR makes use of greedy forwarding strategy to advance the packet at each hop towards the surface sonobuoys and move the void nodes to new depths to adjust the topology. The major advantage of GEDAR protocol over all other protocols proposed for UASNs is in the additional reliability offered by the recovery mode. The recovery mode based on depth adjustment of the void node helps to route the data packet when it gets stuck at a void node. GEDAR aims to mitigate the effects of the acoustic channel, and thus a subset of neighbor nodes is determined to continue forwarding the packet towards some next-hop forwarder set. GEDAR protocol is designed such that it is able to use one or multiple sink nodes on the water's surface to gather the transmitted information. In GEDAR, each node assumes that it has geographic positions of its neighbors. The expected packet progress (EDP) [24] metric is used to generate the candidate set of potential forwarders.

When a source node needs to send data to a sink node located on the surface, it includes its candidate relay set in the header of the data packet. The node then broadcasts the data packet to the network. Every node that receives the data packet checks for its entry in the candidate relay set. If it is the highest priority forwarder in the set, it forwards the node towards the destination. Else it uses a timer-based coordination scheme for deciding on possible retransmissions of the data packet. This strategy is continued till the data packet reaches the destination. GEDAR switches to the recovery mode if the data packet reaches a void node. The node then calculates the new depth to move in order to avoid the void node.

The major issue with GEDAR is the wastage of network bandwidth due to a large number of beacon messages used in obtaining the position information of the nodes. Another limitation of the protocol is that it doesn't consider the energy of the nodes in making routing decisions in the network. This is an important challenge because the best forwarder nodes selected from the candidate set needs to stay alive for forwarding the data packet. Duplicate data transmission due to errors in candidate set coordination process is another major issue with GEDAR.

7.3.1.6 Energy-Scaled and Expanded Vector-Based Forwarding

Energy-scaled and expanded VBF (ESEVBF) is another latest advanced protocol proposed for UASNs [25]. The protocol mainly focuses on energy efficiency in the network. It uses the residual energy of the node to scale and vector pipeline distance ratio to expand the holding time. It is a VBF scheme that forwards data packets to sink using holding time and location information of the sender, forwarder, and sink nodes. The major advantage proposed by ESEVBF is in avoiding multiple forwarding at the intermediate nodes, thus reducing the energy consumption in the network. ESEVBR contributes energy fairness and reduces the packet broadcast by scaling and expanding the holding time with the residual energy. Another unique feature of the protocol is that the expanded proximity closeness ratio of the forwarding candidate nodes towards the virtual pipeline between sender and sink is added in holding time computation to signify the node preference. The major areas of concern with this protocol are the duplicate packet transmissions in the network and the limitations in handling multiple communication voids in the network.

7.3.1.7 Energy-Efficient Cooperative Opportunistic Routing

Energy-efficient cooperative OR (EECOR) protocol is one of the latest protocols proposed for energy efficiency in UASNs [26]. The protocol focuses on enabling data transfer in the network with energy fairness among the deployed sensor nodes. The protocol uses the local position information to generate the candidate relay set. Once the candidate set is generated, a fuzzy logic-based relay selection scheme is used to select the best forwarder to forward the data packet in the network. The algorithm considers the energy consumption ratio and the packet delivery probability of the forwarder in the selection process. EECOR is a source-based protocol where the source node will decide which relay nodes will cooperate in forwarding the packets to the next hop destination. The neighboring relay node in the forwarding relay set will accept the packet, if not the packet will be dropped by the relay node. By taking the advantage of the broadcast nature of wireless communications, the source node locally selects the forwarding relay nodes based on depth information from the embedded depth sensor and thereby links quality between the source node and each neighbor node. The OR concept is used by the EECOR protocol to solve the energy consumption challenge of the acoustic signal propagation and enhance the communication reliability. EECOR protocol improves the network lifetime compared with all previous protocols in UASNs, because the residual energy of the sensor nodes is considered when selecting the best relay to forward the packets. The major issue with EECOR is the limitation in handling communications voids in the network. The protocol mainly focuses on energy efficiency of the sensor nodes and needs to consider the QoS parameters in data transmission. The actual performance of the protocol needs to be tested in a real working environment.

7.3.2 Depth-Based Opportunistic Routing Protocols

Most of the location-based geographic routing protocols proposed for UASNs utilize much energy in obtaining the position information. Many protocols suffer from increased traffic in the network and drainage of energy due to constant beacon messages. The next major category of OR protocols, i.e., depth-based routing (DBR) protocols, does not need information on the position of the sensor nodes. It utilizes the depth and pressure information of each deployed sensor node to make the routing decisions in the network. Node with highest depth sends data to the higher nodes and subsequently the data reaches the surface sink node.

7.3.2.1 Depth-Based Routing

DBR protocol uses the depth information of the nodes as the metric to decide whether to forward the packets or not [12,27]. DBR assumes that each sensor node is attached with a depth sensor to measure the depth of the node. DBR uses the OR concept, where the nodes closer to the water surface are potential candidates for receiving and forwarding the packet. DBR is a receiver-based OR protocol like VBF and HH-VBF, but it uses the depth of the nodes as an OR metric. Moreover, multisink architecture is used in DBR, which helps to increase the packet delivery in the network. One of the objectives of DBR is that it makes the use of timer-based coordination to prevent duplicate transmissions, thereby it allows the candidate that has less depth to wait for a short period of time.

Whenever a sensor node needs to send data in DBR, first it measures the depth, assigns it in the data packet header, and broadcasts it. The received node compares its depth with the one that's included in the data packet header. If a node's depth is lower or closer to the surface of the water than the depth of the previous forwarder, then the receiver is considered as a potential candidate to forward the packet. When the wait time of a candidate expires and if it does not hear the transmission of the same packet, then it will forward the packet. DBR does not consider the energy level of the sensor node as a metric to decide which node should act as a member of the candidate set. A node with less depth and limited remaining energy may choose to forward the packets, while another node with similar depth and greater energy has a rare chance to be in the candidate set. Due to long propagation delays, routing discovery is expensive in UASNs, and the DBR protocols are designed such that its routing discovery cost is almost zero. DBR is a scalable protocol, because it does not need a control packet to obtain depth information. If the depth difference to the previous hop's depth is greater than the threshold, then the node is considered a potential candidate, and it does not have any recovery mechanism when a packet reaches its void node.

The major issue with DBR is the limitation in handling the communication void in the network. Also, the performance of DBR comes down in sparse networks. The larger depth threshold results in fewer nodes in the candidate set, and therefore, this leads to lower PDR. Also if the depth threshold is small, many nodes may be eligible for forwarding the packet, and this leads to duplicate transmissions in the network. DBR does not take any other parameters other than depth, which leads to a few drawbacks.

7.3.2.2 Hydraulic Pressure Based Anycast Routing

Hydraulic pressure based anycast routing (HydroCast) algorithm was proposed to solve the DBR's problem of local maximum [12,13]. Similar to DBR, HydroCast also makes use of the node's depth information to select candidates from the neighboring nodes. The pressure level of the sensor node is used to route the data packets in a greedy multihop fashion to sinks that are deployed on the surface. HydroCast proposes a candidate selection algorithm that considers the EDP as an OR metric. In addition to new candidate selection algorithm in HydroCast, the approach for local maximum recovery is used when a packet is delivered to a void node. The candidate selection and coordination of HydroCast is based on EDP and the heuristic algorithm. A timer-based algorithm is used for candidate coordination, same as the other existing approaches. The major issue with HydroCast is that it fails to consider the energy of sensors in making the routing decisions in the network.

7.3.2.3 Void-Aware Pressure Routing

Void-aware pressure routing (VAPR) protocol is aimed to solve the communication void problem in UASN [12,28]. VAPR is designed with two phases known as the beaconing phase and opportunistic data forwarding phase. In the beaconing phase, each sink in the network starts sending the beacon message along with the depth information, hop count, and direction information of the current node. Whenever a node receives a beacon message, it uses the beacon to store information and later update the packet information by depth and hop count. And later, the node rebroadcasts it. When a node receives a message from another node with smaller depth, then the direction of the node is revised upwards. Otherwise, the direction of the node is revised downwards. VAPR removes void nodes from its search area by using opportunistic forwarding selection. VAPR relies on no recovery phase in the event the packet reaches its void node. Rather, VAPR selects nodes according to the information in each node, such that the packet will not get trapped in a void node. The major issue with VAPR is the increased energy consumption at the nodes. One issue with VAPR is that the protocol is fully based on control messages that increase the delay and overhead. Moreover, the void avoidance technique often fails in different networks.

7.3.2.4 Energy and Depth Variance-Based Opportunistic Void Avoidance

Energy and depth variance-based opportunistic void avoidance (EDOVE) protocol is aimed at energy balancing and void avoidance in the network [29]. EDOVE considers the depth parameter as well as the normalized residual energy of the one-hop node to the normalized depth variance of the second hop neighbors.

The major advantage proposed by this method is in limiting the duplicate packet broadcast and also in achieving good performance even with void areas in the network. EDOVE prioritizes the receiver node to become the potential forwarding candidate if it has maximum residual energy among its one hop neighbors or if it has one or more neighbors with less depth or if it has more number of neighbors with depth less than the receiver node and larger normalized depth variance. The main objective of EDOVE is that it avoids the void region by selecting the forwarder with large residual energy and having multiple neighbors that are largely distributed depthwise. Energy efficiency is achieved by avoiding the packet collision with void avoidance, which results in an increase in the network lifetime. By prioritizing the forwarders with one or more neighbors, communication void avoidance is achieved. The major issue with the protocol is in the limitations with duplicate data transmissions in the network. The delay experienced by the protocol in routing and data transfer is higher compared with the other previous protocols in UASNs (Table 7.2).

Table 7.2 Issues and Challenges with the Major Protocols in UASNs

OR Protocol	*Year*	*OR Type*	*Candidate Coordination*	*Issues and Challenges*
VBF [19]	2006	Location based	Timer based	Low delivery rate in sparse networks. Duplicate data transmissions. Unable to handle communication voids. Unable to handle interferences in the virtual pipe. Lack of energy fairness to sensor nodes
HH-VBF [20]	2007	Location based	Timer based	Duplicate data transmissions. Unable to handle communication voids. Unable to handle interferences in the virtual pipe. Lack of energy fairness to sensor nodes
DBR [27]	2008	Depth based	Timer based	Limitation in handling the communication voids. Performance of DBR comes down in sparse networks. Duplicate transmissions in the network.

(Continued)

Table 7.2 (*Continued*) Issues and Challenges with the Major Protocols in UASNs

OR Protocol	*Year*	*OR Type*	*Candidate Coordination*	*Issues and Challenges*
HydroCast [13]	2010	Depth based	Timer based	Limitation in handling the communication voids in the network. Lack of energy fairness to sensor nodes
VAPR [28]	2013	Depth based	Timer based	Increased energy consumption Increased delay and overhead
GPNC [21]	2015	Location based	Timer based	Limitations in handling communication voids in the network.
OAVR [22]	2016	Location based	Timer based	Error in coordination leads to duplicate messages in the network.
GEDAR [23]	2016	Location based	Timer based	Complexity in execution. Issues with control packet. Wastage of network bandwidth due to large number of beacon messages. Lack of energy fairness to sensor nodes. Duplicate data transmissions due to errors in candidate set coordination.
ESEVBF [25]	2017	Location based	Timer based	Limitations in handling communication void Duplicate data transmissions
EECOR [26]	2017	Location based	Timer based	Limitations in handling communication void Duplicate data transmissions
EDOVE [29]	2017	Depth based	Timer based	Duplicate data transmissions, higher delay

7.4 Performance Evaluation

Working and behavior of the major OR protocols proposed for UASNs are analyzed using simulations in Aqua-Sim [30]. Aqua-Sim is a high-fidelity and flexible packet-level UWSN simulator developed on NS-2 to simulate the impairment of the underwater acoustic channel. Compared with other wireless network simulators, Aqua-Sim offers many distinctive features such as underpinning to discrete event-driven networks, support for mobile and 3D networks, simulation of high-fidelity underwater acoustic channels, and implementation of a complete protocol stack. Six hundred sensor nodes are randomly deployed in a 3D region of 1500×1500×1500 m, where the number of sonbuoys is 45. The data transmission rate of the underwater acoustic medium is 50 kbps. The acoustic signal propagation speed is 1500 m/s. Each sensor node has a transmission range of R=250 m. The most popular OR protocols from both categories VBF and DBR along with the latest GEDAR protocol are taken for performance analysis.

Figure 7.4 shows the comparison of PDR among the three protocols with varying numbers of nodes in the network. It is observed that the location-based protocols gives better data delivery rate in the network compared with the depth-based protocols. The latest proposed protocol GEDAR gives better data delivery rate in the network compared with the other two protocols.

Figure 7.5 shows the comparison of latency among the three protocols with varying number of nodes in the network. Location-based protocols enjoy lower delay in data transmission compared with the pressure-based protocols. The latest

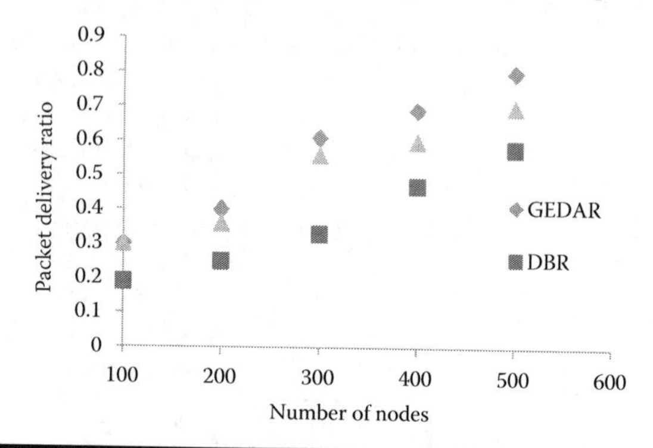

Figure 7.4 Packet delivery ratio vs. number of nodes.

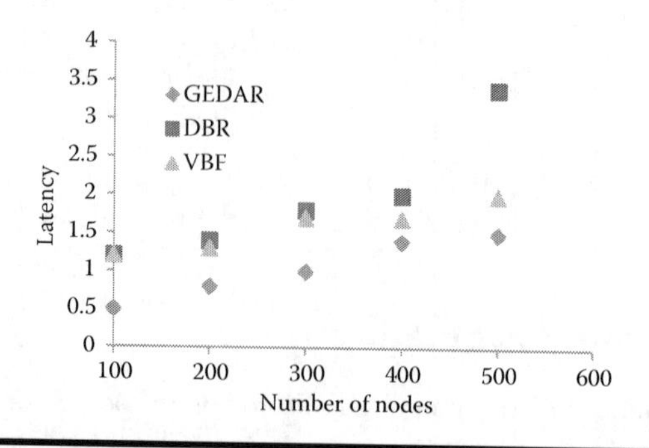

Figure 7.5 Latency vs. number of nodes.

protocol in this category, GEDAR achieves less delay compared with the other two protocols in UASNs. Next section discusses the challenges and open research issues existing in the area of routing in UASNs.

7.5 Challenges and Open Research Issues

A number of issues and challenges still exist in routing of data packets in UASNs. These highlighted areas should be the focus of research in the coming years. Some of the existing open research issues are as follows.

7.5.1 Duplicate Data Transmissions

Most of the OR protocols proposed for UASNs suffer from increased duplicate data transmissions. As most of these protocols use a flooding approach to improve the delivery rate in the network, more and more duplicate transmission occurs. Any error in the coordination scheme among the candidate nodes leads to numerous redundant transmissions in the network. Lack of clarity in the number of required candidate nodes adds to this issue. This is a major issue to be addressed because duplicate transmissions degrade the performance of the protocols considerably in the network leading to low QoS to the application users.

7.5.2 Overhead Incurred

The additional overhead incurred is often unavoidable but affects the performance of the routing protocol adversely. Routing protocols has to minimize the additional overhead used for routing, error correction, and retransmission purposes. The complexity of the protocol needs to limited. Although many proposed protocols give better performance in UASNs, they have higher overhead, leading to limitations in practical usage. How to minimize the overhead is an important research question still not answered.

7.5.3 Number of Nodes in the Candidate Set

Determining the number of nodes to be included in the candidate set is a major confusion among the OR protocols. Maintaining too many candidates can lead to increased retransmissions and storage problems in sensor nodes. Maintaining too less candidates in the set might lead to lack of forwarding nodes in error-prone channels. So it is important to find out the optimum number of nodes to be included in the candidate set.

7.5.4 Simulation-Based Study

A major issue with most of the protocols proposed for UASNs is that the protocols are tested in simulations. The results are obtained based on simulations. Actual working and feasibility of these protocols could only be analyzed with real-time experiments. Researchers should focus on testing the protocols in real UASN environment. Results from these tests would be highly beneficial for future applications compared with the simulation results.

7.5.5 Quality of Service and Energy Limitation of Nodes

A major concern required for the protocols is in maintaining the required QoS considering the limited energy of nodes in the network. Many protocols work on energy efficiency in the network but provide too low QoS to the users. The new protocols should focus on providing good data delivery rate in the network with reduced delay and overhead while considering the residual energy available in the sensor nodes.

7.5.6 Scalability and Communication Voids

A major concern with many protocols proposed for UASNs is that they give good performance only in small networks. As the networks scale, their performance comes down with increased data loss and delay. As the networks become large, communication holes might develop. It is essential for every protocol to have an efficient mechanism to handle the communication void in the network.

7.6 Conclusion

The research paper initially discussed the working of OR in detail and highlighted its advantages over all the previous routing methods. OR utilized the reception of the same broadcasted packet at multiple devices in the network and selected one best forwarder device dynamically from the set of multiple receivers. The paper discussed the working of two major classes of OR protocols proposed for UASNs; LBR and DBR schemes. Working and behavior of the prevalent and latest protocols from the two categories were analyzed, and the issues and challenges existing in each of these protocols were discussed. The performance of three major OR protocols were evaluated using simulations. Further, the existing challenges and open research problems were discussed, which would provide future research direction to researchers worldwide.

References

1. E.M. Sozer, M. Stojanovic and J.G. Proakis, "Underwater acoustic networks", *IEEE Journal of Oceanic Engineering*, vol. 25, no. 1, pp. 72–83, Jan. 2000.
2. Jun-Hong Cui, Jiejun Kong, M. Gerla and Shengli Zhou, "The challenges of building mobile underwater wireless networks for aquatic applications", *IEEE Network*, vol. 20, no. 3, pp. 12–18, May–June 2006.

3. Varun G. Menon and P.M. Joe Prathap, "Comparative analysis of opportunistic routing protocols for underwater acoustic sensor networks", *Proceedings of the IEEE International Conference on Emerging Technological Trends*, Kerala, India, 2016.
4. I. Akyildiz, D. Pompili, and T. Melodia, "Underwater acoustic sensor networks: Research challenges," *Elsevier Ad Hoc Networks*, vol. 3, no. 3, pp. 257–279, 2005.
5. B. Karp and H.T. Kung, "GPSR: Greedy perimeter stateless routing for wireless networks," *Proceedings of the ACM Mobile Computing*, pp. 243–254, 2000.
6. S. Biswas and R. Morris, "ExOR: Opportunistic multi-hop routing for wireless networks," in *Proceedings of the ACM SIGCOMM*, Aug. 2005.
7. E. Rozner, J. Seshadri, Y. Mehta and L. Qiu, "SOAR: Simple opportunistic adaptive routing protocol for wireless mesh networks," *IEEE Transactions Mobile Computing*, vol. 8, no. 12, pp. 1622–1635, Dec. 2009.
8. Varun G. Menon and P.M. Joe Prathap, "Analysing the behaviour and performance of opportunistic routing protocols in highly mobile wireless ad hoc networks", *International Journal of Engineering and Technology*, vol. 8, no. 5, pp. 1916–1924, 2016.
9. Varun G. Menon, P.M. Joe Prathap and A. Vijay, "Eliminating redundant relaying of data packets for efficient opportunistic routing in dynamic wireless ad hoc networks", *Asian Journal of Information Technology*, vol. 12, no. 17, 2016.
10. Varun G. Menon, Joe Prathap Pathrose and Jogi Priya, "Ensuring reliable communication in disaster recovery operations with reliable routing technique," *Mobile Information Systems*, vol. 2016, Article ID 9141329, 10 pages, 2016.
11. G. Han, J. Jiang, N. Bao, L. Wan and M. Guizani, "Routing protocols for underwater wireless sensor networks," *IEEE Communications Magazine*, vol. 53, no. 11, pp. 72–78, 2015.
12. Amir Darehshoorzadeh and Azzedine Boukerche, "Underwater sensor networks: A new challenge for opportunistic routing protocols", *IEEE Communications Magazine*, November 2015.
13. U. Lee et al., "Pressure routing for underwater sensor networks," *Proceedings of the IEEE INFOCOM*, Mar. 2010, pp.1–9.
14. S. Biswas, and R. Morri, "ExOR: Opportunistic multi-hop routing for wireless networks", *Proceedings of the 2005 Conference on Applications, Technologies, Architectures, and Protocols for Computer Communications (SIGCOMM '05)*, ACM, New York, USA, pp. 133–144, 2005.
15. J. Lee, C. Yu, K.G. Shin and Y. Suh, "Maximizing transmission opportunities in wireless multihop networks", *IEEE Transactions on Mobile Computing*, vol. 12, no. 9, pp. 1879–1892, 2013.
16. Z. Zhong and S. Nelakuditit, "On the efficacy of opportunistic routing," in *Proceedings of the IEEE Communications Society Conference on Sensor, Mesh and Ad Hoc Communications and Networks (SECON)*, USA, pp. 441–450, 2007.
17. R. Laufer, H. Dubois-Ferriere and L. Kleinrock, "Multirate anypath routing in wireless mesh networks," *Proceedings of the IEEE Conference on Computer Communications (INFOCOM)*, Rio de Janeiro, Brazil, pp. 37–45, 2009.
18. S. Duquennoy, O. Landsiedel and T. Voigt, "Let the tree bloom: Scalable opportunistic routing with ORPL", *Proceedings of the ACM Conference on Embedded Networked Sensor Systems (SenSys)*, New York, USA, pp. 1–14, 2013.
19. Peng Xie, Jun-Hong Cui and Li Lao, "VBF: Vector-based forwarding protocol for underwater sensor networks", UCONN CSE Technical Report: UbiNet-TR05-03, Last Update: February 2006.

20. N. Nicolaou et al., "Improving the robustness of location-based routing for underwater sensor networks," *OCEANS 2007—Europe*, June 2007, pp. 1–6.
21. K. Hao, Z. Jin, H. Shen and Y. Wang, An efficient and reliable geographic routing protocol based on partial network coding for underwater sensor networks. *Sensors*, vol. 15, pp. 12720–12735, 2015. doi: 10.3390/s150612720.
22. S.M. Ghoreyshi, A. Shahrabi and T. Boutaleb, "A novel cooperative opportunistic routing scheme for underwater sensor networks", *Sensors*, vol. 16, 297 pages, 2016.
23. R.W.L. Coutinho, A. Boukerche, L.F.M. Vieira and A.A.F. Loureiro, "Geographic and opportunistic routing for underwater sensor networks," *IEEE Transactions on Computers*, vol. 65, no. 2, pp. 548–561, Feb. 2016.
24. K. Zeng, W. Lou, J. Yang, and D. Brown, "On geographic collaborative forwarding in wireless ad hocand sensor networks," *Int'l. Conf. Wireless Algorithms, Systems and Applications, 2007*, WASA 2007, Aug. 2007, pp. 11–18.
25. Z. Wadud, S. Hussain, N. Javaid, S.H. Bouk, N. Alrajeh, M.S. Alabed and N. Guizani, "An energy scaled and expanded vector-based forwarding scheme for industrial underwater acoustic sensor networks with sink mobility", *Sensors*, vol. 17, 2251 pages, 2017.
26. M.A. Rahman, Y. Lee and I. Koo, "EECOR: An energy-efficient cooperative opportunistic routing protocol for underwater acoustic sensor networks," *IEEE Access*, vol. 5, pp. 14119–14132, 2017.
27. H. Yan, Z. Shi, and J.-H. Cui, "DBR: Depth-based routing for underwater sensor networks," *Proceedings of the IFIP Networking*, pp. 1–13, 2008.
28. Y. Noh et al., "VAPR: Void-aware pressure routing for underwater sensor networks," *IEEE Transactions Mobile Computing*, vol. 12, no. 5, pp. 895–908, 2013.
29. S.H. Bouk, S.H. Ahmed, K.-J. Park and Y. Eun, "EDOVE: Energy and depth variance-based opportunistic void avoidance scheme for underwater acoustic sensor networks", *Sensors*, vol. 17, 2212 pages, 2017.
30. P. Xie, Z. Zhou, Z. Peng, H. Yan, T. Hu, J.-H. Cui, Z. Shi, Y. Fei, and S. Zhou, "Aqua-sim: An Ns2 based simulator for underwater sensor networks," in *Proceedings of the IEEE OCEANS Conference*, 2009, pp. 1–7.

ACOUSTIC COMMUNICATIONS

Chapter 8

Cooperative Protocol for Medium Access Control in Underwater Acoustic Sensor Networks

Lucas S. Cerqueira and Alex B. Vieira
Universidade Federal de Juiz de Fora

Luiz F. M. Vieira and Marcos A. M. Vieira
Universidade Federal de Minas Gerais

José Augusto M. Nacif
Universidade Federal de Viçosa

Contents

8.1 Introduction to Underwater Sensor Networks

Underwater sensor network (UWSN) is an important research field, and its applications are increasing. The use of such networks in military, commercial, and research applications turns a number of previously challenging tasks feasible. For instance, by employing UWSNs, it is possible to efficiently monitor aquatic life and water quality, explore and monitor natural resources such as oil and gas, detect mines and submarine vehicles, prevent disasters such as oil spills, and alert about natural disasters such as tsunamis.

The conditions of the aquatic environment impose great difficulties to communication. Below the water surface, electromagnetic and optical waves suffer high attenuation, being absorbed in a few meters [1]. Therefore, radio frequency based technologies are unsuitable for the deployment of networks in an aquatic environment.

In this sense, acoustic waves have been adopted by both industry and academia as a standard to underwater communication [2]. However, acoustic communication is still challenging, and is characterized by three major issues: (i) the limited and distance-dependent bandwidth, (ii) the multipath fading varied by time, and (iii) the low speed of sound in water when compared with radio frequency. In addition, UWSNs have limited power. As expected, these three major issues enhance well-known sensor network problems and make energy consumption a crucial factor in the development of new technologies in this area.

To improve communication quality and overcome UWSN challenges, many techniques have been studied from physical to network layer. In fact, we are aware about a number of studies involving the development of acoustic modems and effective use of communication channels [3–5], multiple access of the communication channel (detailed in Section 8.2), and routing data among sensors [6–14].

One of these techniques is the cooperative transmission of messages that takes advantage of the broadcast nature of wireless transmissions [15]. Each node on the network can overhear other nodes transmissions and act as relays, realizing spatial diversity. This form of space diversity is referred to as cooperative diversity or cooperative communication [15,16]. Additionally, by means of cooperation, it is possible to achieve time diversity by transmitting the same signal that was overheard through a broadcast transmission at a different time instant.

More precisely, aquatic nodes broadcast messages to communicate with each other. As expected, a broadcasted message may be well listened by a number of nodes, while the final destination may not receive it properly. Intermediate nodes that listen to a failed transmission can serve as relays and retransmit this message. Thus, the node that originated the message may try to go ahead, leaving the retransmission task to its cooperation partners. Moreover, both can try to retransmit the same message, increasing the diversity of transmission paths. While the first approach mainly increases the throughput, the second reduces the error rate.

Another method to ensure reliable information delivery is the use of Automatic Repeat reQuest (ARQ) protocols [17]. Through these techniques, the receiver manages incoming packets and informs the transmitter about packets that require retransmission. Although there are some diverse ARQ schemes, the stop-and-wait (S&W) is the most common due its simplicity and the half duplexing nature of acoustic modems [18]. In brief, in the S&W scheme, the transmitter sends a packet and waits for the acknowledgement (ACK) before sending the next packet. A retransmission occurs when no ACK is received after a timeout period or a negative-ACK (NACK) is received. The key S&W schemes problem is the long propagation delays combined with high bit error rate (BER) of the acoustic channel, making high-throughput efficiency hard to achieve [19].

Go-back-N is another well-known ARQ scheme in which the source transmits packets in a window of size N without waiting for any ACK. The receiver sends back an ACK for each packet received, containing the number of the acknowledged packet. When the source does not receive an ACK, it retransmits beginning from the unacknowledged packet. In this process, the destination accepts packets in order, eliminating the need for buffering. However, when a single packet is lost, it may result in the whole window of packets being retransmitted, even if they were already delivered. This behavior implies a waste of energy and increase in latency, which is not desirable in UWSNs.

The selective repeat (SR) scheme allows the receiver to acknowledge out-of-order packets, sending to the transmitter only information about packets with errors, using NACK packets. While SR and go-back-N can significantly improve channel utilization, they require full-duplex operation, which is not the case of the acoustic channel. However, full-duplex can be emulated using channel access methods such as time or frequency division multiplexing [20].

In this chapter, we discuss cooperative communication on medium access protocols for underwater acoustic sensor networks. We first overview UWSNs from underwater acoustic communication concepts to modeling. We review works on cooperative communication and ARQ schemes and how they deal with medium access control. We then present a model that describes the packet error rate of underwater communications based on ambient properties and acoustic modem characteristics. Finally, we present our approach for cooperative underwater communication, detail the simulations, and discuss the results.

We propose COoperative Protocol for PERvasive Underwater Acoustic Networks (COPPER) as a cooperative UWSN communication protocol that takes into account both sublayers of the data link layer: the media access control and the logical link control. In summary, COPPER works on top of time division multiple access (TDMA) method combined with an ARQ scheme based on SR technique. COPPER explores node idleness to enhance space diversity, or more specifically, cooperative diversity. Different from existing approaches [18–22], our protocol is designed to efficiently integrate cooperative transmission and an ARQ scheme in a collision-free medium access control protocol.

We have evaluated COPPER through ns-3 simulations [23], in a scenario where various underwater sensors attempt to communicate to a sink node. When compared with a noncooperative protocol, the results show that COPPER enhances overall network performance metrics. For example, on the best-case scenario, the packet loss rate is reduced by 48.53%, goodput increases by 21.86%, and the relative energy consumption decreases by 17.65%.

The reminder of this chapter is organized as follows. First, in Section 8.2, we present a brief review of the literature on cooperation in underwater acoustic sensor networks. Then, in Section 8.3, we present the mathematical model used in the simulations. In Section 8.4, we detail the operation of the proposed cooperative protocol. In Section 8.5, we first describe the protocol evaluation and then our simulation results. Finally, our conclusions are presented in Section 8.6.

8.2 Related Work

In this section, we briefly review the literature on cooperation in underwater acoustic sensor networks. The earliest works on cooperation in underwater acoustic sensor networks have emerged as extensions of cooperation work on terrestrial wireless networks [24–29]. In most cases, authors focus on showing that cooperative transmission, through one or more relays, can be effectively applied to underwater networks.

For instance, Carbonelli et al. [25] use a decode-and-forward scheme to show that multihop cooperation schemes are energy efficient. Vajapeyam et al. [28] presented a time-reversal distributed space-time block coding scheme in which the relays use amplify-and-forward protocols. Authors have shown that their scheme can be extended to any number of relays. Again, the experimental and simulation results show that the performance of cooperation scheme is superior to direct transmission. Han et al. [26] also propose a cooperation scheme with amplify-and-forward and show that, even with amplification of the relay signal with noise, the quality of transmissions is improved with cooperation.

Amplify-and-forward, decode-and-forward, and estimate-and-forward schemes have been evaluated by Han et al. [27]. Moreover, authors presented a new scheme named wave cooperative, which is similar to amplify-and-forward. In a glance, authors compared schemes with each other in both cooperative and noncooperative scenarios. Results show that the wave cooperative is superior in relation to channel capacity. Carbonelli et al. [24] also analyzed error propagation with decode-and-forward in cooperative and multihop scenarios. Their results show that the gains are significant even when the error propagation is considered.

Finally, Wang et al. [29] propose an asynchronous cooperation scheme, applicable in scenarios with large and variable propagation delay. The authors have compared amplify-and-forward, decode-and-forward, and direct transmission schemes. Their results show that each scheme performs better depending on the signal-to-noise ratio (SNR) conditions. In short, for optimum SNR conditions, direct transmission has a better result, whereas for bad SNR conditions, the amplify-and-forward scheme is superior.

A number of works jointly uses ARQ schemes and cooperative transmissions [19,21,22]. For example, Lee et al. [21] propose a cooperative S&W ARQ scheme in a single-hop acoustic channel. The protocol uses ACKs and NACKs to inform message reception. When the destination node receives an erroneous packet, it requests the cooperative nodes for retransmissions one by one until retransmission succeeds, recruiting closest nodes first. Despite the benefits, authors assume each node knows the internode distance for its neighbor nodes. The destination node selects the relay, based on distance only. Lee et al. [22] also have proposed the use of the cooperative protocol in a multihop scenario. Authors observed gains while using the cooperation scheme that provides spatial diversity. Moreover, additional gains were achieved by replacing ACK messages with overhearing the next hop transmission in the relaying process.

Ghosh et al. [19] propose a protocol that combines cooperative ARQ and Hybrid ARQ, called C-HARQ, using the S&W scheme. Hybrid ARQ is a combination of an ARQ scheme with forward error correction (FEC), where original data are encoded with an FEC code, and redundant FEC bits are transmitted along with the data or requested by destination when errors are detected. C-HARQ works by using the relays to send the correction codes, which are smaller messages, but still providing the opportunity to correct erroneous messages. However, there is still the possibility of not being able to recover a failed message and a full retransmission from the source will be required.

These three works [19,21,22] consider simple scenarios, where only one node originates messages. Also, media access control issues are either assumed as resolved or ignored.

A few works propose a joint approach to media access and error control schemes [18,20]. They consider a time division and random access Medium Access Control (MAC) protocols while proposing an SR ARQ scheme [18,20]. However, these works do not address system node cooperation.

Kim et al. [30] consider a handshake-based MAC protocol with a cooperative ARQ scheme. The handshaking process is based on the request-to-send and clear-to-send mechanism, and the cooperation information is shared during the handshaking process. Their results show a better performance in terms of throughput when compared with the MACAU-ACK protocol, which uses S&W ARQ. The major shortcoming in their approach is the amount of control messages exchanged, which may cause more collisions and also extend the duration of the handshaking process, decreasing the throughput.

In a glance, the cooperative UWSN communication protocol we propose, namely COPPER, uses an ARQ scheme that takes into account the issues of media access. Our approach effectively coordinates the retransmission of erroneous messages when more than one node shares the medium, using TDMA.

8.3 Underwater Acoustic Modeling

Acoustic communication in the aquatic environment presents high error rate, low throughput, and low bandwidth. These combined characteristics are unique to the

UWSN [1] domain. In this section, we present widely used mathematical models to simulate underwater acoustic communication and its error rate.

In an aquatic environment, we can estimate the packet delivery probability based on the transmission frequency f, the modulation of the acoustic wave, the distance d between the transmitter and receiver nodes, the transmitting power SL, and the amount of ambient noise. First, we calculate the path loss due to large-scale fading, which is a combination of spreading loss and absorption loss. The spreading loss results from the geometrical spreading of the acoustic signal and is given by

$$PL_{\text{spreading}}(d) = k \cdot 10\log(d), \tag{8.1}$$

where k is the spreading factor, which, for a practical scenario, is given by $k = 1.5$. The absorption loss results from energy loss in the form of heat as the acoustic signal propagates and is given by

$$PL_{\text{absorption}}(f, d) = 10\log a(f) \cdot d, \tag{8.2}$$

where $a(f)$ is the absorption coefficient given by Thorp's formula [31]:

$$10\log a(f) = 0.11\frac{f^2}{f^2+1} + 44\frac{f^2}{f^2+4100} + 2.75 \cdot 10^{-4} f^2 + 0.003. \tag{8.3}$$

Finally, the path loss is the combination of equations (1.1) and (1.2) and is given by [32]:

$$A(d, f) = k \cdot 10\log d + 10\log a(f) \cdot d \cdot 10^{-3}. \tag{8.4}$$

Figure 8.1 shows how the spreading and absorption affects path loss. The absorption loss $PL_{\text{absorption}}$ is shown for frequencies 4, 10, 20, and 60 kHz, while spreading loss $PL_{\text{spreading}}$ is shown using $k = 1.5$. For lower distances, the spreading factor has a greater impact on path loss, being dominant even for higher frequencies [33].

We then calculate the SNR using the passive sonar equation [32]:

$$\gamma(d) = SL - TL - NL + DI, \tag{8.5}$$

where SL is the source level of the transmitted signal, TL is the transmission loss given by Equation (8.4), NL is the ambient noise given by the Wenz equation [34], and DI is the directing factor for omnidirectional hydrophones (e.g., acoustic modems) $DI = 0$ [35]. For a given frequency f, the Wenz equation is given by

$$NL(f) = N_t(f) + N_s(f) + N_w(f) + N_{th}(f), \tag{8.6}$$

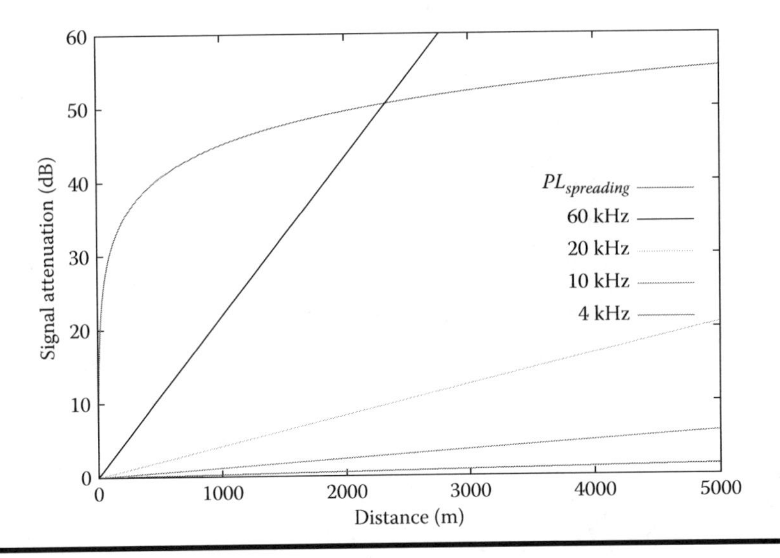

Figure 8.1 Path loss components: Spreading loss and absorption loss.

where N_t is the turbulence noise, N_s is the shipping noise, N_w is the wind noise, and N_{th} is the thermal noise. The power spectral density of each noise components is given by

$$\begin{aligned}
&10 \log N_t(f) = 17 - 30 \log f \\
&10 \log N_s(f) = 40 + 20(s - 0.5) + 26 \log f - 60 \log(f + 0.03) \\
&10 \log N_w(f) = 50 + 7.5w^{1/2} + 20 \log f - 40 \log(f + 0.4) \\
&10 \log N_{th}(f) = -15 + 20 \log f.
\end{aligned} \tag{8.7}$$

The shipping factor s from the shipping noise N_s expresses naval activity and varies from 0, indicating no naval traffic, to 1, indicating heavy naval traffic. In the wind noise N_w, w is the wind speed in m/s. Among the four noise components, the wind noise is the dominating noise in the frequency region 100 Hz–100 kHz, which is the frequency region used by most modern acoustic modems. In this work, we use $w = 1.5$ for wind speed and $s = 0.5$ (moderate shipping) for shipping factor.

We can then describe the probability of error in the bit, for the BPSK modulation in a Rayleigh fading channel [36], given a distance d as

$$p_e(d) = \frac{1}{2}\left(1 - \sqrt{\frac{\Gamma(d)}{1 + \Gamma(d)}}\right), \tag{8.8}$$

where $\Gamma(d)$ is given by

$$\Gamma(d)=10^{\frac{\gamma(d)}{10}}. \tag{8.9}$$

Finally, the delivery probability of a packet of m bits is given by [37]

$$p_p=(1-p_e(d))^m \tag{8.10}$$

8.4 Cooperative Communication

In this work, we consider a single-hop sensor network containing N nodes and a single sink node. In this network, all generated data packets are addressed to the sink node. We also consider a TDMA. Typically, in the TDMA protocol, each node is assigned to a time period, called slot, in which it exclusively accesses the communication medium.

A node is said to be idle when it does not have any data to transmit during its time slot. During these idleness periods, a node may cooperate to the system, aiding other nodes to retransmit previously failed messages. Note that, we consider that any node can act as a relay, which will depend on its status (idle or not). Therefore, in case a node is generating a high burst of packets, it will not be able to cooperate. On the other hand, if the same node is experiencing an idle period, it may cooperate with other nodes, retransmitting their packets.

Without loss of generality, we propose a TDMA modification where each frame has a signaling period (SP) followed by a data period (DP). The SP is used for the exchange of cooperation control messages, and during the DP, the exchange of messages occurs as in a traditional TDMA scheme. As cooperative messages are synchronized, we call this protocol synchronized COPPER.

In our scheme, a node is able to transmit a single packet at a single frame in its time slot. Additionally, the acoustic communication follows a broadcast-like pattern. Then, when a node successfully receives a data packet that is not addressed to it, this node stores this message in a cooperation buffer. By the end of the frame, the sink node transmits a single NACK to signal which packets of the current frame have failed. Upon receiving this message, each node can know which packets have failed. Nodes can then manage their buffers, and, in case they have any packet that the sink node did not correctly receive, they can try to cooperate during the next frame.

In the SP, each node that is idle can cooperate by sending a want to cooperate message (WTC), indicating it wants to cooperate to a given node. The WTC messages indicate to other nodes which packets will be retransmitted, avoiding additional retransmissions of the same packet. In addition, it allows the node that originated the packet to continue to its next packet transmission. By the end of the SP, each node already knows which packet it will transmit in the current frame. Thus, the cooperation buffer can be cleared or overwritten. Consequently, the cooperative buffer size of each node is $n-1$, where n is the number of nodes in the network.

Figure 8.2 depicts a scenario containing three nodes: a source (O), a relay (R), and a sink (S). During the frame i, node O transmits message m_1, which is successfully received by node R, but fails to reach S. By the end of the frame, the sink signals the failed message by transmitting a NACK packet, which is received by both nodes. On the next frame, R is idle and has the message m_1, so it signals—sending the WTC message—that it will cooperate during its next data slot. Thus, node O can transmit its next message m_2 in the same frame as R retransmits m_1. Note that if S had to retransmit its own message m_1 on frame $i + 1$, it would only be able to transmit m_2 on the next frame $i + 2$.

Figure 8.3 represents a scenario with four transmitter nodes and one sink node. In the first DP i, represented in Figure 8.3a, all nodes have messages to transmit. Messages P1 and P2 are delivered, but messages P3 and P4, from nodes 3 and 4, respectively, are not delivered correctly to sink. The sink then signals with a NACK message containing the identification of nodes 3 and 4. Upon receiving the NACK, nodes 1 and 2 check that they have missed messages, and in the next SP $i + 1$ sends the WTC cooperation message. When node 2 receives the WTC message of 1, it receives the information that node 1 will cooperate with node 3, leaving node 2 as the only option to cooperate with node 4. Upon receiving the cooperation message, nodes 3 and 4 know they can transmit their next messages P5 and P6. In the next DP $i + 1$, represented in Figure 8.3b, cooperative messages P3 and P4 are sent and received correctly. Likewise, the message P6 of node 4 is also delivered successfully. However, the message P5 of node 3 fails, causing a NACK message to be sent by the sink node. In this opportunistic process, a node becomes a cooperator when (i) it is

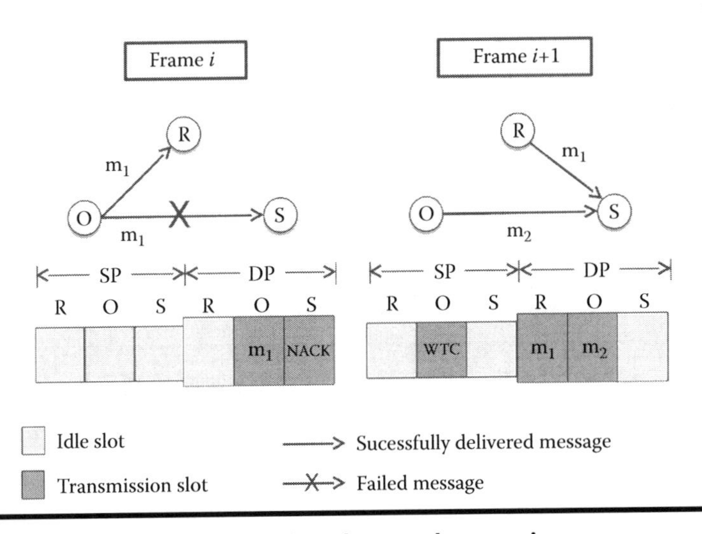

Figure 8.2 Synchronous COPPER in a three-node scenario.

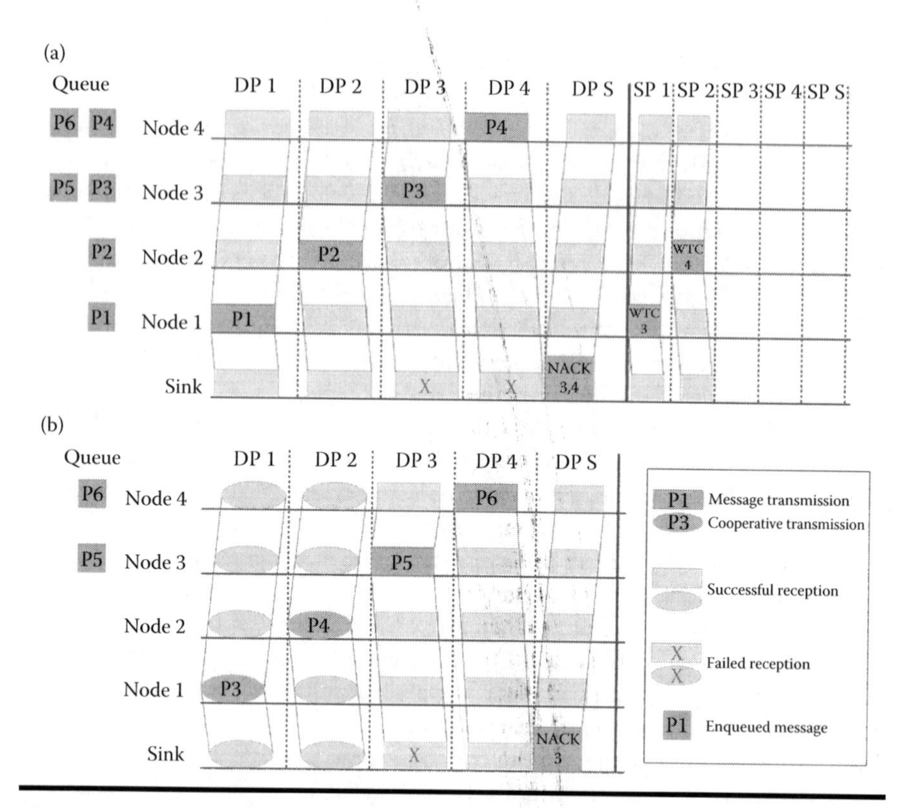

Figure 8.3 Synchronous COPPER in a five-node scenario. (a) DP *i* and SP *i* + 1. (b) DP *i* + 1.

idle, (ii) receives a message correctly that has failed to reach the sink, (iii) sends a WTC message signaling that it will cooperate.

In addition, we have developed a COPPER modification where nodes do not need to signalize to cooperate. In this scheme, the SP is removed. As a consequence, nodes do not have the information of which messages already have a cooperating node and which do not. So, when a node has the opportunity to cooperate, acting as a relay, it simply draws a random message it owns. This draw takes into account only the messages that were answered with a NACK by the sink node. In this approach, the same failed message can be retransmitted by multiple nodes, which may lead to a waste of energy. On the other hand, it may increase the number of transmission paths, and consequently, reduces the packet error rate. We name this approach as asynchronous COPPER.

Figure 8.4 shows the asynchronous protocol in the same scenario of Figure 8.3: Four nodes transmitting messages to the sink node over two frames. Note that each frame contains only the DP, and no longer both the SP and DP. In the first frame, each node has a message to transmit. However, messages P3 and P4, from

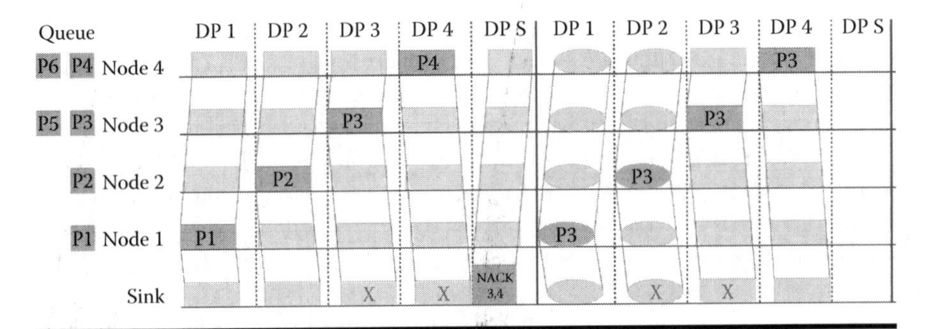

Figure 8.4 Two frames of asynchronous COPPER in a five-node scenario.

nodes 3 and 4 respectively, are not delivered correctly. At the end of the frame, the sink signals with a NACK that messages from nodes 3 and 4 have failed. In the next frame, nodes 1 and 2 have no messages to transmit, being available to cooperate. Since both received the NACK and the P3 and P4 messages correctly, they randomly draw one of them cooperate. In this example, both end up selecting the same P3 message. Node 3 also retransmits its own failed message, totaling three possible paths for P3.

The relay selection is crucial to the success of the cooperation. Intuitively, the closer the relay is to the sink, the better will be the chances of a successfully retransmission. Indeed, the closest nodes to the sink node will present a lower packet error rate than the more distant nodes as the packet error rate is proportional to the distance, according to equation (8.10). In sum, we order the nodes time slots in relation to their distance from the sink node, similar to what was done in [38]. In other words, the first slot of each frame will belong to the closest node to the sink. This way, in the synchronous COPPER, this node will present a greater opportunity to act as a relay, when it announces itself as a relay, by sending a WTC message. The major shortcoming of this approach is related to energy consumption balancing. Nodes closer to the sink may transmit more cooperative messages than other nodes, causing unequal power consumption. We highlight that different relay selection policies may be used. For example, policies may consider node energy level.

8.5 Evaluating COPPER

8.5.1 Evaluation Methodology

To evaluate the proposed protocol, we use ns-3 [23], a discrete event network simulator. We implemented in ns-3 the proposed protocols and a noncooperative TDMA protocol. For both, we follow the bit error probability shown in equation (8.10).

The simulation scenario consists of 16 nodes, including the sink node. Regular nodes are responsible for generating data packets and transmitting them to the sink node. Each regular node can act as a relay node when idle. The sink node only receives the packets and positively/negatively confirms transmission, either by an ACK or NACK message. In this work, we do not consider node mobility. In other words, nodes are static and are randomly distributed in a star topology inside a 200 m side square area and 70 m depth. Nodes are repositioned at each simulation execution, except the sink node, which is always positioned in the center of the square, on the surface.

To generate random values, ns-3 implements an algorithm based on streams and substreams. Each stream generates a set of substreams that does not overlap [39]. Thus, to produce multiple independent runs, we fix the stream by choosing a value for the seed and change only the substream for each run. In this work, we use 138 as seed and i for the ith execution ($i \in \mathbb{N}^*$).

Data traffic is randomly generated according to a uniform random variable. A node may generate a packet according to a probability L at the beginning of each frame. We call this probability L network load. In other words, when network load $L = 0$, no packet is generated in the whole simulation, and when $L = 1$, all nodes will always have a packet to transmit, at every frame. We have varied L to evaluate the performance of the protocol with different network loads, from 10% to 100%.

We have executed each protocol simulation 25 times, for 10 different network loads, totaling 250 executions per protocol. Each execution simulates the operation of the network during 3,600 s. The size of data packets is set to 540 bytes, WTC packets to 3 bytes and NACK packets to 5 bytes, which are in the same order of magnitude as [19,20]. The transducer settings are based on the UNET-2 [40] acoustic modem: data rate of 2,400 bps, center frequency of 4,000 Hz and BPSK modulation, bandwidth of 2,000 Hz, transmitting power of 138 dB, power consumption for packet transmission of 50 W, power consumption for reception of packets of 158 mW, and power consumption in idle mode of 158 mW. Data slots present 2 s duration, with 1.8 s for data transmission and 0.2 s as guard time to avoid packet collision. In the SP, the control slots last for 0.2 s, which represents only 10% of the data slot time.

In this work, we assume the network can keep the time synchronized among nodes, as it can be achieved in a practical scenario. Indeed, the slot synchronization of all nodes of the network is of extreme importance for the operation of the protocol [41]. If one node is not synchronized, it can lead to collisions in the sink node, negatively impacting network performance. To avoid this problem, time synchronization can be achieved using the modem's own functionality [42]. The modem's ranging feature is used to get the clock offset between the nodes and adjust it accordingly, so the time slots can also be synchronized.

Finally, we evaluate synchronous and asynchronous COPPER and a system without any cooperation, according to a set of three metrics: (i) the goodput that corresponds to the amount of data packets bytes which sink node receives, in relation to the total simulation period; (ii) a packet loss rate which represents the percentage of data packets that are not properly received by sink node; (iii) energy spent in the simulation per packet delivered, which we call relative energy spent. Unless we tell otherwise, results we present are mean values, and are plotted with the standard error of the mean*.

8.5.2 Results

Figure 8.5 presents the mean packet error rate (and confidence interval) for each network load we simulated. For lower network loads, the synchronous COPPER protocol has a much lower packet loss rate when compared with the noncooperative protocol. For example, at a 10% network load, the synchronous COPPER is almost 50% better than the noncooperative TDMA. In fact, while the first presents about 16% of mean packet error rate, the former presents more than 31%.

Despite the notable improvement for packet error rate, for lower network load scenarios, all three protocols tend to present similar results in high-loaded scenarios. In fact, in high-loaded scenarios, all nodes are transmitting data practically all the time. In this case, nodes are not idle and do not have the opportunity to cooperate. Conversely, for lower network loads, the cooperative protocol achieves better results. The lower the network load, the more nodes are idle, and, consequently, the greater the chance of some nodes to cooperate.

Asynchronous COPPER also presents better goodput, when compared with synchronous and noncooperative TDMA. For instance, according to Figure 8.6, asynchronous COPPER is about 36% better than the noncooperative TDMA.

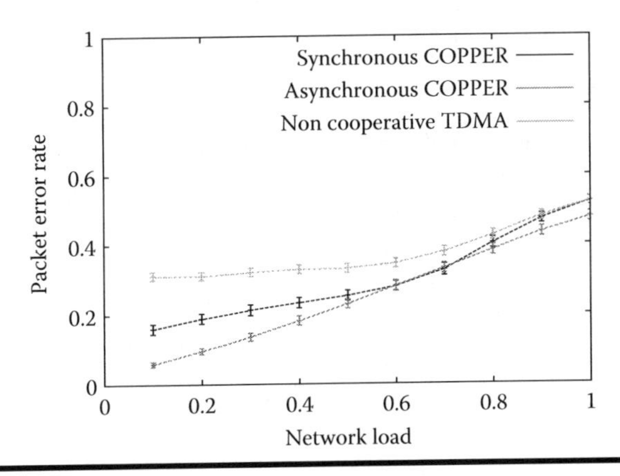

Figure 8.5 Package error rate.

* $\mathrm{SEM}(x) = \mathrm{SD}(x) \cdot \sqrt{\mathrm{length}(x)}^{-1}$

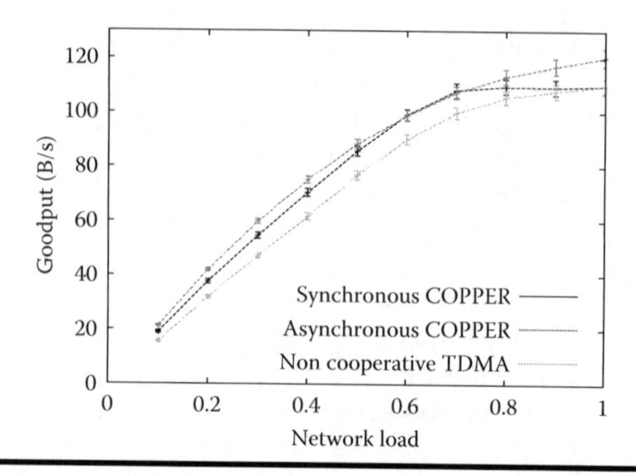

Figure 8.6 Goodput.

Despite the gains in goodput and packet error rate, asynchronous COPPER protocol demands more energy for its proper functioning. Indeed, when multiple nodes try to cooperate, without any synchronism, they may waste energy. According to our results, at a 10% network load, the asynchronous COPPER spends almost 38% more energy, when compared with the synchronous COPPER. In this scenario of low network load, synchronous COPPER presents a 21.44% better energy consumption when compared with the noncooperative TDMA. In this case, the use of a synchronized protocol that will organize relays presents much lower impact than the waste of energy that retransmissions imposes to the sensor network.

8.6 Conclusions

In this chapter, we discuss cooperative communication on medium access layer for underwater acoustic sensor networks. We also present a cooperative MAC protocol for UWSNs. Our cooperation technique is based on an SR ARQ scheme and incorporates the error signaling and retransmission of messages into medium access control. The retransmission of failed messages is made by nodes that would, otherwise, be idle. Our simulation results show an improvement in the packet loss rate, relative energy spent, and goodput metrics. More specifically, it achieves a reduction of 48.53% in packet loss rate in the best-case scenario. The proposed synchronous protocol performs better at mid-to-lower network loads, which corresponds to most UWSN applications [43]. Nevertheless, it still retains gains at higher network loads. Future work includes the scalability evaluation of the proposed scheme as well as the cooperation using other MAC protocols, besides implementing them in a real environment.

References

1. Luiz FM Vieira, Antonio AF Loureiro, Antonio Fernandes, and Mario Campos. *Redes de Sensores Aquáticas. XXVIII Simpósio Brasileiro de Redes de Computadores e Sistemas Distribuídos*, 1:199–240, 2010.
2. John Heidemann, Milica Stojanovic, and Michele Zorzi. Underwater sensor networks: applications, advances and challenges. *Philosophical Transactions of the Royal Society A*, 370(1958):158–175, 2012.
3. Emrecan Demirors, George Sklivanitis, G Enrico Santagati, Tommaso Melodia, and Stella N Batalama. Design of a software-defined underwater acoustic modem with real-time physical layer adaptation capabilities. In *Proceedings of the International Conference on Underwater Networks & Systems*, page 25. ACM, 2014.
4. David Pinto, Sadraque S Viana, José Augusto M Nacif, Luiz FM Vieira, Marcos AM Vieira, Alex B Vieira, and Antônio O Fernandes. Hydronode: a low cost, energy efficient, multi purpose node for underwater sensor networks. In *Local Computer Networks (LCN), 2012 IEEE 37th Conference on*, pages 148–151. IEEE, 2012.
5. Christian Renner and Alexander J Golkowski. Acoustic modem for micro AUVs: design and practical evaluation. In *Proceedings of the 11th ACM International Conference on Underwater Networks & Systems*, page 2. ACM, 2016.
6. Stefano Basagni, Chiara Petrioli, Roberto Petroccia, and Daniele Spaccini. CARP: A channel-aware routing protocol for underwater acoustic wireless networks. *Ad Hoc Networks*, 34:92–104, 2015.
7. Rodolfo WL Coutinho, Azzedine Boukerche, Luiz FM Vieira, and Antonio AF Loureiro. GEDAR: Geographic and opportunistic routing protocol with depth adjustment for mobile underwater sensor networks. In *2014 IEEE International Conference on Communications (ICC)*, pages 251–256. IEEE, 2014.
8. Rodolfo WL Coutinho, Azzedine Boukerche, Luiz FM Vieira, and Antonio AF Loureiro. Design guidelines for opportunistic routing in underwater networks. *IEEE Communications Magazine*, 54(2):40–48, 2016.
9. Rodolfo WL Coutinho, Azzedine Boukerche, Luiz FM Vieira, and Antonio AF Loureiro. Modeling and analysis of opportunistic routing in low duty-cycle underwater sensor networks. In *Proceedings of the 18th ACM International Conference on Modeling, Analysis and Simulation of Wireless and Mobile Systems*, pages 125–132. ACM, 2015.
10. Rodolfo WL Coutinho, Azzedine Boukerche, Luiz FM Vieira, and Antonio AF Loureiro. Geographic and opportunistic routing for underwater sensor networks. *IEEE Transactions on Computer*, 65(2):548–561, 2016.
11. Rodolfo WL Coutinho, Luiz FM Vieira, and Antonio AF Loureiro. : Depth-controlled routing protocol for underwater sensor networks. In *2013 IEEE Symposium on Computers and Communications (ISCC)*, pages 453–458. IEEE, 2013.
12. Uichin Lee, Paul Wang, Youngtae Noh, Luiz Filipe M Vieira, Mario Gerla, and Jun-Hong Cui. Pressure routing for underwater sensor networks. In *INFOCOM 2010. The 29th Conference on Computer Communications*, pages 1676–1684. IEEE, 2010.
13. Luiz Filipe M Vieira. Performance and trade-offs of opportunistic routing in underwater networks. In *2012 IEEE Wireless Communications and Networking Conference (WCNC)*, pages 2911–2915. IEEE, 2012.

14. Z Zhou, B Yao, R Xing, L Shu, and S Bu. E-CARP: An energy efficient routing protocol for UWSNs in the internet of underwater things. *IEEE Sensor Journal*, 16(11):4072–4082. IEEE, 2015.
15. J Nicholas Laneman, David NC Tse, and Gregory W Wornell. Cooperative diversity in wireless networks: Efficient protocols and outage behavior. *IEEE Transactions on Information Theory*, 50(12):3062–3080, 2004.
16. Rana Azeem M Khan and Holger Karl. MAC protocols for cooperative diversity in wireless lans and wireless sensor networks. *IEEE Communications Surveys & Tutorials*, 16(1):46–63, 2014.
17. Ethem M Sozer, Milica Stojanovic, and John G Proakis. Underwater acoustic networks. *IEEE Journal of Oceanic Engineering*, 25(1):72–83, 2000.
18. Saiful Azad, Paolo Casari, Federico Guerra, and Michele Zorzi. On ARQ strategies over random access protocols in underwater acoustic networks. In *OCEANS, 2011 IEEE-Spain*, pages 1–7. IEEE, 2011.
19. Arindam Ghosh, Jae-Won Lee, and Ho-Shin Cho. Throughput and energy efficiency of a cooperative hybrid arq protocol for underwater acoustic sensor networks. *Sensors*, 13(11):15385–15408, 2013.
20. Saiful Azad, Paolo Casari, and Michele Zorzi. The underwater selective repeat error control protocol for multiuser acoustic networks: Design and parameter optimization. *IEEE Transactions on Wireless Communications*, 12(10):4866–4877, 2013.
21. Jae Won Lee, Jin Yong Cheon, and Ho-Shin Cho. A cooperative ARQ scheme in underwater acoustic sensor networks. In *OCEANS 2010 IEEE-Sydney*, pages 1–5. IEEE, 2010.
22. Jae Won Lee and Ho-Shin Cho. A cooperative ARQ scheme for multi-hop underwater acoustic sensor networks. In *Underwater Technology (UT), 2011 IEEE Symposium on and 2011 Workshop on Scientific Use of Submarine Cables and Related Technologies (SSC)*, pages 1–4. IEEE, 2011.
23. ns-3. https://www.nsnam.org/. Accessed: 2017-09-27.
24. Cecilia Carbonelli, Shiou-Hung Chen, and Urbashi Mitra. Error propagation analysis for underwater cooperative multi-hop communications. *Ad Hoc Networks*, 7(4):759–769, 2009.
25. Cecilia Carbonelli and Urbashi Mitra. Cooperative multihop communication for underwater acoustic networks. In *Proceedings of the 1st ACM international workshop on Underwater networks*, pages 97–100. ACM, 2006.
26. Jung-Woo Han, Hyung-Jun Ju, Ki-Man Kim, Seung-Yong Chun, and Kyoung-Cheol Dho. A study on the cooperative diversity technique with amplify and forward for underwater wireless communication. In *OCEANS 2008-MTS/IEEE Kobe Techno-Ocean*, pages 1–3. IEEE, 2008.
27. Zhu Han, Yan Lindsay Sun, and Hongyuan Shi. Cooperative transmission for underwater acoustic communications. In *Communications, 2008. ICC'08. IEEE International Conference on*, pages 2028–2032. IEEE, 2008.
28. Madhavan Vajapeyam, Satish Vedantam, Urbashi Mitra, James C Preisig, and Milica Stojanovic. Distributed space–time cooperative schemes for underwater acoustic communications. *IEEE Journal of Oceanic Engineering*, 33(4):489–501, 2008.
29. Ping Wang, Wei Feng, Lin Zhang, and Victor OK Li. Asynchronous cooperative transmission in underwater acoustic networks. In *Underwater Technology (UT), 2011 IEEE Symposium on and 2011 Workshop on Scientific Use of Submarine Cables and Related Technologies (SSC)*, pages 1–8. IEEE, 2011.

30. Hee-won Kim and Ho-Shin Cho. A cooperative ARQ-based MAC protocol for underwater wireless sensor networks. In *Proceedings of the 11th ACM International Conference on Underwater Networks & Systems*, page 28. ACM, 2016.
31. Leonid M Brekhovskikh, Yu P Lysanov, and Robert T Beyer. *Fundamentals of Ocean Acoustics*. Springer Science & Business Media, 2003.
32. Robert J Urick. *Principles of Underwater Sound*, 2nd ed., McGraw-Hill, 1975.
33. Gunilla Burrowes and Jamil Y Khan. Short-range underwater acoustic communication networks. In *Autonomous Underwater Vehicles*. InTech, 2011.
34. Gordon M Wenz. Acoustic ambient noise in the ocean: Spectra and sources. *Journal of the Acoustical Society of America*, 34(12):1936–1956, 1962.
35. Hao Wang, Shilian Wang, Eryang Zhang, and Jianbin Zou. A network coding based hybrid ARQ protocol for underwater acoustic sensor networks. *Sensors*, 16(9):1444, 2016.
36. Theodore S Rappaport et al., volume 2. Prentice Hall PTR New Jersey, 1996.
37. Rodolfo WL Coutinho, Azzedine FM Boukerche, Luiz Vieira, and Antonio Loureiro. A novel centrality metric for topology control in underwater sensor networks. In *Proceedings of the 19th ACM International Conference on Modeling, Analysis and Simulation of Wireless and Mobile Systems*, pages 205–212. ACM, 2016.
38. Mari Carmen Domingo. A distributed energy-aware routing protocol for underwater wireless sensor networks. *Wireless Personal Communications*, 57(4):607–627, 2011.
39. Pierre L'ecuyer, Richard Simard, E Jack Chen, and W David Kelton. An object-oriented random-number package with many long streams and substreams. *Operations Research*, 50(6):1073–1075, 2002.
40. Mandar Chitre, Iulian Topor, and Teong-Beng Koay. The UNET-2 modem—an extensible tool for underwater networking research. In *OCEANS, 2012-Yeosu*, pages 1–7. IEEE, 2012.
41. Sadraque S Viana, Luiz FM Vieira, Marcos AM Vieira, José Augusto M Nacif, and Alex B Vieira. Survey on the design of underwater sensor nodes., pages 1–20, 2015.
42. Prasad Anjangi and Mandar Chitre. Design and implementation of super-TDMA: A mac protocol exploiting large propagation delays for underwater acoustic networks. In *Proceedings of the 10th International Conference on Underwater Networks & Systems*, page 1. ACM, 2015.
43. Beatrice Tomasi, Paolo Casari, Leonardo Badia, and Michele Zorzi. Cross-layer analysis via Markov models of incremental redundancy hybrid ARQ over underwater acoustic channels. *Ad Hoc Networks*, 34:62–74, 2015.

Chapter 9

Hybrid Infrastructure for AUV Operations

Seyedmohammad Salehi and Chien-Chung Shen
University of Delaware

Aijun Song
The University of Alabama

Contents

9.1 Introduction

The oceans cover more than 70% of the surface of our planet, forming one of the most critical physical systems to life. To support ocean monitoring and exploration missions, the prevailing strategies use either seafloor fiber-optic cables, e.g., ocean observatories around the globe [1–4], or satellite-linked stationary in-water moorings [5] as backbones for communications and networking. The sea-floor observatories often have enormous price tags for development and maintenance. Further, the seafloor infrastructures are inflexible to relocate or to accommodate evolving societal needs, although supporting invaluable long-term ocean observations. In addition, although a dense array of satellite-linked stationary moorings may cover a relatively large area, this static solution results in high costs as well as operational difficulties for deployment and recovery.

In recent decades, autonomous underwater vehicles (AUVs; including underwater gliders) have emerged as effective and versatile tools to respond to vital needs in the oceans, lakes, and estuaries [6]. Successful applications enabled by AUVs include, just to name a few, adaptive environmental monitoring, geological surveys, ocean observations, and national defense. In these applications, AUVs may gather orders of magnitude more measurements than the traditional ship-based surveys, at much lower cost, and/or in hazardous conditions (e.g., underwater during hurricanes). In addition, the ability to retrieve imagery and scientific data from AUVs via a communication network will greatly enhance human–vehicle interactions and real-time decision making [7], thus supporting critical real-time underwater missions, e.g., disaster responses.

Fleets of coordinated AUVs operating together facilitate applications of distributed sampling and exploration [8], including (1) tracking marine life to understand the life cycles of sharks, jellyfish, lobsters, etc. [9]; (2) monitoring and tracking fast-evolving plumes, algae, or other fast-evolving features [10]; and (3) mine detection and other national defense applications [11]. In addition, several trends have driven the need to establish motion coordination and team behaviors [12,13]. For instance, distributed real-time measurements are critical to sparse sampling in vast oceans or great lakes. Coordinated AUV fleets are poised to perform sophisticated missions in highly dynamic oceans, and AUV fleets can greatly reduce the sensory and capability requirements on individual members, thus reducing the overall mission cost.

These applications demand reliable communication and networking among the participating AUVs, and between AUVs and their external monitoring, control, and human decision making. However, it is well known that wireless communications in the underwater realm is an intractable challenge. In field operations, scientists have been adopting the concepts of delay-tolerant networking to cope with intermittent underwater communications [14]. In such context, *encounters* are used as the main opportunities to communicate.

When considering underwater wireless communications over ranges beyond tens of meters, acoustic communications should be used, because both electromagnetic and optical waves suffer strong attenuation in the aquatic environment. At the same time, the unique characteristics of acoustics further challenge underwater communications. The fundamental difficulty lies in the limited bandwidth, with a maximum of only tens of kilohertz. In addition, due to the highly dynamic ocean environment, the acoustic communication channel suffers large dispersion in both time and frequency domains (i.e., time-varying multipath), constraining spectral efficiency. Further, underwater sound speed, 1500 m/s, is five orders of magnitude slower than that of electromagnetic waves in air. The resulting long propagation delay introduces spatiotemporal uncertainty [15], which seriously limits the efficiency of networking protocols.

The mobility of AUV fleets introduces additional challenges. First, AUVs often experience high uncertainties in localization and time synchronization due to the lack of global positioning system (GPS) signals underwater. Second, AUVs may be sparsely deployed over a large aquatic region, so that network connectivity becomes intermittent. Third, the network topology of a fleet of AUVs is in constant change, leading to variable and long propagation delay. Mobility also creates variation on data rates in different geographical locations of the network, as the achievable data rate decreases with the increase of communication range.

We propose to use low-cost autonomous surface vehicles (ASVs) equipped with both acoustic and radio-frequency (RF) modems to support underwater missions, as depicted in Figure 9.1. The ASVs form a connected and adaptive backbone via RF links above the water surface while connecting AUVs via underwater acoustic links. The connected backbone is maintained by a swarming-based ASV navigation strategy for enhanced data rates and much reduced end-to-end latency.

Figure 9.2 illustrates the functional architecture of the proposed mission-defined hybrid infrastructure, which (1) directly addresses the communication and network challenges and (2) allows seamless integration with autonomy and control. The hybrid infrastructure complements existing AUV autonomy middleware and behavior architecture, such as Mission Oriented Operating Suite-Interval Programming (MOOS-IvP) [16], and trilevel hybrid control architecture of mission planning and executive [17], as an efficient and reliable communication infrastructure among AUVs. Using the defined mission from mission planning as inputs, the ASV-based hybrid RF-acoustic infrastructure facilitates networking among AUVs and to the outside world by *optimizing* the navigation of ASVs to *jointly* (1) trail respective AUVs to maintain short range and close to "vertical" acoustic links for improved data rates, reduced propagation delay, and enhanced

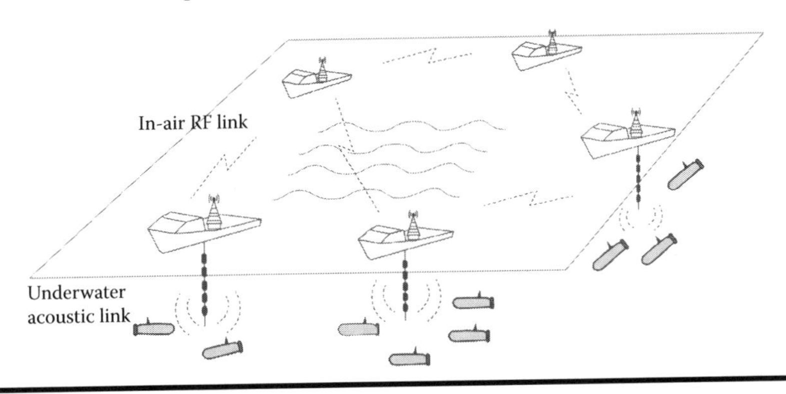

Figure 9.1 Scenario of hybrid RF-acoustic networking and ASV navigation.

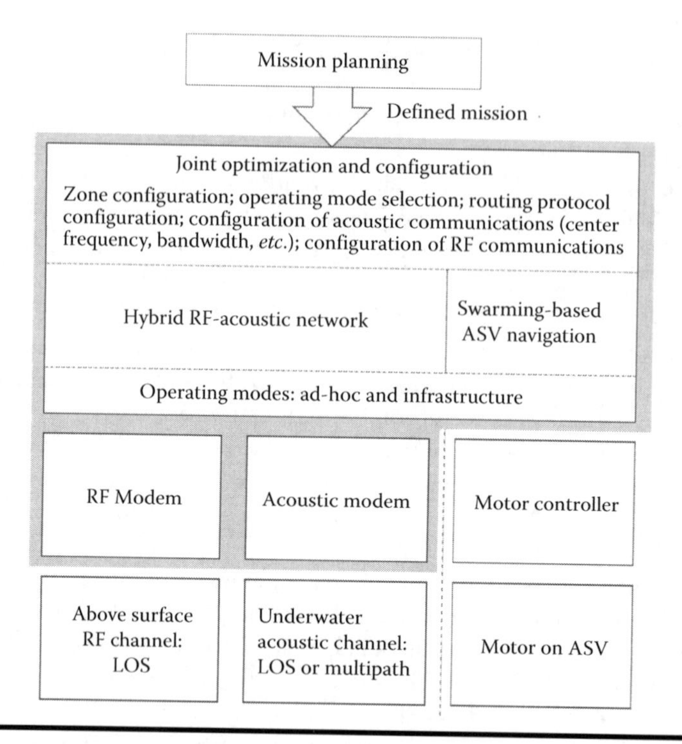

Figure 9.2 Functional architecture of hybrid infrastructure.

reliability and (2) form an adaptive and connected RF "backbone" above water surface to support high data rate and reliable communications. The hybrid short range underwater acoustic links and low-latency in-air RF links create much improved network throughput, efficiency, and reliability. To sustain a connected ASV backbone, AUVs may be instructed not to move away from associated ASVs so as to be connected with other AUVs within the same mission.

Such a hybrid networking infrastructure represents a new network, where two communication constituents differ greatly in their data rates, link performance dynamics, power efficiency, and network coverage. Further, the five orders of magnitude difference in wave propagation speed leads to a large disparity in network latency between subsurface and in-air subnetworks. One critical issue is to guarantee reliable connectivity among AUVs through navigation of ASVs, in the presence of aquatic dynamics (ocean currents, surface waves) and location uncertainty of AUVs.

The chapter proceeds to review related work in Section 9.2. Hybrid RF-acoustic networking among AUVs via ASVs is introduced in Section 9.3. Swarming-based ASV navigation is briefly described in Section 9.4. Parameters used for the simulation are discussed in Section 9.5. Simulation results of hybrid RF-acoustic communications between AUVs are presented in Section 9.6. Section 9.7 concludes the chapter with future research directions.

9.2 Related Work

It is well recognized that acoustic communications alone cannot meet the needs of data telemetry in underwater missions. To address the issues, a number of hybrid schemes have been proposed: acoustics combined with fiber-optic cabled sea-floor stations [Ocean Observatories Initiative (OOI) projects], acoustics with satellite links, and RF-acoustic method that is used in a centralized network to collect sensory information of underwater nodes and to control them.

Mobility of AUVs has been used to assist routing among drifting sensors [18–20] or in data muling and encounter-based connectivity [21,22]. Some of these schemes used only acoustic communications [23], while others used a combination of optical and acoustic methods for communications [24].

ASVs are low-cost, easy-to-operate, and versatile platforms [25–27]. Being on the surface, ASVs have several advantages: (1) access to GPS and RF communications [26], (2) more cargo space and possible long endurance in the ocean, (3) access to solar energy [28] and different propulsion solutions. In addition, ASVs can continuously provide GPS information to assist AUVs with more accurate and precise localization [29–31].

The use of ASVs has also been reported in various scientific field experiment efforts since 2000, for example, in cooperative marine autonomy [32], ocean remote sensing [33], and hydrographic survey [34]. As reported in [35–37], *individual* ASVs were used as communication gateways for underwater platforms. A single semisubmersible ASV was used to support AUV communication and positioning [38]. Large-scale experiments in [39,40] also reported the use of individual ASVs as communication gateways to control centers or satellites. To our best knowledge, there are no reported efforts on using multiple ASVs to form a hybrid network or even an RF network above the sea surface.

As a communication platform, although ASVs face several challenges, solutions exist. First, the stability of these ASVs are subject to the dynamics of surface waves. Therefore, they are more suitable to operate in relatively calm sea water surfaces. One solution is to use semisubmersibles. Second, close to the surface, the acoustic receiving array may not have good reception when the ocean is downward-refracting for acoustic waves. One solution is to use relatively long cables for reception as well as transmission. Third, the RF modems above the water surface often rely on line-of-sight for reliable communications. To cover large areas, ASVs need to install elevated RF antennas that can be accomplished with bigger vessels.

9.3 Hybrid RF-Acoustic Networking among AUVs via ASVs

The proposed hybrid infrastructure consists of two complementary components: hybrid RF-acoustic networking of ASVs and AUVs and swarming-based ASV

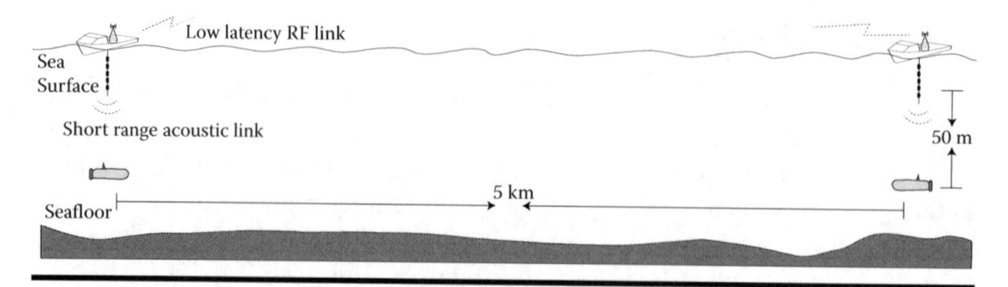

Figure 9.3 Simple scenario of hybrid RF-acoustic networking.

navigation. The benefits of hybrid RF-acoustic networking can be illustrated by a simple scenario depicted in Figure 9.3, where two AUVs, separated some distance apart, navigate collaboratively to sample the ocean. Using conventional schemes, the two AUVs communicate via the direct acoustic link over a *horizontal* channel. Due to the slow underwater sound speed, the communication latency is high. In addition, due to the long distance between the two AUVs, the acoustic link can only support lower data rates with limited reliability subject to multipath and ocean fluctuations.

In contrast to a single long delay and unreliable acoustic link, the central idea of hybrid RF-acoustic networking is to use ASVs to *trail* AUVs by a short distance so as to bridge the two short-range underwater acoustic communications (between two pairs of AUV and ASV) with high speed, low latency RF communications (between the two ASVs above the sea surface). Having short-range underwater acoustic communications between a pair of AUV and ASV not only reduces the latency of acoustic communications but also makes the acoustic communications closer to *vertical* to mitigate refraction* and multipath. Overall, end-to-end communications between two AUVs over a hybrid RF-acoustic network achieve lower latency, higher bandwidth, and improved reliability.

Figure 9.4 compares the latency of transmitting one data packet between the two AUVs in Figure 9.3. In this illustrative comparison, the data packet has W kbits. The two AUVs are separated by distance D_{HA} ranging from 2 to 10 km. Distance D_{VA} between an AUV and its trailing ASV is 50 m. Underwater sound speed c_A is 1500 m/s. It is commonly believed that the achievable data rate R_{HA} over a horizontal acoustic channel decreases with the increase of communication distance, so that the achievable *rate-range product* is a constant, say K kbps × km (i.e., $R_{HA} \cdot D_{HA} = K$). Over the vertical acoustic channel, the data rate is largely limited by the available bandwidth. Based on these two principles, we assume that the rate-range product $R_{HA} \cdot D_{HA}$ is 20 kbps × km for the direct horizontal acoustic link between the two AUVs. We assume that the vertical acoustic channel supports data rate R_{VA} of 40 kbps. These data rates are realistic and have been demonstrated via

* Because water is much more stratified in the vertical than in the horizontal.

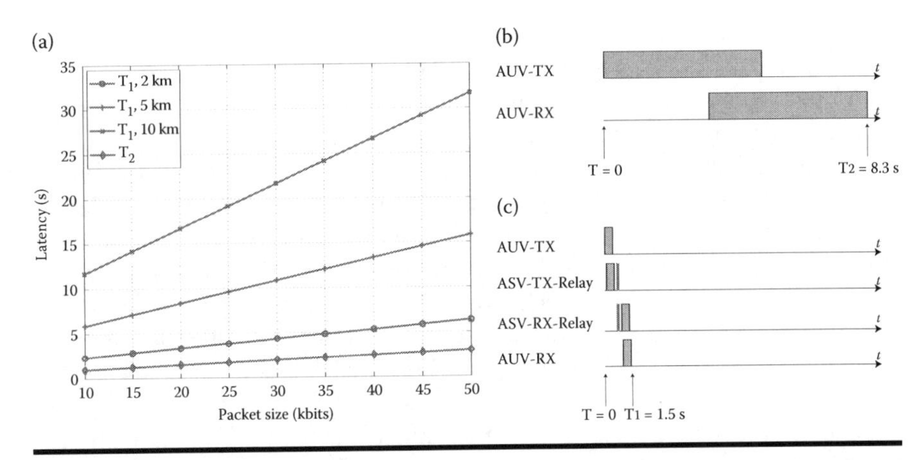

Figure 9.4 (a) Comparison of packet delivery latency between the traditional and proposed schemes for different AUV-AUV ranges. Timing diagram comparison between the traditional scheme, shown in (b), and our hybrid scheme, shown in (c), for $W = 20$ kb and $D_{HA} = 5$ km. In (b), $T_1 = 8.3$ s while $T_2 = 1.5$ s in (c).

different commercial products. Using the traditional schemes, the latency associated with packet delivery is: $T_1 \simeq \frac{W}{R_{HA}} + \frac{D_{HA}}{c_A}$.

Using the hybrid scheme, the same data packet traverses two (short-range) acoustic links and one (long-range) RF link, and ASVs need to translate the data packet between the acoustic and RF links. We assume there is a delay, T_δ, associated with such translations. We assign $T_\delta = 0.2$ s to allow the conversion between acoustic and RF signals and the forwarding decisions for data packets across the two constituent networks. The RF link, with a propagation speed of $c_{EM} = 3 \cdot 10^8$ m/s, can support much higher data rates than the acoustic links, for example 500–800 kbps. Therefore, not only is RF link's propagation latency negligible when compared with that of acoustic links, but RF link's packet transmission latency is also very small (T_ϵ). Therefore, the packet delivery latency in the hybrid scheme is $T_2 \simeq 2\left(\frac{W}{R_{VA}} + \frac{D_{VA}}{c_A} + T_\delta\right) + T_{(\varepsilon)}$.

For different communication ranges (i.e., the AUV-to-AUV distance), latency does not vary in the hybrid scheme, where the RF link is used to address the range above the surface. In the traditional pure acoustic solution, the communication range matters in two ways. First, it increases the acoustic propagation delay. Second, the range reduces the allowed acoustic data rates. At a 2 km range, the traditional scheme uses 50–100 percent of extra time to deliver the same packet, compared with the proposed scheme. When the range increases to 5 or 10 km, the advantage of hybrid scheme becomes significant. The traditional scheme uses about 5.8 and 11.6 s to deliver a 10 kb packet at 5 and 10 km, which are 3.7 and 7.4

folds of the latency in the hybrid scheme, respectively. Timing diagrams for transmission of a data packet of $W = 20$ kb are shown in Figures 9.4b, c for direct and hybrid schemes, respectively. The latency values in the direct AUV-AUV link and the hybrid network are $T_1 = 8.3$ and $T_2 = 1.5$ s, respectively.

When the packet size (PS) increases, the traditional scheme lags behind more than the hybrid scheme. We neglect the physical layer (PHY) receiver decoding delay, which is often small compared with packet duration. If we take into account the link reliability, we will see further advantage of the hybrid scheme. The short-range acoustic links are much more reliable than the long-range horizontal acoustic link, especially in the dynamic ocean environment. Often in the traditional scheme, high packet loss in the long horizontal channels leads to excessive retransmission and even network failure. Furthermore, in the hybrid scheme, there are two segments of short-range acoustic links in the end-to-end path between two AUVs, which may be far apart to form two different contention domains so that respective acoustic transmissions do not interfere with each other. This allows concurrent acoustic transmissions to further reduce packet delivery latency.

9.4 Swarming-Based ASV Navigation

Given a defined mission for AUVs (such as waypoints, destination, etc.), the hybrid infrastructure is to navigate ASVs by jointly (1) trailing respective AUVs to maintain local acoustic links and (2) forming a connected and adaptive RF backbone to support inter-AUV communications. However, given the dynamic nature of the aquatic environment (current, wind, etc.), a fully decentralized scheme is deemed necessary to "coordinate" the navigation of ASVs so that all the AUVs move toward the common goal to complete the defined mission, stay connected during the mission, and avoid potential collision. To accomplish this objective, we propose *swarming-based ASV navigation* based on the three-zone swarming model [41].

In general, swarming is a collective behavior exhibited by entities, particularly animals, of similar size which aggregate together, perhaps milling about the same spot or perhaps moving *en masse* or migrating in some direction. Swarming is typically defined by a set of rules which a group of nodes follow to interact *locally* with other proximal nodes without any centralized control.

In ASV swarming, the perceptual field of each ASV, as defined by its RF communication range, is divided into zone of repulsion (ZOR), zone of orientation (ZOO), and zone of attraction (ZOA), as depicted in Figure 9.5. Given a distribution of N ASVs, to coordinate with neighboring ASVs in different zones, an ASV will move *away* from its neighboring ASVs in ZOR or move *along* with its neighboring ASVs in ZOO while moving *towards* its neighboring ASVs in ZOA, as depicted in Figure 9.6.

Let ASV i be located at position vector P_i and pointing in direction D_i. We define three decision vectors,

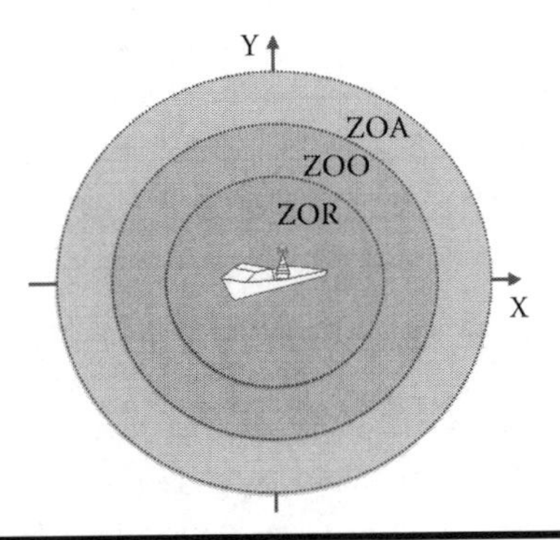

Figure 9.5 Move away, move along, and move towards.

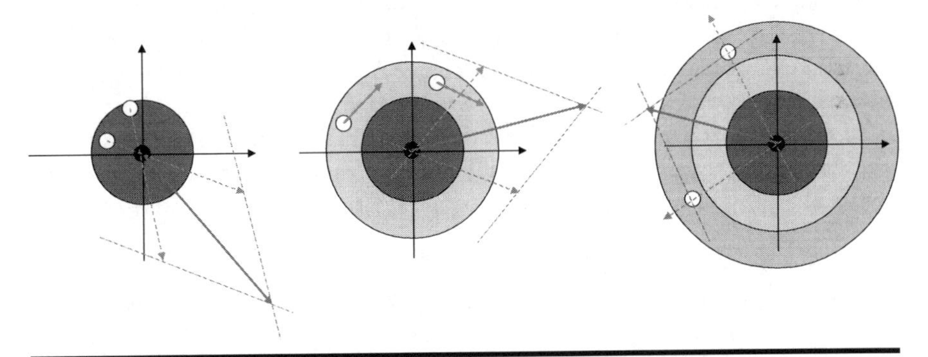

Figure 9.6 Simulation model of Figure 9.3 in ns-3.

$$DR_i = \sum_{j \in S_{ZOR}} \frac{R_{ij}}{|R_{ij}|} \quad DO_i = \sum_{j \in S_{ZOO}} \frac{D_j}{|D_j|} \quad DA_i = \sum_{j \in S_{ZOA}} \frac{R_{ji}}{|R_{ji}|} \tag{9.1}$$

where $R_{ij}=P_i-P_j$ is a displacement vector between ASV i and ASV j, and S_{ZOR}, S_{ZOO}, and S_{ZOA} are the sets of indices of ASVs in the ZOR, ZOO, and ZOA, respectively.

Let $|S_Z|$ denote the number of ASVs in zone Z. Assuming that decision vectors are normalized as unit vectors, a generic ASV swarming algorithm can be summarized as follows.

1. If $|S_{ZOR}| \neq 0$ then $V_i = DR_i$, Break;
2. If $|S_{ZOO}| \neq 0$ and $|S_{ZOA}| = 0$ then $V_i = DO_i \,/\, |DO_i|$, Break;
3. If $|S_{ZOO}| = 0$ and $|S_{ZOA}| \neq 0$ then $V_i = DA_i \,/\, |DA_i|$, Break;
4. If $|S_{ZOO}| \neq 0$ and $|S_{ZOA}| \neq 0$ then $V_i = \alpha \times DO_i \,/\, |DO_i| + (1-\alpha) \times DA_i \,/\, |DA_i|$,

where α is an optimization variable between 0 and 1. Changing the relative sizes of the zones in this model resulted in different swarming behavior [42], e.g., milling or migrating in some direction. In the context of swarming-based ASV navigation, defined mission, such as desired destination, represents extra information. In this case, let $F_i = P_d - P_i$ point to the desired destination, where P_d is the position vector of the desired destination. Each ASV i then sets its new orientation to be $D_i = \beta \times V_i + (1 - \beta) \times F_i$, where β is another optimization variable between 0 and 1.

9.5 Evaluation of Hybrid AUV-AUV Communications

We simulate different scenarios in the ns-3 simulator, where underwater acoustic network modules are available. To create hybrid network simulations, we integrate multiple components of acoustic and RF networks (PHY and medium access control (MAC) layers, channel models, and net devices) on ns-3 node objects. On the ASV nodes, we install the Internet stack that is used by both acoustic and RF networks. The detailed hybrid network structure in ns-3 is depicted in Figure 9.7. In the hybrid method, application packets of source AUV are encapsulated in the UDP protocol, sent to its associated ASV, routed to the destination AUV's associated ASV, and finally received by the destination AUV. Since each source ASV knows the IP address of destination ASV to successfully route the packets, we use a mapping class at each ASV to transform the IP address of destination AUV to its associated ASV and vice versa.

Parameters used for simulations are depicted in Table 9.1. In all simulations, since the traffic load of RF links is less than that of acoustic links, we use the request to send/clear to send (RTS/CTS) mechanism to reserve the channel. In the hybrid scheme, the data rate is 40 kbps for the acoustic link and 800 kbps for the RF link [43], the AUV transmission power level is 177 dB re 1 μPa, the acoustic carrier frequency is 200 kHz, and the symbol rate is 40 kHz. In direct acoustic AUV-AUV communications, the AUV transmission power level is 187 dB re 1 μPa, the acoustic carrier frequency is 12 kHz, and the symbol rate is 4 kHz. In both schemes, for the acoustic links, binary phase-shift keying modulation is used. An acoustic attenuation model, the Thorp approximation in ns-3, is used [44] to characterize the path loss. Therefore, multipath is not simulated for either of the schemes. We simulated four PSs varying between 500 and 2000 bytes. We assume

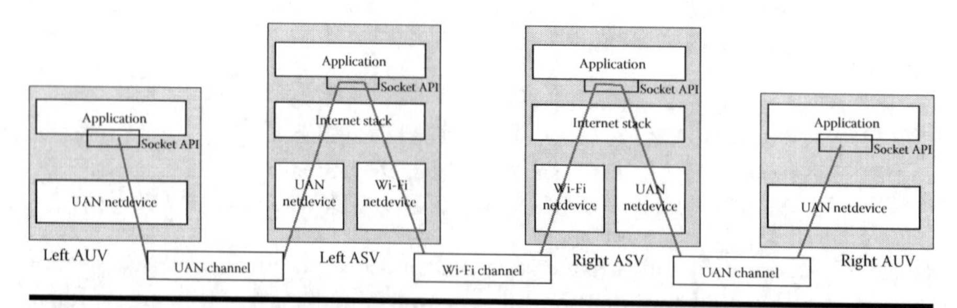

Figure 9.7 Three-zone ASV swarming model.

Table 9.1 Parameters for the Simulations

Parameter	*Direct AUV-AUV*	*Hybrid AUV-AUV*
PHY rate (sps)	4000	40,000
Data rate (kbps)	range-rate product (20 km × kbps)	40,800
PS (bytes)	500, 1000, 1500, 2000	500, 1000, 1500, 2000
Center frequency(kHz)	12	200
AUV TX power(dB re 1 μPa)	187	177
Acoustic model	Thorp	Thorp
Simulation stop time	400	400

an oracle to compute optimal intervals for traffic generations. Packet generation rate has an inverse relationship with the next packet transmission (Next TX) time. In other words, a higher packet generation rate leads to a shorter transmission interval, which is computed by the following formula:

$$\text{Next TX} = \text{Packet Size}(\text{bit}) / \text{Packet Generation Rate} \tag{9.2}$$

Throughput is computed as the total received bits divided by the time it takes from transmission of the first packet (by a sending AUV) to the reception of the last packet (by a receiving AUV). Simulations are chosen to assess the maximum achievable throughput (at the application layer) for a single AUV per each AUV-ASV pair in the hybrid acoustic-RF and cross-layer MAC-routing protocols. Hence, to build a collision-free schedule, the Aloha MAC protocol is used in the simulations. Our simulation results show that higher throughputs are achieved with the hybrid method, which also remain intact even with increasing distance. As an example, for two AUVs located 5 km apart and transmitting PSs of 2 kB, direct acoustic link achieves a maximum one-way throughput of 3960.6 bps. In contrast, the hybrid network achieves a maximum of 39847.5 bps, 10-fold of the direct acoustic link's maximum throughput. Further, the end-to-end delay between two AUVs (5 km away from each other) is 0.1 s in the ns-3 simulations for the hybrid network. The delay includes both propagation delay and PHY/MAC algorithm processing delay (T_δ is not added). In comparison, the delay is 3.53 s in direct acoustic AUV-AUV communications.

We choose three scenarios to simulate in ns-3 with variable AUV-AUV distances of 1, 2, 5, and 10 km. The first scenario simulates one-way and two-way (bidirectional) communications among two AUVs. The second scenario simulates a network of four AUV nodes with four application flows running on them. The third scenario simulates underwater infrastructure-based networks. Each scenario compares the achieved throughputs between the hybrid and direct methods. Variable PSs and different distances are examined.

9.5.1 First Scenario: One-Way and Bidirectional AUV-AUV Communications

The deployment scenario is shown in Figure 9.3. Results for one-way application from AUV-1 (left) to AUV-2 (right) are shown in Figure 9.8, where four clusters of bars denote throughput for different PSs. Each cluster has eight thin bars for four different distances of the hybrid and direct schemes. In the hybrid scheme, with 40 kbps AUV-ASV link data rate, the throughput obtained is not affected by the PS or AUV-AUV distance owing to contention-free links in both the acoustic and RF domains. Thus, we observe a throughput of 40 kbps. In the direct scheme, the acoustic link data rate is computed from the range-rate product of 20 kbps × km, *that is,* 20 kbps at 1 km distance. Thus, this results in higher data rate (hence throughput) for closer AUVs. The throughput for the direct one-way application flow from AUV-1 to AUV-2 is also unaffected by the PS in the direct scheme.

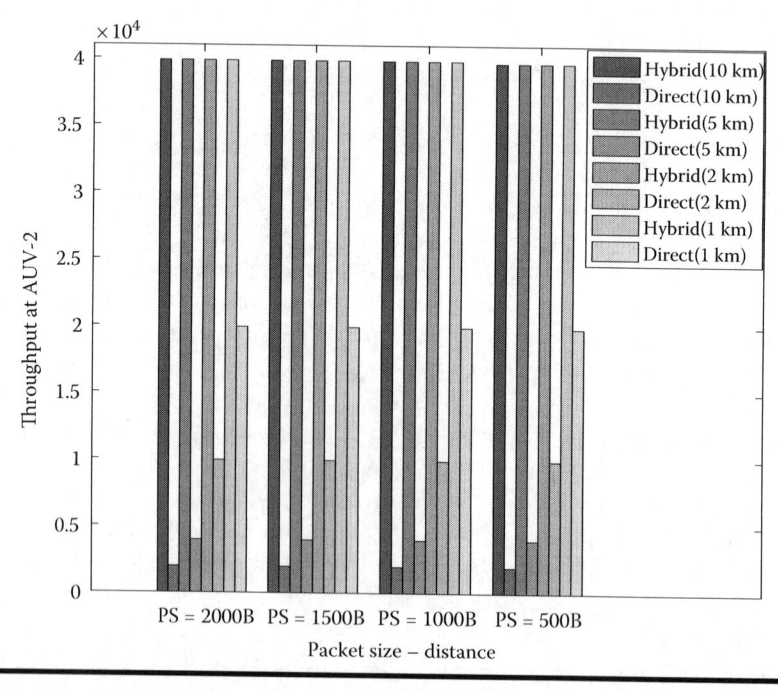

Figure 9.8 PS vs. throughput for one-way flow from AUV-1 to AUV-2 using hybrid RF-acoustic and direct method (only acoustic). Throughput is computed at AUV-2.

Throughput results in bidirectional hybrid/direct AUV-AUV communications for variable distances/PSs are shown in Figure 9.9. In the direct scheme, if the application start times for both AUV1 and AUV2 are the same, the PS of 2 kB cannot be used. This is due to TX-RX interference, since in half-duplex communication, a node cannot send and receive packets at the same time using the same frequency. When the data rate is 2 kbps (in the case of 10 km AUV-AUV distance), 8s packet transmission delay (PTD) and 6.6s propagation delay cause a collision and packet loss. To transmit packets of 2 kB size, packet scheduling is used, in which an AUV transmits right after reception of a packet. For other PSs, no scheduling is used.

We notice from Figure 9.9 that the aggregated throughput of the network is on average 10% lower than the one-way results for PS of 2 kB and 30% for PS of 500 B. Since we use no pipelining and packet generation is interval-based, larger packets are preferred. For example, PS = 1500 B reaches a throughput of 10 kbps per AUV. If we use pipelining and scheduling for other PSs (not presented in this chapter), the attainable throughput reaches half of the acoustic link data rate for each AUV. In the hybrid method, with the increase in the number of packets when smaller PSs are used, collision probability of RF transmissions also increases. This is due to more transmissions of control packets to reserve the channel, and, hence, larger PSs are preferred.

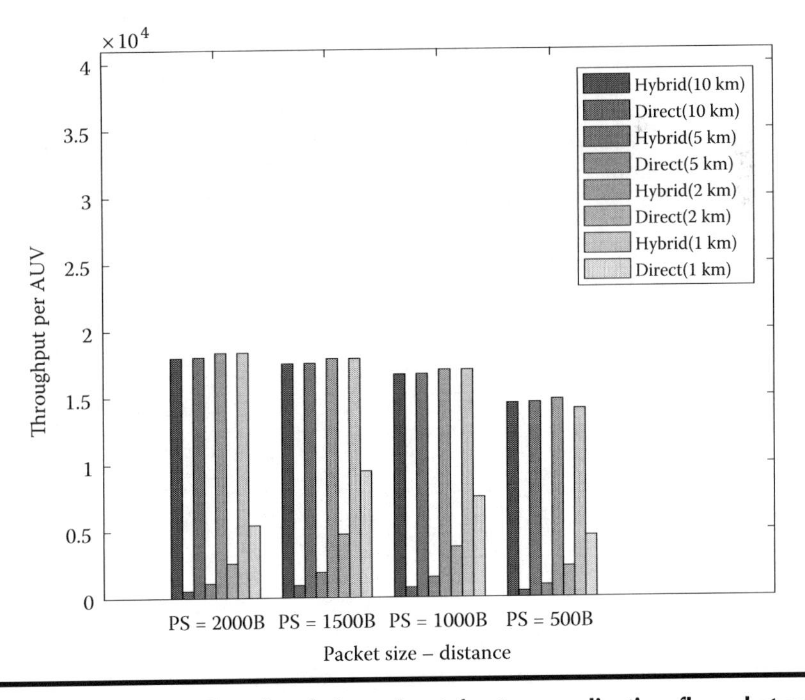

Figure 9.9 PS vs. bidirectional throughput for two application flows between the two AUVs in the hybrid and direct scheme.

In the simulations of direct scheme and bidirectional communications, except for the PS of 2 kB, no scheduling is used. Further, in the RF domain, both ASVs send and receive packets, which limit the achievable throughput. In the second scenario, we relax this constraint by four one-way flows in a network of four AUVs along with an optimal scheduler [45].

9.5.2 Second Scenario: Network of Four Nodes with Four Application Flows

This scenario has four ASV-AUV pairs located at the edges of D_{HA} by D_{HA} grid as shown in Figure 9.10. There are two flows from AUV-1 to AUV-2 and from AUV-3 to AUV-4 on the sides of the grid. There are two diagonal flows from AUV-1 to AUV-4 and AUV-3 to AUV-2. In direct AUV-AUV communications, packets of the diagonal flows traverse longer distances $\sqrt{2} \cdot D_{HA}$, and, therefore, have a slightly lower data rate, as we assumed a constant rate-range product of 20 kbps×km.

In the direct method, if applications start at the same time, we experience high collision rate. To deal with this issue, we use scheduling in the direct scheme, where application start times have a lag to avoid collisions. As shown in Figure 9.11, the aggregate throughput of the direct scheme is far less than that of the hybrid scheme. This is due to lower acoustic link rates in the direct scheme, which causes higher PTDs. In the hybrid scheme, using the RTS/CTS mechanism for the RF links does not affect the network efficiency because of the higher data rates of the RF links.

This simulation highlights the difference between the hybrid and direct AUV-AUV communications in a network of four AUVs. As mentioned earlier, since the traffic load of RF links is far less than that of the acoustic links, the RTS/CTS mechanism is suitable for hybrid networks.

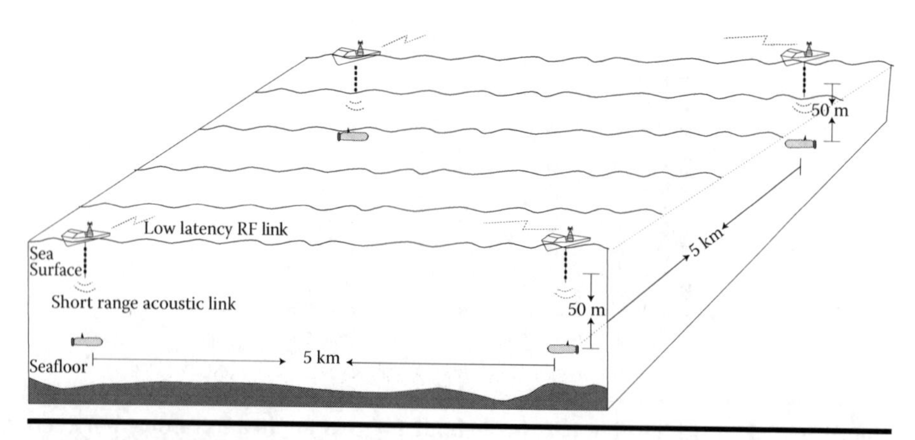

Figure 9.10 Network of four AUVs.

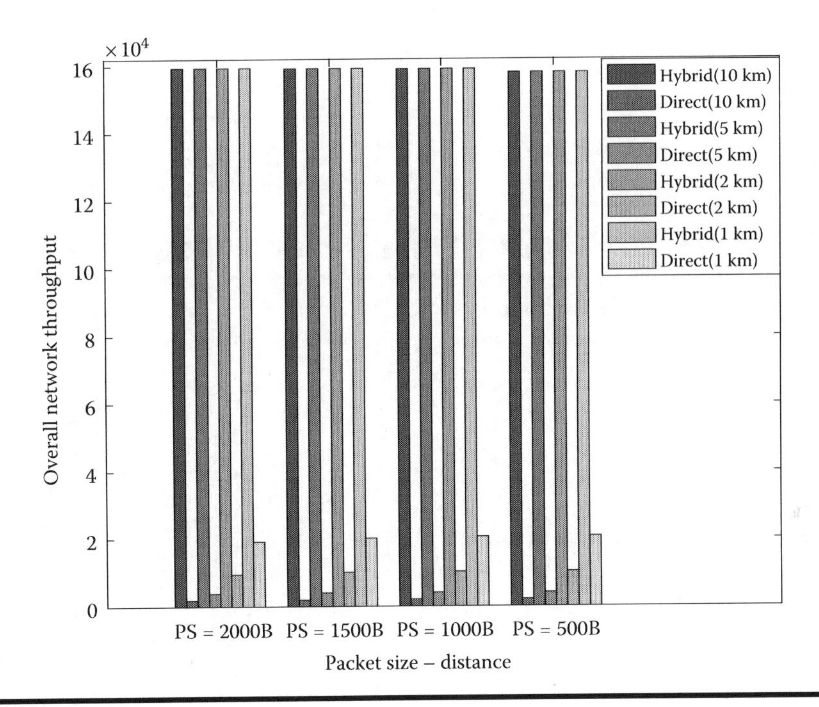

Figure 9.11 Throughput results of four AUVs.

9.5.3 Third Scenario: Infrastructure-Based Networks

Underwater networks are often constructed with a centralized (infrastructure-based) structure, where all the nodes transmit their sensory information to a centralized entity, or an access point (AP). An AP will have another antenna to communicate with a sink node by using RF link. This scenario simulates either two or four AUVs communicating with one AP using the hybrid or the direct scheme. The AP is located equidistant from the AUVs. In the hybrid method, ASVs transfer the information collected from the AUVs to the AP using RF links. Figure 9.12 illustrates two AUVs 5 km apart from each other. The AP is located halfway between the two AUVs at water surface and AUVs navigate at a depth of 50 m similar to previous scenarios.

The results of aggregate network throughput for two and four nodes are presented in Figure 9.13. In the direct scheme, we use optimal packet scheduling (pipelining), where the AP always receives packets and an AUV transmits a packet with a lag. The lag is referred to as the PTD plus interframe space (IFS), which is 0.01 s in our simulations. For instance, if two AUVs are 2 km apart and the PS is 2 kB, PTD is 0.8 s. If the first AUV transmits at time $T = 0$ s, the second AUV transmits at time $T = \text{PTD} + \text{IFS} = 0.81$ s, the third AUV at time $T = 2 \times (\text{PTD} + \text{IFS}) = 0.162$ s, and so on. The acoustic link data rate is also computed based on the same rate-range

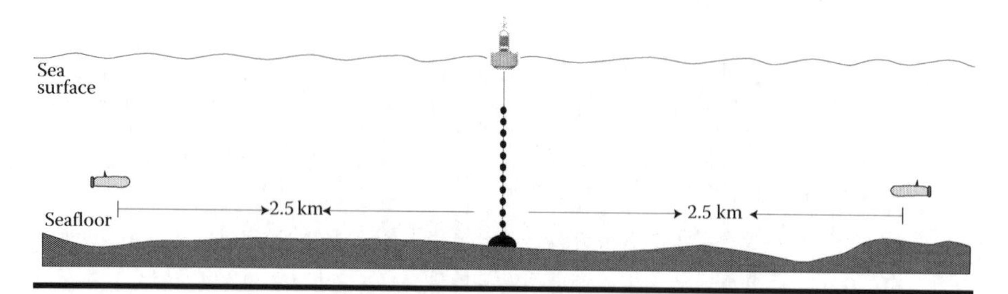

Figure 9.12 Infrastructure mode.

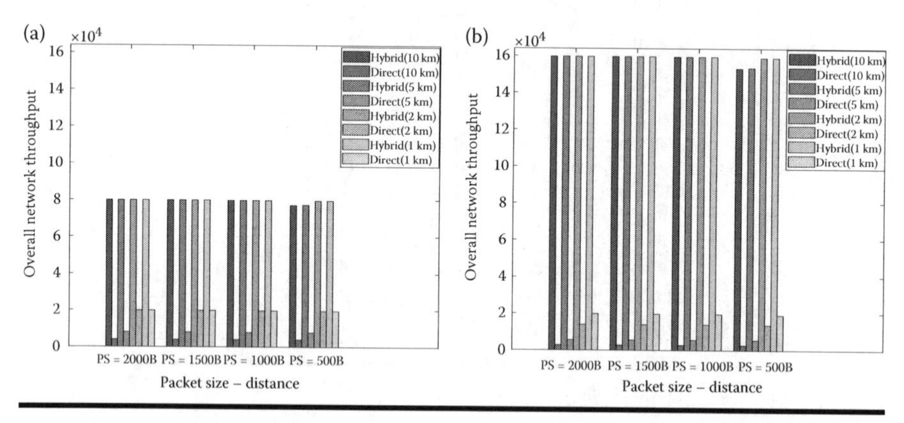

Figure 9.13 Aggregate throughput of variable PSs for (a) two and (b) four AUVs communicating with an AP equidistant from the AUVs.

product as before, 20 kbps × km. In case of two AUVs separated by 2 km, the acoustic link data rate is 20 kbps for the AUV-AP link.

In the hybrid scheme, higher throughputs are achieved for larger PSs owing to smaller amount of control packets (RTS/CTS/ACK). Advantages of the hybrid scheme are more remarkable with increase in the network size or AUV-AUV distance.

9.6 Conclusion and Future Work

We proposed a functional architecture for autonomous underwater operations with swarming-based ASV navigation and hybrid RF-acoustic communications as its constituents. The connected RF backbone is maintained by a swarm of ASVs for enhanced data rates and much reduced end-to-end latency for AUV-AUV communications. As a first step, we evaluated, via ns-3 simulation, the performance of hybrid RF-acoustic communications in comparison to direct underwater AUV-AUV acoustic communications.

In the simulations, we have an AUV associated with one ASV. In practice, there might be multiple AUVs or even hierarchical clusters of AUVs (at different depths) associated with one ASV, which demands effective MAC protocols to coordinate AUV and ASV transmissions. We intend to design cross-layer protocols to facilitate such communications. Mission-defined joint optimization of ASV navigation and adaptive RF backbone will be further developed.

Acknowledgement

This work was supported in part by the NSF Grants (CNS-1704097 and CNS-1704076).

References

1. US Ocean Observatories Initiative (US-OOI). http://oceanobservatories.org/observatories/, 2017. Accessed: Mar 4, 2017.
2. MBARI. "Monterey Accelerated Research System (MARS)." http://www.mbari.org/at-sea/cabled-observatory/, 2017. Accessed: Mar 4, 2017.
3. EMSO. http://www.emso-eu.org/, 2017. Accessed: Mar 4, 2017.
4. Ocean Networks Canada. Neptune. http://www.oceannetworks.ca/observatories/pacific, 2017. Accessed: Mar 4, 2017.
5. Mark Chaffey, Larry Bird, Jon Erickson, John Graybeal, Andy Hamilton, Kent Headley, Mike Kelley, Lance McBride, Ed Mellinger, Tim Meese, Tom O'Reilly, Walter Paul, Mike Risi, and Wayne Radochonski. MBARI's buoy based seafloor observatory design. In *OCEANS'04. MTTS/IEEE TECHNO-OCEAN'04*, volume 4, pages 1975–1984. IEEE, 2004.
6. Fumin Zhang, Giacomo Marani, Ryan N Smith, and Hyun Taek Choi. Future trends in marine robotics. *IEEE Robotics & Automation Magazine*, 22(1):14–122, 2015.
7. J. Borges De Sousa and G. Andrade Gonçalves. Unmanned vehicles for environmental data collection. *Clean Technologies and Environmental Policy*, 13(2):369–380, 2011.
8. Signe Redfield. Cooperation between underwater vehicles. In *Marine Robot Autonomy*, pages 257–286. Springer, 2013.
9. Brooks Reed, Josh Leighton, Milica Stojanovic, and Franz Hover. Multi-vehicle dynamic pursuit using underwater acoustics. In *Robotics Research*, pages 79–94. Springer, 2016.
10. Stephanie Petillo, Arjuna Balasuriya, and Henrik Schmidt. Autonomous adaptive environmental assessment and feature tracking via autonomous underwater vehicles. In *OCEANS 2010 IEEE-Sydney*, pages 1–9. IEEE, 2010.
11. Francesco Maurelli, Pedro Patrón, Joel Cartwright, Jamil Sawas, Yvan Petillot, and David Lane. Integrated MCM missions using heterogeneous fleets of AUVs. In *OCEANS, 2012-Yeosu*, pages 1–7. IEEE, 2012.
12. Naomi Ehrich Leonard. Cooperative vehicle environmental monitoring. In *Springer Handbook of Ocean Engineering*, pages 441–458. Springer, 2016.
13. Pierre FJ Lermusiaux, Tapovan Lolla, Patrick J Haley Jr, Konuralp Yigit, Mattheus P Ueckermann, Thomas Sondergaard, and Wayne G Leslie. Science of autonomy: time-optimal path planning and adaptive sampling for swarms of ocean vehicles. In *Springer Handbook of Ocean Engineering*, pages 481–498. Springer, 2016.
14. RH Rahman and MR Frater. Delay-tolerant networks (DTNs) for underwater communications. Sawston, U.K.: Woodhead Publishing, 2015.

15. Affan Syed, Wei Ye, Bhaskar Krishnamachari, and John Heidemann. Understanding spatio-temporal uncertainty in medium access with aloha protocols. In *Proceedings of the WUWNet*, pages 41–48, 2007.
16. MOOS-IvP. http://oceanai.mit.edu/moos-ivp/pmwiki/pmwiki.php, 2017. Accessed: Mar 4, 2017.
17. Christopher C Sotzing and David M Lane. Improving the coordination efficiency of limited-communication multi-autonomous underwater vehicle operations using a multiagent architecture. *Journal of Field Robotics*, 27(4):412–429, 2010.
18. Zheng Guo, Bing Wang, and Jun-Hong Cui. Prediction assisted single-copy routing in underwater delay tolerant networks. In *Global Telecommunications Conference (GLOBECOM 2010), 2010 IEEE*, Miami, Florida, 2010.
19. Xiaoyan Hong, Meng Kuai, and Wenhua Hu. Routing with bridging nodes for drifting mobility. In *WUWNet '12 Proceedings of the Seventh ACM International Conference on Underwater Networks and Systems*, Los Angeles, CA, 2012.
20. Seokhoon Yoon, Abul K. Azad, Hoon Oh, and Sunghwan Kim. Aurp: An AUV-aided underwater routing protocol for underwater acoustic sensor networks. *Sensors*, 12(2):1827–1845, 2012.
21. Ian Katz. A delay-tolerant networking framework for mobile underwater acoustic networks. 2007.
22. Fabricio JL Ribeiro, Aloysio de CP Pedroza, and Luis H.M.K. Costa. Deepwater monitoring system using logistic-support vessels in underwater sensor networks. *IEEE Latin America Transactions*, 10(1):1324–1331, 2012.
23. Jun-Hong Cui, Jiejun Kong, Mario Gerla, and Sengli Zhou. The challenges of building mobile underwater wireless networks for aquatic applications. *IEEE Network*, 20(3):12–18, 2006.
24. Iuliu Vasilescu, Keith Kotay, Daniela Rus, Matthew Dunbabin, and Peter Corke. Data collection, storage, and retrieval with an underwater sensor network. In *Proceedings of the 3rd International Conference on Embedded Networked Sensor Systems*, pages 154–165. ACM, 2005.
25. Atish P Shirodkar and Samarth Borkar. Autonomous surface craft and a significant operating system–an overview. 2014.
26. Stefano Brizzolara and Robert A Brizzolara. Autonomous sea surface vehicles. In *Springer Handbook of Ocean Engineering*, pages 323–340. Springer, 2016.
27. Zhixiang Liu, Youmin Zhang, Xiang Yu, and Chi Yuan. Unmanned surface vehicles: An overview of developments and challenges. *Annual Reviews in Control*, 41:71–93, 2016.
28. Francisco García-Córdova and Antonio Guerrero-González. Intelligent navigation for a solar powered unmanned underwater vehicle. *International Journal of Advanced Robotic Systems*, 10(4):185, 2013.
29. Alexander Bahr, John J Leonard, and Maurice F Fallon. Cooperative localization for autonomous underwater vehicles. *The International Journal of Robotics Research*, 28(6):714–728, 2009.
30. Maurice F Fallon, Georgios Papadopoulos, and John J Leonard. Cooperative AUV navigation using a single surface craft. In *Field and Service Robotics*, Eds. Kelly, A., Iagnemma, K., Howard, A. pages 331–340. Berlin, Heidelberg: Springer, 2010.
31. Sarah E Webster, Louis L Whitcomb, and Ryan M Eustice. Preliminary results in decentralized estimation for single-beacon acoustic underwater navigation. *Robotics: Science and Systems VI*, pages 1–8, 2010.

32. Joseph Curcio, John Leonard, and Andrew Patrikalakis. Scout: A low cost autonomous surface platform for research in cooperative autonomy. In *OCEANS, 2005. Proceedings of MTS/IEEE*, pages 725–729. IEEE, 2005.
33. Elgar Desa, Pramod Kumar Maurya, Arvind Pereira, António M Pascoal, RG Prabhudesai, Antonio Mascarenhas, R Madhan, SGP Matondkar, G Navelkar, S Prabhudesai, and S Afzulpurkar. A small autonomous surface vehicle for ocean color remote sensing. *IEEE Journal of Oceanic Engineering*, 32(2):353–364, 2007.
34. Damian Manda, May-Win Thein, Andrew D'Amore, and Andrew Armstrong. A low cost system for autonomous surface vehicle based hydrographic survey, U.S. Hydrographic Conference, National Harbor, MD, March 16–19, 2015.
35. Thomas C O'Reilly, Brian Kieft, and Mark Chaffey. Communications relay and autonomous tracking applications for wave glider. In *OCEANS 2015-Genova*, pages 1–6. IEEE, 2015.
36. João Borges de Sousa, João Pereira, João Alves, Madaleno Galocha, Baptista Pereira, Claro Lourenço, and Marinha Portuguesa. Experiments in multi-vehicle operations: The rapid environmental picture atlantic exercise 2014. In *OCEANS 2015-Genova*, pages 1–7. IEEE, 2015.
37. Tawfiq Taher, Vinothkumar Viswanathan, Tony Varghese, Hongchuan Jiang, Nicholas Patrikalakis, and Audren Cloitre. Multi-domain autonomous mobile network for sensing. In *OCEANS 2016 MTS/IEEE Monterey*, pages 1–6. IEEE, 2016.
38. Masahiko Sasano, Shogo Inaba, Akihiro Okamoto, Takahiro Seta, Kenkichi Tamura, Tamaki Ura, Shinichi Sawada, and Taku Suto. Development of a regional underwater positioning and communication system for control of multiple autonomous underwater vehicles. In *Autonomous Underwater Vehicles (AUV), 2016 IEEE/OES*, pages 431–434. IEEE, 2016.
39. Michael R. Benjamin, Henrik Schmidt, Paul M. Newman, and John J. Leonard. Nested autonomy for unmanned marine vehicles with MOOS-IVP. *Journal of Field Robotics*, 27(6):834–875, 2010.
40. Toby Schneider and Henrik Schmidt. Unified command and control for heterogeneous marine sensing networks. *Journal of Field Robotics*, 27(6):876–889, 2010.
41. Iain Couzin, Jens Krause, Nigel Franks, and Simon Levin. Effective leadership and decision-making in animal groups on the move. *Nature*, 433 (7025):513–516, 2005.
42. Iain Couzin, Jens Krause, Richard James, Graeme Ruxton, and Nigel Franks. Collective memory and spatial sorting in animal groups. *Journal of Theoretical Biology*, 218 (1):1–11, 2002.
43. FreeWave. http://www.freewave.com/wp-content/uploads/LDS0001HTE_Rev_A_HT_PE.pdf.
44. Albert F Harris III and Michele Zorzi. Modeling the underwater acoustic channel in ns2. In *Proceedings of the 2nd International Conference on Performance Evaluation Methodologies and Tools*, page 18. ICST (Institute for Computer Sciences, Social-Informatics and Telecommunications Engineering), 2007.
45. Yang Guan, Chien-Chung Shen, and Justin Yackoski. MAC scheduling for high throughput underwater acoustic networks. In *Wireless Communications and Networking Conference (WCNC), 2011 IEEE*, pages 197–202. IEEE, 2011.

Chapter 10

Adaptive Underwater Acoustic OFDM

Zhijing Ye, Yuehai Zhou, Qiang Fu, and Aijun Song
The University of Alabama

Contents

10.1 Introduction

Over the past several decades, a large number of efforts have been pursued in the area of underwater acoustic communications. Higher data rates have been achieved through spectrally efficient modulation schemes, spatial diversity, and high-performance equalizers [1–11]. For instance, data rates up to several tens of kbps have been demonstrated by multichannel decision-feedback equalization, time reversal methods, Orthogonal frequency-division multiplexing (OFDM), and multi-input/multi-output (MIMO) techniques. However, the demonstrated reliability and the achieved spectral efficiency are still inadequate, often failing to meet the communication needs in the ocean.

The underwater acoustic channel is a challenging environment for data communications. One of the commonly cited difficulties is the time-varying multipath propagation. The variability of the ocean parameters, mainly the water temperature profiles and the sea surface condition, has been studied to understand their impact on the performance of acoustic communications at high frequencies, greater than 10 kHz. Both the long-term (in hours) changes in the ocean volume and the fast fluctuations (in seconds) of the sea surface generate direct impacts on acoustic communication receiver performance [12,13].

The time-varying characteristic severely constrains the performance of acoustic communication systems. Most underwater acoustic systems adopt a fixed set of transmission parameters, such as power level, modulation order, or coding rate. The power level is often set adequately high, and the code rate and modulation order are usually set low to achieve reliable communications to cope with extreme channel conditions. Apparently, power and bandwidth resource are wasted when the channel condition turns to a favorable one [14].

Close-loop adaptive transmissions can achieve both reliability and spectral efficiency of the acoustic systems in time-varying underwater acoustic channel conditions. In adaptive transmission schemes, the transmission power, modulation mode, and coding rate are adjusted based on the current channel state information. When the channel condition is favorable, the power level is reduced for energy efficiency or high-level modulation orders and high coding rates are chosen to make full use of the spectrum. When the channel condition deteriorates, the system adopts high source levels, low modulation orders, or low coding rate methods to ensure reliability.

Single-carrier and multicarrier systems often differ in adaptive strategies. For instance, in source power level adaptation, the total power level is adjusted in a single-carrier system. In the OFDM transceiver, the power levels at individual subcarriers are adjusted, in addition to the total source power level. Another adaptive option in OFDM is to adjust mapping indexes for individual subcarriers. The receiver adjusts the mapping schemes for individual subcarriers based on channel conditions. The selected mapping indexes for individual subcarriers are sent back to the transmitter. In this chapter, we focus on adaptation schemes of underwater acoustic OFDM systems.

In adaptive OFDM systems, the challenges are multifold. First, the combination of fast channel fluctuations and slow sound speed limits the value of the receive feedback. When the receiver feedback gets to the transmitter, the channel state information may well be obsolete. Channel prediction is often needed. Second, the limited bandwidth of underwater acoustic systems restricts how much information can be sent back to the transmitter. This is especially limiting for the single-input/multiple-output (SIMO) system, where a large number of parameters are needed to describe multiple impulse responses. Third, it is still unknown to the community what best channel metrics should be used for selection of modulation order or power level in the dynamic ocean environment.

A number of papers in the literature investigated adaptive acoustic communications, especially in OFDM systems. For example, adaptive OFDM systems were used to support high spectral efficiency over extended periods of time [15]. The predicted impulse responses were used to select power levels and modulation orders for OFDM subcarriers. An adaptive modulation and coding scheme in an OFDM system was demonstrated in at-sea experiments [16]. In [17], an adaptive MIMO OFDM system based on partial knowledge of channel state information was studied. An adaptive modulation and coding technique for phase-shift keying modulation in underwater acoustic (UWA) channels was considered [18]. Adaptive bit and power loading were used to maximize the data rate in OFDM systems [19]. In [20], the forward link utilized the coded OFDM transmission, while the feedback link used a binary chirp spread-spectrum modulation.

Here we propose to investigate adaptive modulation based on time-reversed OFDM (TR-OFDM). Multiple advantages exist to use TR-OFDM [21] for adaptive systems. First, the time reversal processing generates a compact equivalent impulse response, namely q-function, between the transmitter and the receiver array. This greatly reduces the number of channel parameters for feedback. The cyclic prefix (CP) length can also be reduced for OFDM communications. Second, the compact q-function is more stable than the individual impulse responses. Frequency of receiver feedback can be reduced. Channel prediction requirements may also be relaxed. Third, for the compact, stable q-function, it is easier to identify a channel metric for modulation order selection.

We investigate three different types of receiver feedback strategies, all taking advantage of the compact and stable characteristics of the q-function $q(n)$. The first option is to use effective signal-to-noise ratio (SNR) after OFDM demodulation, as in [16]. The second option is to send mapping order for individual subcarriers back to the transmitter, after the receiver performs power allocation. The third option is to use a truncated time domain q-function as feedback. The q-function can be calculated based on estimated impulse responses. These three options have their respective advantages. For example, the effective SNR uses minimum communication bandwidth. Usage of selected mapping index as feedback means selection is made at the receiver, where the demodulation performance is immediately known. The receiver can adjust the predetermined modulation selection scheme based on the demodulation results. Also, sending q-function back allows the transmitter to perform power allocation and modulation selection at individual subcarriers.

We simulate all three feedback strategies in the TR-OFDM system. Impulse responses obtained from local river tests are used in communication simulations. Communication performance, in terms of data rates and bit error rate (BER), is obtained for three feedback options. For the first and third feedback options, we also investigate system performance based on adaptive modulation at individual subcarriers. The rest of the chapter is organized as follows. In section 10.2, the system model of the adaptive TR-OFDM is introduced. Three different adaptive structures are outlined. Section 10.3 presents the performance metrics and results of the computer simulations. Section 10.4 draws the conclusion. We use the following notations: superscripts $(\cdot)^T$, $(\cdot)^*$, and $(\cdot)^H$ stand for transpose, conjugate, and conjugate transpose, respectively.

10.2 System Description and Feedback Strategies

In this section, we present the structure of our adaptive TR-OFDM system. We also describe three types of receiver feedback strategies.

Figure 10.1 illustrates our adaptive TR-OFDM system, which has one transmitting element and M receiving elements. At the transmitter side, the source signal $S(k)$ is the frequency domain data that is fed to an OFDM modulator. The time domain signal $s(n)$ is the output from the OFDM modulator. After the transmission, the transmitted data arrives at multiple receiving elements after being distorted by the multipath propagation, described by the impulse response $c_j(n)$, and contaminated by the ambient noise, $w_j(n)$, where j is the receiving element index. At the receiver side, a TR-OFDM receiver is used to demodulate the transmitted symbols. Specifically, based on the received signal $y_j(n)$, Doppler correction and channel estimation are performed. Then, the received signals from the multiple receiving elements are filtered with $c_j^*(-n)$ and the outputs are combined. This is the time reversal processing, which produces a single composite signal, $r(n)$. After that, regular OFDM demodulation is applied on the composite signal $r(n)$.

In the system in Figure 10.1, we assume K be the number of subcarriers and $L_d = K + L_{CP}$, L_{CP} be the prefix length. The source information block is

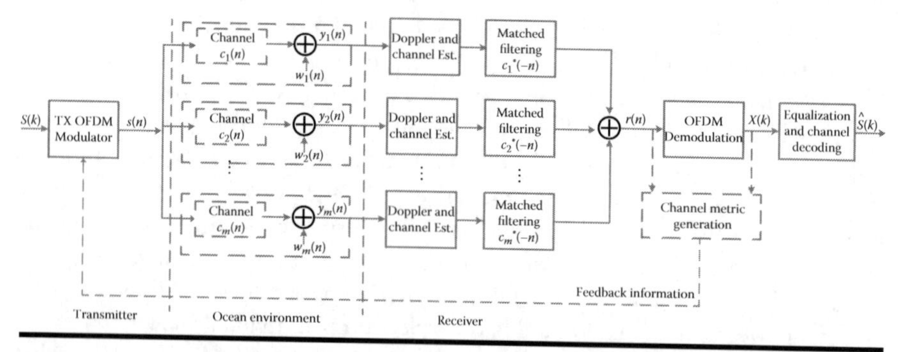

Figure 10.1 Adaptive acoustic communication system based on TR-OFDM.

$\mathbf{S} \triangleq [S(0),\ldots,S(K-1)]^T$. The corresponding transmitted time domain block $\mathbf{s} \triangleq [s(0),\ldots,s(L_d-1)]^T$ is

$$\mathbf{s} = \mathbf{T}_{CP}\mathbf{F}_k^H\mathbf{S}. \tag{10.1}$$

In equation (10.1), $\mathbf{T}_{CP}$ is a $L_d \times K$ matrix $\mathbf{T}_{CP} \triangleq \begin{bmatrix} \mathbf{I}_{CP}^T & \mathbf{I}_K^T \end{bmatrix}^T$, $\mathbf{I}_{CP}$ be the last L_{CP} rows of the $K \times K$ identity matrix $\mathbf{I}_K$. The variable $\mathbf{F}_K$ is a $K \times K$ matrix with the (p,q)-th entry as $[\mathbf{F}_K]_{p,q} = \left(1/\sqrt{K}\right)e^{-2i\pi(p-1)(q-1)/K}$. The matrix $\mathbf{F}_K$ represents the operations of CP insertion and K-point FFT.

At the TR-OFDM receiver, utilizing the passive time reversal (TR) leads to a composite signal $r(n)$. The equivalent impulse response between the transmitter and receiving array is

$$q(n) \triangleq \sum_{j=1}^{M} c_j(n) \circledast c_j^*(-n), \tag{10.2}$$

where $\circledast$ represents the convolution operation. Then, the composite signal $r(n)$ can be denoted as

$$r(n) = \sum_{p=-L}^{L} q(n)s(n-p) + z(n), \tag{10.3}$$

where $z(n) \triangleq \sum_{j=1}^{M} w_j(n) \circledast c_j^*(-n)$ is the noise component.

The OFDM demodulator output $X(k)$ is related to $r(n)$ via a relationship in a matrix/vector form as

$$\mathbf{X} = \mathbf{F}_K\mathbf{R}_{CP}\mathbf{r}, \tag{10.4}$$

where $\mathbf{X} \triangleq [X(0),\ldots,X(K-1)]^T$, $\mathbf{r} \triangleq [r(0),\ldots,r(L_d-1)]^T$ and the $K \times L_d$ matrix $\mathbf{R}_{CP} \triangleq \begin{bmatrix} \mathbf{0}_{K\times L_{CP}} & \mathbf{I}_K \end{bmatrix}$ represents the operation of discarding the prefix.

In the TR-OFDM system, the equivalent impulse response, or the q-function $q(n)$, is compact and stable. Therefore, feedback based on the q-function can greatly reduce the number of channel parameters for receiver feedback. The system model assumes that the residual Doppler effects are negligible after proper initial motion compensation, that is resampling by a nominal Doppler factor and removal of the frequency offset. There is also an assumption that the channel is static (linear time invariant) at least over the transmission interval of one OFDM symbol.

10.2.1 Feedback Strategies

The receiver has knowledge of the channel frequency response for each subcarrier and the corresponding channel impulse responses. To perform adaptive modulation, some forms of the channel state information needs to be sent back to the transmitter.

10.2.1.1 Effective SNRs

The first option is to send back the effective SNR, which is more reliable in capturing channel quality than raw SNRs. Note that the effective SNR is calculated after OFDM demodulation, thus accounting for channel estimation errors.

With TR-OFDM, the equivalent channel input–output relationship is single-input/single-output. The OFDM demodulator output is

$$\hat{S}_{eq}(k) = \frac{\hat{Q}^*(k)X(k)}{\sqrt{\hat{Q}^*(k)\hat{Q}(k)}}, \tag{10.5}$$

where $\hat{Q}(k)$ is the discrete Fourier transform of the estimated q-function. Define $\hat{Q}_{eq}(k) = \sqrt{\hat{Q}^*(k)\hat{Q}(k)}$ as the equivalent channel gain. The effective SNR can be calculated as

$$\rho_e = \frac{E_{k\in D}[|\hat{Q}_{eq}(k)S(k)|^2]}{E_{k\in D}[|\hat{S}_{eq}(k) - \hat{Q}_{eq}(k)S(k)|^2]}. \tag{10.6}$$

As in equation (10.6), the effective SNR is obtained via averaging of all the data subcarriers, $D \triangleq \{0,\ldots,K-1\}$. When we use effective SNR, all subcarriers use the same mapping order. The following methods allow adjustment of mapping orders at individual subcarriers.

10.2.1.2 Mapping Indexes for Subcarriers

In the second option, the receiver adjusts the mapping schemes for individual subcarriers based on the channel state information, which is available at the receiver. The selected mapping indexes for individual subcarriers are sent back to the transmitter.

The noise in the frequency domain for each subcarrier is calculated as

$$Z(k) = \sum_{n=0}^{K-1}\left(z(n)e^{-i2\pi\frac{nk}{K}}\right). \tag{10.7}$$

Based on the noise $Z(k)$ and estimated q-function $\hat{Q}(k)$ in the frequency domain, the receiver can adjust mapping orders for each subcarrier. We assume that the source information $S(k)$ is independent across subcarriers. The candidate mapping schemes are binary phase-shift keying (BPSK), quadrature phase-shift keying (QPSK), 8 phase-shift keying (8PSK), and 16-quadrature amplitude modulation (16QAM). For any $k \in D$, the modulation order $W_k \in \{2,4,8,16\}$ is selected based on $Z(k)$ and $\hat{Q}(k)$. The objective is to maximize the data rate while maintaining a target average BER. Hence, the optimization criterion is

$$\begin{aligned} &\max_{W_0,\ldots,W_{K-1}} \sum_{k=0}^{K-1} \log_2 W_k \\ &\text{subject to } \frac{1}{K}\sum_{k=0}^{K-1} P_{e,k} \leq P_b, \end{aligned} \tag{10.8}$$

where $P_{e,k}$ is the average BER for the kth subcarrier and P_b is the target average BER. To solve this optimization problem, the SNRs at subcarriers are divided into four ranges. Each SNR range is assigned to one of the candidate mapping schemes, from BPSK to 16QAM.

10.2.1.3 Truncated Time Domain Q-Function

The third feedback strategy uses a truncated time domain q-function as feedback. The truncated time domain q-function allows adaptive mapping for the OFDM symbol or individual subcarriers. Knowing that the q-function is compact, we use the central "impulse-like" part as the feedback instead of the entire function. The q-function $q(n)$ has a time support of $-L < n < L$, with a length of $L_q = 2L - 1$. To reduce the amount of feedback information, we choose the central part of the q-function over the indexes $[-L_\tau, L - L_\tau]$ for feedback.

In addition to the q-function, the transmitter needs to know the noise information, $Z(k)$, to adjust mapping schemes. Again to limit feedback information, we use the average noise power σ_z of all subcarriers for feedback.

10.3 Experiments and Simulations

In this section, we first show the TR-OFDM performance from an at-sea experiment. Then, we show our computer simulation results for adaptive TR-OFDM systems.

Figure 10.2 shows effective SNR variations from an at-sea experiment. The experiment was conducted at the 20 m isobath in the northern Gulf of Mexico, August 2016. A five-element hydrophone array was deployed in a stationary mooring. The receiving array covered 9–16 m of the water column, with an element spacing of 1.75 m. We tested four source–receiver ranges, 250 m, 500 m, 1000 m,

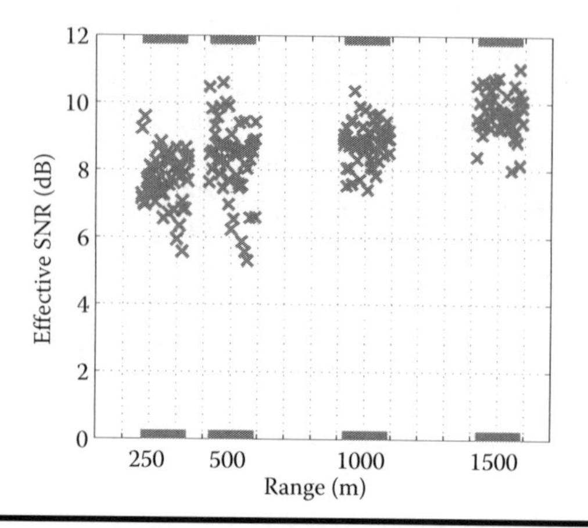

Figure 10.2 Effective SNR based on TR-OFDM demodulation.

and 1500 m, using a 85 kHz transducer hung from a research vessel. The transducer depth was 12 m. At each of the four ranges, three OFDM packages were transmitted from the transducer. Each package has 18 OFDM symbols, and the bandwidth of the OFDM transmission was 8.5 kHz.

As shown, the effective SNR showed significant fluctuations at the same range or across the ranges. For instance, at the distance of 500 m, the effective SNR varied from about 5 to 11 dB. In light of such channel fluctuations, it was necessary to use the adaptive system to adjust the modulation parameters.

10.3.1 Adaptive Simulations

Our simulation used impulse responses extracted from an acoustic experiment at the local river, the Black Warrior River, close to the University of Alabama campus. At the experimental site, the water depth was about 9 m. We used an 85 kHz transducer for OFDM transmission, deployed at 3 m below the water surface. The receiver array had six hydrophones, which were towed by a kayak. The hydrophone array covered the water depth of 2.5–5 m. Its element spacing was 0.5 m. One example of impulse response over 32 OFDM symbols is shown in Figure 10.3. The OFDM transmission parameters are listed in Table 10.1. The length of the estimated impulse response was 20 ms. During the OFDM packet, the impulse responses showed a modest level of fluctuations, as shown in Figure 10.3. We used the impulse responses from this OFDM packet in adaptive TR-OFDM simulations. The OFDM transmission parameters from Table 10.1 were used. During the experiment, the source–receiver range was short, less than several hundred meters. Therefore, we assumed a small propagation delay of 0.5 s in feedback simulations.

Four different cases were compared in each feedback strategy, including ideal case, feedback case, fixed BPSK, and fixed QPSK. In the ideal case, the transmitter

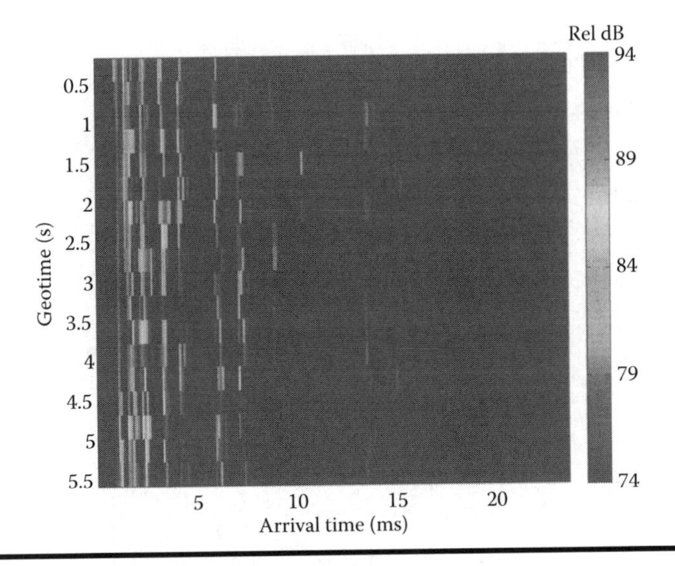

Figure 10.3 Example impulse response.

Table 10.1 TR-OFDM Parameters

Parameters	*Numbers*
Bandwidth B	10 kHz
Symbol duration T	204.80 ms
Cyclic prefix duration T_{CP}	51.20 ms
Data subcarriers K	2048
Number of hydrophones	6

knew the complete channel state information without delay. In the feedback case, two OFDM symbol's delay time was assumed during feedback. The fixed BPSK case used BPSK mapping for all OFDM subcarriers without adaptation. Similarly, the fixed QPSK case used QPSK for all subcarriers.

10.3.2 Performance Comparison

10.3.2.1 Effective SNRs as Feedback

To use effective SNRs for adaptive modulation, we first calculated effective SNR thresholds for candidate mapping scheme. Figure 10.4 shows the BER curves of four different mapping schemes. Note here, all OFDM subcarrier used the same mapping schemes. Impulse responses were extracted from four ranges of the at-sea

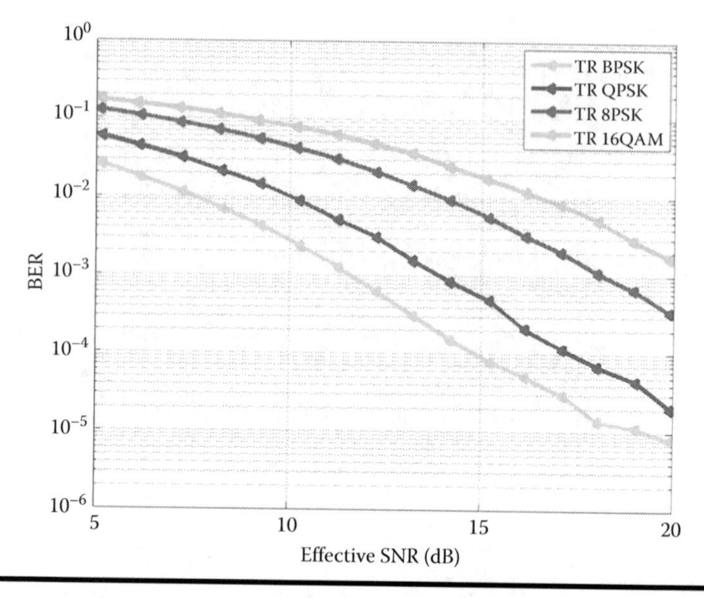

Figure 10.4 BER versus effective SNR for four transmission modes.

experiment. The BER curves were obtained as an average result over all impulse responses. On the basis of performance results, we set the effective SNR thresholds for the four transmission modes as in Table 10.2, with the target BER 10^{-3}, plus a 2 dB protection margin.

Figure 10.5 shows selected mapping orders over 32 OFDM symbols for two feedback cases: ideal feedback and delayed feedback. The impulse responses from Figure 10.3 were used in this and later TR-OFDM simulations. Both cases started with the BPSK mapping. The ideal case quickly adjusted the transmitter to 8-PSK mapping. Compared with the ideal case, the delayed feedback case used less of the higher-order mapping schemes, which resulted in some loss in data rates. Table 10.3 summarizes such results. The ideal feedback case supported 19.5 kbps, 2.4 times of the fixed BPSK rate, and 1.2 times of the fixed QPSK rate. The delayed feedback resulted in 18.0 kbps, still higher than the fixed BPSK or QPSK mapping schemes.

Table 10.2 Effective SNR Thresholds for Four Transmission Modes

Mapping Methods	*Effective SNR (dB)*
BPSK	<12
QPSK	12–16
8 PSK	16–20
16 QAM	>20

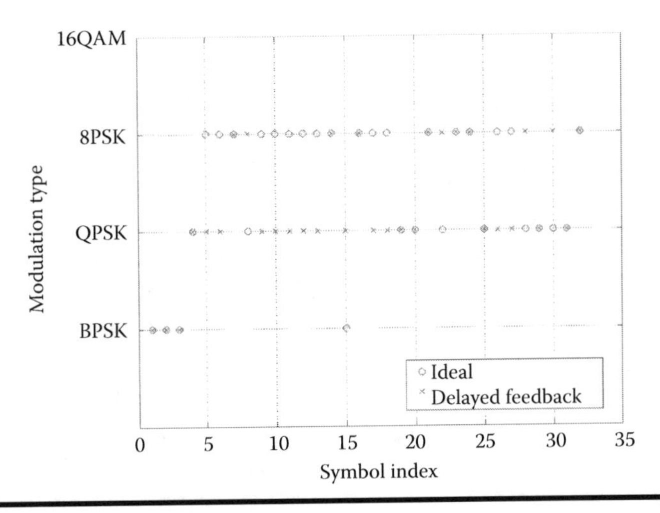

Figure 10.5 Effective SNR's modulation types over all OFDM symbols. The ideal case is compared with the delayed case.

Table 10.3 Communication Performance of the First Feedback Strategy: Effective SNR

Parameters	*Four Cases*			
	Ideal	*Feedback*	*BPSK*	*QPSK*
Average data rate (bps)	19,500	18,000	8,000	16,000
Average ESNR (dB)	14.78	14.83	14.39	14.93
Average BER (10^{-3})	4.81	5.01	3.00	4.40

All four cases had similar BERs or effective SNRs. This means that the first feedback strategy resulted in high data rates without loss of BER performance, through the use of effective SNRs.

10.3.2.2 Mapping Indexes of Subcarriers as Feedback

Figure 10.6 shows the selected mapping orders for all subcarriers, when the adaptive modulation was performed at the receiver for individual subcarriers, for just one OFDM symbol. Due to frequency selectivity of the q-function, some subcarriers used 8PSK or even 16-QAM. In comparison, the delayed feedback case used primarily BPSK and QPSK mapping schemes. Despite some differences, the two cases used the same mapping schemes for a significant portion of the subcarriers. It means that the frequency responses of the q-function did not change significantly.

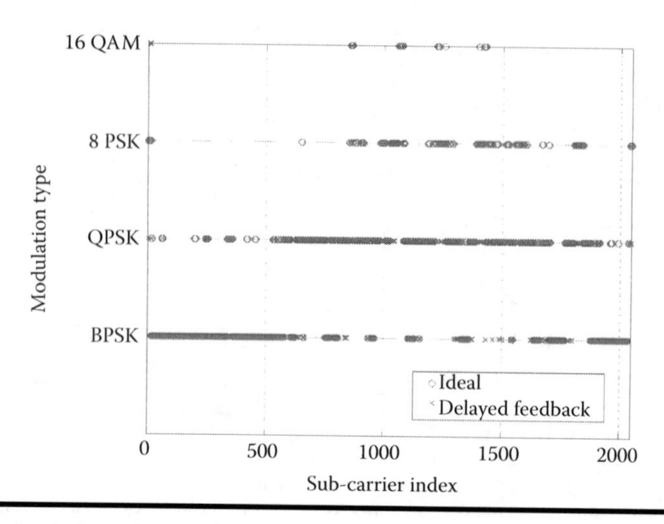

Figure 10.6 Selected mapping indexes over all subcarriers for one OFDM symbol.

Table 10.4 Receiver Performance of Feeding Back Mapping Index for Subcarriers

Parameters	*Four Cases*			
	Ideal	*Feedback*	*BPSK*	*QPSK*
Average data rate (bps)	22,114	21,317	8,000	16,000
Average ESNR (dB)	15.07	14.83	14.39	14.93
Average BER (10^{-3})	6.93	6.51	3.00	4.40

Table 10.4 summarizes communication results for the second feedback strategy. The ideal case of this strategy further increased the data rate to about 22.1 kbps. The delayed feedback case supported a data rate of about 21.3 kbps, higher than the data rate of the first feedback strategy.

10.3.2.3 Truncated Q-Function as Feedback

We first show a comparison between an impulse response and its q-function in Figure 10.7, for one OFDM symbol in Figure 10.3. As shown, q-function had most of its significant value at its center. The delay spread of this impulse response was 20 ms, which corresponded to a channel length of L = 200. Because of the impulse-like structure of the q-function, we were able to use a truncated q-function for feedback. Multiple values of L_τ, L_τ = 50,30,15, were examined. Because of the shape of the q-function, these three values led to similar data rates. When L_τ = 50,

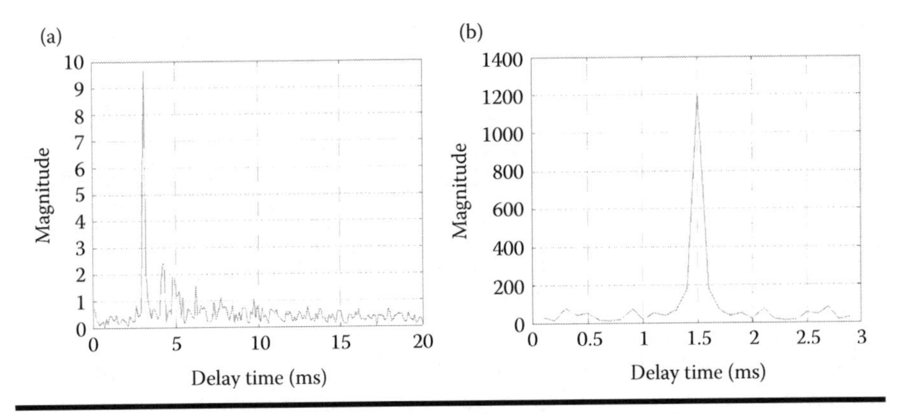

Figure 10.7 (a) Snapshot of channel impulse response and (b) the time domain q-function.

the OFDM data rate achieved in the simulations was 20.87 kbps. The cases with $L_\tau = 30$ and $L_\tau = 15$ generated data rates of 20.63 and 20.32 kbps, respectively. Because the length of the truncated q-function was $2L_\tau - 1$, decrease of L_τ led to a smaller number of parameters for feedback. When $L_\tau = 15$, the delay spread of the truncated time domain q-function was 3 ms. The amount of feedback information was greatly reduced when compared with direct feedback of multiple impulse responses of the receiver array.

Figure 10.8 shows the mapping indexes of the OFDM subcarriers for one OFDM symbol when the third feedback strategy was used. Results from both the ideal case and the delayed feedback cases are both shown. Here, the ideal case

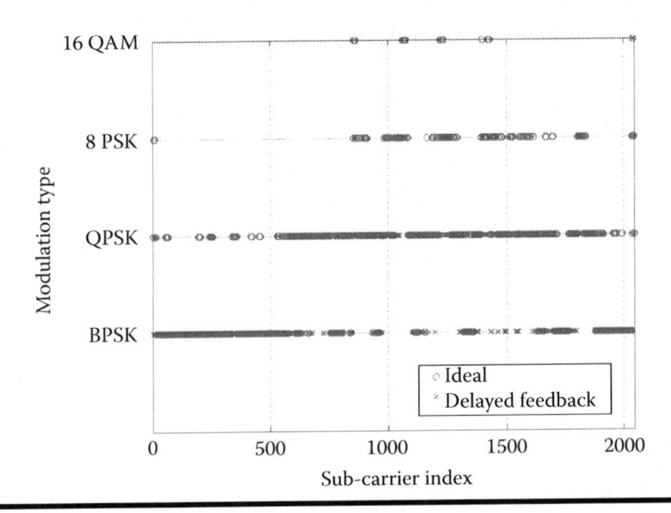

Figure 10.8 Mapping indexes for all subcarriers for one OFDM symbol for two cases: Ideal and delayed feedback, when the third feedback strategy was used.

Table 10.5 Communication Performance of the Third Feedback Strategy

Parameters	*Four Cases*			
	Ideal	*Feedback*	*BPSK*	*QPSK*
Average data rate (bps)	21,324	20,322	8,000	16,000
Average ESNR (dB)	14.97	14.66	14.39	14.93
Average BER (10^{-3})	9.51	7.71	3.00	4.40

refers to the condition that the transmitter knows the truncated q-function once it is available at the receiver. Due to frequency selectivity of the q-function, the ideal case used all four mapping schemes, with a goal to maximize the data rate. The mapping indexes of the delay feedback had a significant overlap. The reason was that the q-function varied slowly, especially over a couple of OFDM symbols.

Table 10.5 shows the communication performance of the third feedback strategy. The ideal cases supported a data rate of 21.3 kbps, slightly below the ideal case of the second strategy. That is because feedback via q-function brought a minimum reduction in communication data rates. The delayed feedback case had a data rate of 20.3 kbps. Therefore, using truncated time domain q-function as feedback had similar performance as sending back mapping indexes for individual subcarrier.

10.3.3 Feedback Cost

In addition to the communication data rate, we need to consider the feedback cost and the amount of information that is sent back to the transmitter, for the sake of spectral efficiency. Table 10.6 lists the feedback information bits of three different strategies.

In the first feedback strategy, the effective SNR per OFDM symbol is the information for feedback. If we use a single byte to represent the effective SNR number, the number of feedback information bits is 8. In the second strategy, we need to send back mapping indexes for all the OFDM subcarriers. We have four mapping schemes, which are indexable by two bits. Therefore, with 2048 subcarrier numbers in each OFDM symbol, the number of feedback information bits is 4096. In the third strategy, we

Table 10.6 Feedback Information Bits for Three Strategies

Feedback Strategy	*Feedback Information Bits*
Effective SNR	8
Mapping order	4096
Truncated q-function	960

send back the truncated q-function and average noise power. The q-function has a length of $2L_\tau - 1$. If we use 32 bits to represent a complex number, either q-function values or the noise power, the number of feedback information bits is $64L_\tau$. If $L_\tau = 50$, that number is 3200, which is at the same order as the second feedback strategy. When $L_\tau = 15$, we only need to send back 960 information bits as feedback.

10.4 Summary and Further Research

We proposed an adaptive TR-OFDM modulation system for underwater acoustic communications. Effective SNRs, mapping indexes, and truncated time domain q-function were examined as three different feedback parameters. Simulation results showed the benefits of adaptive schemes versus the nonadaptive one. All three feedback strategy provided significant data rate increases compared with fixed BPSK mapping in TR-OFDM. When compared with fixed QPSK mapping, the data rate of three adaptive schemes increase was marginal, as we need to consider the feedback cost. We argue that the use of truncated q-function can generate further data rate increase, while only requiring a modest level of feedback information.

As the underwater mobile platform gains increasing popularity [22,23], one of the research needs is to address acoustic communication challenges on mobile platforms. In moving platforms, the location and track of AUVs can be adjusted to ensure quality communication performance [24–27]. Moreover, the location information can be utilized to adjust communication protocols. In [28], a model-based adaptive scheme was proposed, where the AUV path was optimized to ensure connectivities. Geometric information, such as locations of the transmitter and receiver, communication range, and source/receiver depth, were utilized for AUV path optimization. In [26], a composite adaptive algorithm for moving platform was put forward. A position estimator, a data transmission performance tracker, and a rate selector were its three components. An online adaptive transmission rate selector was used to learn from the past communication performance, rather than relying on the channel metric. One of the promising future directions is to use learning-based algorithms instead of the acoustic model based prediction, for moving platforms of adaptive transmissions. The learning strategies should incorporate a multitude of system information for adaptation: platform location and relative velocity, ocean physical parameters, and communication performance. With an interface with platform control algorithms, such learning strategies may result in jointly optimized communications and navigation.

References

1. M. Chitre, S. Shahabudeen, L. Freitag, and M. Stojanovic. Recent advances in underwater acoustic communications networking. In *OCEANS 2008*, 2008.

2. S. Climent, A. Sanchez, J. Capella, N. Meratnia, and J. Serrano. Underwater acoustic wireless sensor networks: Advances and future trends in physical, MAC and routing layers. *Sensors*, 14(1):795, 2014.
3. G. F. Edelmann, H. C. Song, S. Kim, W. S. Hodgkiss, W. A. Kuperman, and T. Akal. Underwater acoustic communications using time reversal. *IEEE Journal of Oceanic Engineering*, 30(4):852–864, 2005.
4. M. Fink, D. Cassereau, A. Derode, C. Prada, P. Roux, M. Tanter, J. L. Thomas, and F. Wu. Time-reversed acoustics. *Reports on Progress in Physics*, 63(12): 1933, 2000.
5. T. H. Liew, B. L. Yeap, C. H. Wong, and L. Hanzo. Turbo-coded adaptive modulation versus space-time trellis codes for transmission over dispersive channels. *IEEE Transactions on Wireless Communications*, 3(6): 2019–2029, 2004.
6. S. Roy, T. M. Duman, V. McDonald, and J. G. Proakis. High-rate communication for underwater acoustic channels using multiple transmitters and space-time coding: Receiver structures and experimental results. *IEEE Journal of Oceanic Engineering*, 32(3):663–688, 2007.
7. A. Song, M. Badiey, A. E. Newhall, J. F. Lynch, H. A. DeFerrari, and B. G. Katsnelson. Passive time reversal acoustic communications through shallow-water internal waves. *IEEE Journal of Oceanic Engineering*, 35(4):756–765, 2010.
8. M. Stojanovic. Recent advances in high-speed underwater acoustic communications. *IEEE Journal of Oceanic Engineering*, 21(2):125–136, 1996.
9. M. Stojanovic, J. Catipovic, and J. G. Proakis. Adaptive multichannel combining and equalization for underwater acoustic communications. *The Journal of the Acoustical Society of America*, 94(3): 1621–1631, 1993.
10. M. Stojanovic, J. A. Catipovic, and J. G. Proakis. Phase-coherent digital communications for underwater acoustic channels. *IEEE Journal of Oceanic Engineering*, 19(1):100–111, 1994.
11. S. Zhou and Z. Wang. *OFDM for Underwater Acoustic Communications*. West Sussex: John Wiley & Sons, 1st edition, 2014.
12. A. Song, M. Badiey, H. C. Song, and W. Hodgkiss. Ocean variability effects on high-frequency coherent acoustic communications. *The Journal of the Acoustical Society of America*, 125(4): 2580–2580, 2009.
13. A. Song, M. Badiey, H. C. Song, and W. S. Hodgkiss. Impact of source depth on coherent underwater acoustic communications. *The Journal of the Acoustical Society of America*, 128(2):555–558, 2010.
14. P. Qarabaqi and M. Stojanovic. Adaptive power control for underwater acoustic communications. In *OCEANS 2011*, pages 1–7, June 2011.
15. A. Radosevic, R. Ahmed, T. M. Duman, J. G. Proakis, and M. Stojanovic. Adaptive OFDM modulation for underwater acoustic communications: Design considerations and experimental results. *IEEE Journal of Oceanic Engineering*, 39(2):357–370, 2014.
16. L. Wan, H. Zhou, X. Xu, Y. Huang, S. Zhou, Z. Shi, and J. Cui. Adaptive modulation and coding for underwater acoustic OFDM. *IEEE Journal of Oceanic Engineering*, 40(2):327–336, 2015.
17. P. Xia, S. Zhou, and G. B. Giannakis. Adaptive MIMO-OFDM based on partial channel state information. *IEEE Transactions on Signal Processing*, 52(1):202–213, January 2004.

18. S. Mani, T. M. Duman, and P. Hursky. Adaptive coding/modulation for shallow-water UWA communications. In *7th European Conference on Noise Control 2008, EURONOISE 2008*, Proceedings—European Conference on Noise Control, pages 4255–4260, 2008.
19. J. Huang, Z. Wang, S. Zhou, and Z. Wang. Turbo equalization for OFDM modulated physical layer network coding. In *2011 IEEE 12th International Workshop on Signal Processing Advances in Wireless Communications,* IEEE Workshop on Signal Processing Advances in Wireless Communications, SPAWC, pages 291–295. Institute of Electrical and Electronics Engineers Inc.
20. E. Demirors, G. Sklivanitis, G. E. Santagati, T. Melodia, and S. N. Batalama. Design of a software-defined underwater acoustic modem with real-time physical layer adaptation capabilities. In *Proceedings of the International Conference on Underwater Networks & Systems*, WUWNET '14, pages 25:1–25:8, New York, 2014. ACM.
21. Z. Liu and T. C. Yang. On the design of cyclic prefix length for time-reversed OFDM. *IEEE Transactions on Wireless Communications*, 11(10):3723–3733, October 2012.
22. J. A. Farrell, S. Pang, and W. Li. Chemical plume tracing via an autonomous underwater vehicle. *IEEE Journal of Oceanic Engineering*, 30(2):428–442, 2005.
23. A. Pottier, F. X. Socheleau, and C. Laot. Robust noncooperative spectrum sharing game in underwater acoustic interference channels. *IEEE Journal of Oceanic Engineering*, PP(99):1–16, 2017.
24. B. Chen, P. C. Hickey, and D. Pompili. Trajectory-aware communication solution for underwater gliders using WHOI micro-modems. In *Sensor Mesh and Ad Hoc Communications and Networks (SECON), 2010 7th Annual IEEE Communications Society Conference on*, pages 1–9. IEEE.
25. D. Nams. Online bandwidth adaptation and data tracking of underwater acoustic transmission to optimize collaborative AUV missions. Thesis, 2014.
26. D. Nams, M. L. Seto, and J. J. Leonard. On-line adaptation of underwater acoustic transmission rates to optimize communications for collaborative AUV missions. In *OCEANS'15 MTS/IEEE Washington*, pages 1–8. IEEE, 2015.
27. B. Reed, J. Leighton, M. Stojanovic, and F. Hover. *Multi-vehicle Dynamic Pursuit Using Underwater Acoustics*, pages 79–94. Springer International Publishing, Cham, Switzerland, 2016.
28. T. Schneider and H. Schmidt. Unified command and control for heterogeneous marine sensing networks. *Journal of Field Robotics*, 27(6):876–889, 2010.

Chapter 11

Signal Classification Using Feature Extraction Techniques and Artificial Neural Network in Underwater Acoustic Environment

Mehdi Shadloo-Jahromi
Petropars Operation and Maintenance Company

Mohammad Reza Khosravi
Shiraz University of Technology

Habib Rostami
Persian Gulf University

Contents

11.1 Introduction to Sonar

The term "SONAR" is defined as the method or equipment that uses underwater sound propagation to explore the presence, location, or nature of objects in the sea. It is an acronym for "*SOund NAvigation and Ranging*" and uses underwater sound propagation to navigate, communicate, or detect other vessels or targets of interest. Since its introduction during the early half of the 20th century, various evolutionary stages has been performed and is one of the priority areas of research in scientific societies. Sonar systems are classified into two categories:

- Passive sonar, where an acoustic noise source is radiated by the target, and the sonar only listen to and record the acoustic signals in order to analyze them (Figure 11.1). Such devices can be used to detect seismic occurrences, early warning of ships, submarines, torpedoes, etc., and marine creatures that emit characteristics of sounds.
- Active sonar, where the sonar itself transmits an acoustic signal and waits for the echoes of the emitted sound reflected by remote objects (Figure 11.2).

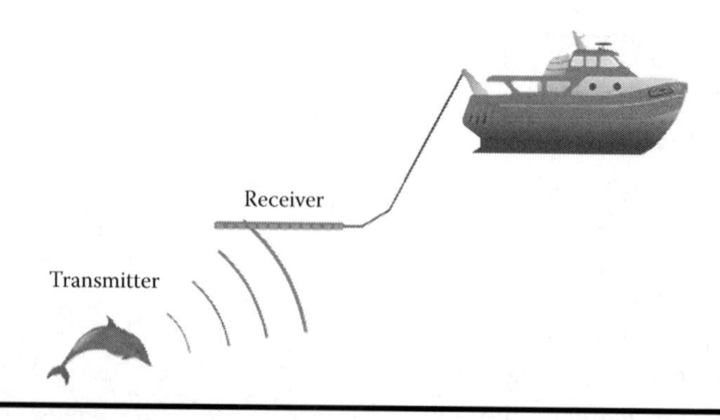

Figure 11.1 Passive sonar.

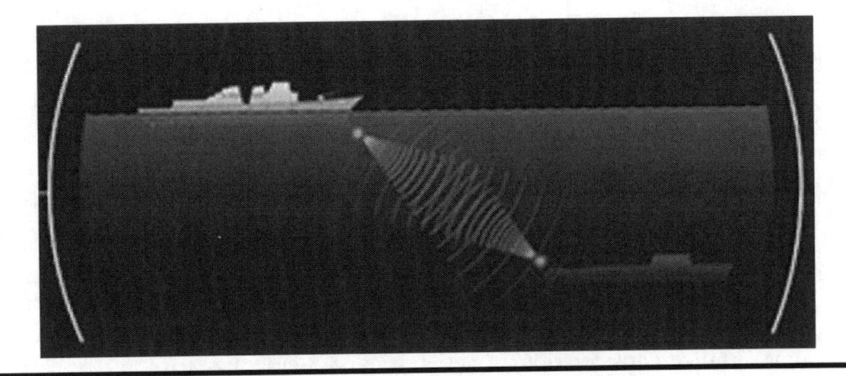

Figure 11.2 Active sonar.

Depending on the characteristics of the interested target, passive or active sonar is used. Although active sonar was mainly used in the World War II, with the advent of noisy nuclear submarines, passive sonar was preferred for early detection and warning applications. Compare with active sonar, passive sonar is generally capable of detecting targets in greater range, which helps in performing the identification of targets.

11.2 Underwater Ambient Noise

To select the discriminating features and proper classification algorithms, understanding the characteristics of underwater ambient noise and the acoustic radiated noise of vessels is very important. There are a wide variety of noise sources present in the underwater environment that are different in frequency range and other inherit parameters. Generally, from the sonar signal processing perspective, they can be classified into four different groups. These are ambient noise, radiated noise, self-noise, and reverberation noise.

- Ambient noise, unlike other sources, does not come from a particular direction or source. The noise level is the same everywhere in the local area. The ambient noise level is the intensity (in dB) of the ambient noise. The major sources for ambient noises are marine mammals, wind, waves, shipping, rainfall, etc.
- Radiated noise constitutes the acoustic output of surfaced or submerged vessels, weapons, or machinery which can reveal the details of the noise sources, and as such, this type of noise is important in target classification/identification scenarios. Generation of radiated noise can be attributed to propellers, machinery, and hydrodynamic effects present in the system.
- The self-noise mainly gets generated from the vibration of structural parts induced by water flow and rotating as well as reciprocating machinery.

Those sources, viz. propellers, machinery, and hydrodynamic effects, which contribute to the radiated noise, also contribute to self-noise. Analysis has revealed that the spectrum of self-noise has both line spectra as well as continuous spectra and has a steep negative slope at high frequencies.

- Another source of noise in bathymetric systems is that caused by reverberation. Reverberation can be defined as reradiations of sound by the sum of all unwanted scattering contributions from all scatterers not pertinent to bathymetric measurement. The reverberation that occurs can come from the water surface, reradiation from a body of marine life scatterers (volume reverberation) in the sea, or from the bottom. Bottom reverberation is a significant noise source consideration in active sonar systems. For bathymetric applications, bottom reverberation is considered only when the application is to discriminate some feature that extends above or is buried at the bottom and might be obscured.

11.3 Sonar Signal Processing

Sonar signal processing comprises of a large number of signal processing algorithms, implemented for achieving various sonar functions like target detection, localization, and classification. One of the important necessities and objectives of sonar is to extract the discriminant features from the acoustic received signals and expound this information precisely and clearly, without having any ambiguity to the end user. The extraction of the discriminant features and the selection of a proper classifier are considered as two important issues related to the recognition and classification of sonar signals.

The detection and classification problems generally addressed by the sonar systems are used to detect the presence of targets, by comparing the level of certain statistics with the assumed or estimated statistics, and to classify the targets adopting the joint concepts of estimation, localization, and tracking. A signal can be represented as a function of time, which shows how the signal magnitude changes over time. Time representation is the most natural description of a signal since almost all physical signals are obtained by receivers recording variations with time. However, in passive sonar, targets are usually categorized according to the components present in the frequency spectrum of the received signals radiated from the targets using spectral estimation techniques. Fourier transform is a powerful way to describe a signal because the concept of frequency is seen in many domains where periodic events occur. Fourier transform of a signal $x(t)$ is defined as follows.

$$X(f)=\int_{-\infty}^{+\infty} x(t)e^{-j2\pi ft}dt. \tag{11.1}$$

Underlying a great deal of traditional signal processing theory is the notion of a sinusoidal wave. With the advent of modern computing and Fast Fourier Transform (FFT), the use and interest in frequency domain signal processing have increased dramatically. However, in a wide range of previous research, the automatic classification of the noise radiated by ships and submarines includes the features extracted from the frequency domain using the FFT of the power spectrum [1–8].The algorithm presented in [9] applies six parameters extracted from the power spectrum. In this method, a standard feature vector is extracted for each class of ships. After that, the weighted distance between the standard vector and the feature vector extracted from the test data is calculated. At last, the classification is made. The discriminant features presented in [8] utilize the features of the power spectrum radiated from four different ships for data classification. In this chapter, the background noise was estimated by applying two-pass split window (TPSW) and then by subtracting it from the total power spectral density, the signal-to-noise ratio was improved, leading to more accurate extraction of tonal components. Many natural and man-made signals have spectral characteristics that vary with time. A signal can be represented as a function of time, which shows how the signal magnitude changes over time. Short-time Fourier transform (STFT) is known to be the first time-frequency method (TFM) that was applied in practical systems like order tracking, speech processing systems, and Inverse Synthetic Aperture Radar (ISAR) imaging.

11.4 Short-Time Fourier Transform

Fourier analysis becomes inadequate when the signal contains nonstationary or transitory characteristics like transients, trends, and so on. To introduce time-dependency in the Fourier transform, a simple and intuitive solution consists in prewindowing the signal to be analyzed $x(t)$ around a particular time t calculating its Fourier transform, and doing that for each time instant t. The resulting transform called the STFT, is therefore defined as follows.

$$\mathrm{STFT}\left(t,f\right)=\int_{-\infty}^{+\infty} x\left(u\right)h^{*}\left(u-t\right)e^{-j2\pi fu}du. \tag{11.2}$$

Here $h(t)$ is a short-time analysis window, localized around $t = 0$ and $f = 0$. Because multiplication by the relatively short window $h^*(u{-}t)$ effectively suppresses the signal outside a neighborhood around the analysis time point $u = t$, the STFT is a local spectrum of the signal $x(t)$. This relation expresses that the total signal can be decomposed as a weighted sum of elementary waveforms $h_{t,f}(u) = h(u - t)e^{j\pi fu}$. These waveforms are obtained from the window $h(t)$ by a translation in time and a translation in frequency (modulation). While the STFT's compromise between time and frequency information can be useful, the drawback is that once a particular size is chosen for the time window, it remains the same for all frequencies. The time resolution of the STFT is proportional to the effective duration of the analysis window $h(t)$. Similarly, the frequency resolution of the STFT is proportional to the effective bandwidth of the analysis window $h(t)$, consequently, for the STFT, we have a tradeoff between the time and frequency resolutions. On one hand, a good time resolution requires a short window $h(t)$. On the other hand, a good frequency resolution requires a narrowband filter, i.e., a long window $h(t)$. Therefore, the width of the windowing function relates to how the signal is represented. A wide window gives better frequency resolution but poor time resolution. A narrower window gives good time resolution but poor frequency resolution. These are called narrowband and wideband transforms, respectively. This is the major drawback of STFT.

In this work, to improve the classification efficiency of STFT method, variable parameters in this method have been evaluated, and consequently, a new feature vector has been introduced using feature extraction (FE) and feature selection technique.

11.4.1 STFT in Sonar Systems

Sonar systems make use of the acoustic emission for detection, communication, and navigation in underwater environments. The main purpose of these systems is to analyze the acoustic waves received by the sensors to detect and classify the acoustic wave emitters. In passive sonar systems, the classification is mainly made by sonar operators, which are trained to perform this task. The acoustic signals radiated from a target are received by the sensors and processed to provide the auditory and visual information for sonar operators. Sonar operators investigate the compiled information to make decisions. The form of the radiated noise in the underwater environment depends on the source of the sound radiator. Submarines, ships, torpedoes, and marine mammals are some examples of noise sources. The propulsion components (turbines, engines, gear trains, propellers, and pumps) are the principal sources of the noise emitted by marine vessels. The purpose of passive sonars is to implement an automatic system to exploit the radiated noises and distinguish them from similar interfering noises, e.g., the ambient noise, array self-noise, and noises from platforms not of interest. The automatic system to detect the emitted signals from targets reduces the operator workload and improves their decision-making accuracy.

In the last 20 years, using the time–frequency signal analysis and processing has been considered as an effective way to remove the noises and extract the discriminant features [10–16]. The STFT analysis and discrete wavelet transform [17–20] are such methods. Since the passive sonar signals are nonstationary signals, by applying the short temporal windows on the nonstationary signal in the STFT method, the main signal is converted into a signal with a minor period with respect to the stationary signal assumption. The main shortage of the STFT time–frequency processing method is that only the stationary content of sonar signal is used for data classification. In the following sections, STFT method is implemented for a passive sonar signal classification system.

11.4.1.1 Preprocessing

Acoustic signals are received and recorded by underwater microphones, called hydrophones. These signals were digitized by an 8-bit Analog to Digital (A/D) converter at a sampling rate of 22,050 Hz. Since the main passive sonar signal information is at frequencies below 3 kHz [1], the digitized acoustic signal was filtered by an eight-order type-I low-pass Chebyshev filter with a cut-off frequency of 2.87 kHz, which provides an attenuation of 72 dB in the transition region (Figure 11.3). Since the primary signal sampling frequency is 22,050 Hz, the filtered signal is decimated with no aliasing and the sampling frequency is reduced to 7350 Hz. In signal processing, most often, Hanning and Hamming windows are used for window processing [14]. So, the window with a smaller duration than the whole duration of the acoustic signal is used. The window starts to move from the beginning of the signal and shifts to the right to reach the end of the signal. Here, two quantities must be determined: the window width and the shift between consecutive windows. A window width of about 278 ms (2048 samples) has been selected, which corresponds to an average duration that is essential for the stationary assumption to hold [1]. Due to nonsignificant changes in surface vessel spectrum during the test, no overlap between successive windows is used [1].

11.4.1.2 FFT Power Spectrum

The signals multiplied by a Hanning window, called windowed signal, then transformed to the frequency domain by using the FFT. For each obtained spectrum, 800 frequency bins, corresponding to an alias-free frequency range from DC to 2.87 kHz, were retained for further analysis (Figure 11.4).

11.4.1.3 Classification Based on Artificial Neural Network

As shown in Figure 11.5, an RBF neural network has three layers [15]. The first layer is an input layer; the second layer is a hidden layer that includes some radial basis

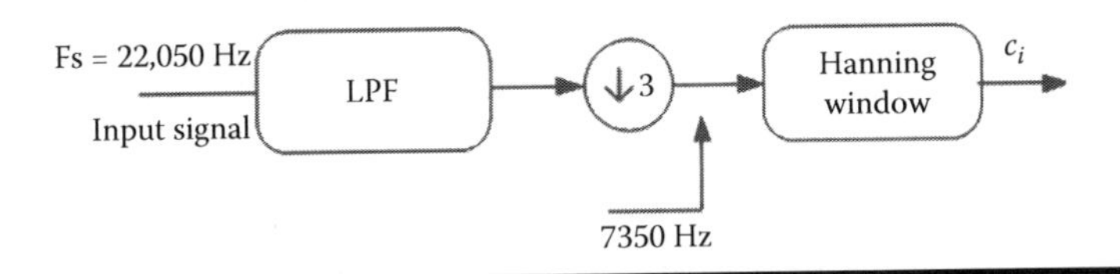

Figure 11.3 Signal preprocessing on received sonar signal.

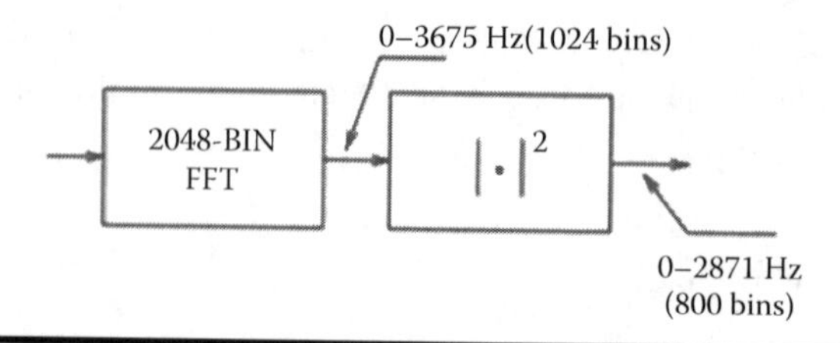

Figure 11.4 An FFT power spectrum calculation [1].

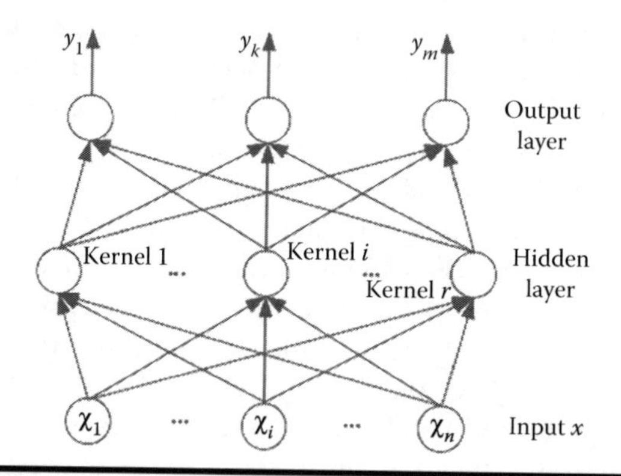

Figure 11.5 An RBF neural network.

functions, also known as hidden kernels; and the third layer is the output layer. An RBF neural network can be considered as a mapping of input domain X onto the output domain Y.

$$y_k(\vec{x}) = \sum_{i=1}^{N} w_{ki} G(\| \vec{x} - \vec{t}_i \|) + b_k, \quad i = 1, 2, \ldots, N; k = 1, 2, \ldots, K \tag{11.3}$$

where $\| . \|$ exhibit the Euclidean norm, k is the number of outputs, N is the number of hidden kernels, $y_k(\vec{x})$ is output k corresponding to the input $\vec{x}$, $\vec{t}_i$ is the center of kernel i, w_{ki} is the weight between kernel i and output k, b_k is the bias on output k, and $G(\| \vec{x} - \vec{t}_i \|)$is the kernel function. The most commonly used kernel function for RBF neural networks is Gaussian kernel function as follows:

$$G(\| \vec{x} - \vec{t}_i \|) = \exp\left(-\frac{\| \vec{x} - \vec{t}_i^{\,2} \|}{2\sigma_i^2} \right), \tag{11.4}$$

where σ_i is the radius of the kernel i. The main steps to construct an RBF neural network include (a) determining the positions of all the kernels $\vec{t}_i$, (b) determining the radius of each kernel, and (c) calculating the weights between the kernels and the output nodes.

In this work, we use an RBF neural network as the classifier and the back-propagation algorithm as the network-training algorithm. Kernel function that has been used in the RBF network is Gaussian function and the error criterion is mean square error. Ten-fold cross-validation is used to validate the classifier.

11.4.2 STFT Performance Evaluation

In this section, we change variable parameters in STFT process to test various configurations of usual STFT to achieve maximum classification efficiency that STFT can ensure.

11.4.2.1 Dataset

It is a time-consuming and expensive job to compile real underwater sounds of either various targets or ambient noises. Therefore, it is common to use simulated data to evaluate different algorithms. In this work, we use real recorded sounds include five different classes of surface vessels as Table 11.1. Vessels are classified based on their dimension and classes of the engine. Data is recorded by an omnidirectional hydrophone and an 8-bit A/D convertor with a sample rate of 22,050 Hz in shallow water. Table 11.1 shows data details about number of records and time duration per class.

11.4.2.2 STFT Different Configurations

Using STFT as a signal processing analysis (Figure 11.6) is one of the most common time–frequency analyses in passive sonar target detection and classification. It provides multiprocessing blocks to process sonar signals spectrum and shows the STFT spectrum processing method which we name it STFT configuration block (Figure 11.7). Number of Bins (Block F), background noise correction

Table 11.1 Dataset Details

Class	*Number of Records*	*Duration of Each Record (s)*	*Total (s)*	*%*
V1	6	130	780	22.6
V2	6	129	774	22.5
V3	6	127	762	22.1
V4	6	125	750	21.7
V5	3	129	384	11.1
Total	27		3450	100

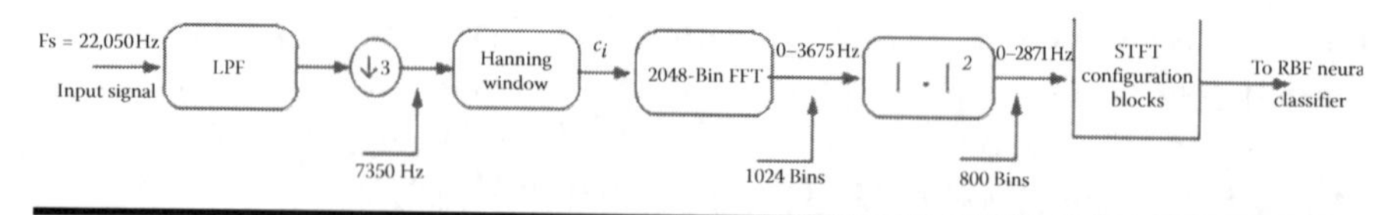

Figure 11.6 Passive sonar signal classification using STFT signal processing analysis.

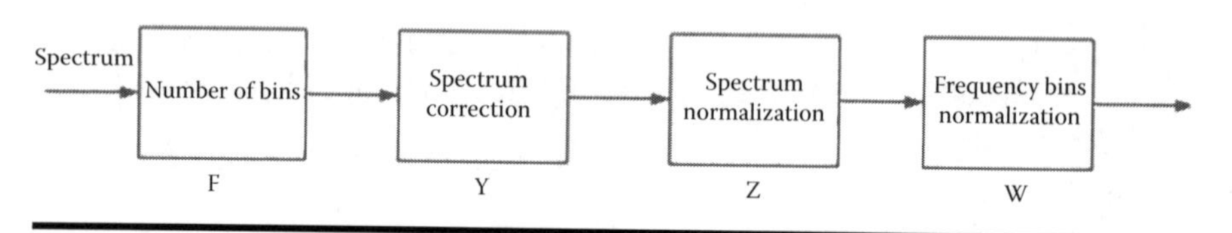

Figure 11.7 STFT spectrum processing blocks.

(Block Y), spectrum normalization (Block Z), and bin normalization (Block W) are the spectrum processing blocks, which methods we use to implement can affect the final classification accuracy [1].

11.4.2.2.1 Frequency (F)

Separation of two nearby frequency lines depends on the frequency bin width used in spectrum calculation. It may become more complicated to separate such lines if frequency bin width gets worse. To evaluate the results of changes in frequency bin width on the classification efficiency of the classifier, spectra with 200, 400, or 800 bins have been considered. Respectively, this corresponds to frequency bin widths of 14.36, 7.18, and 3.59 Hz. The spectra with 800-bin was filtered using an impulse response $h = [0.5\ 0.5]$, and was decimated by 2 to achieve the 400-bin spectrum. Moreover, the resulting 400-bin spectrum was decimated using the same impulse response to obtain the corresponding 200-bin spectrum (Figure 8).

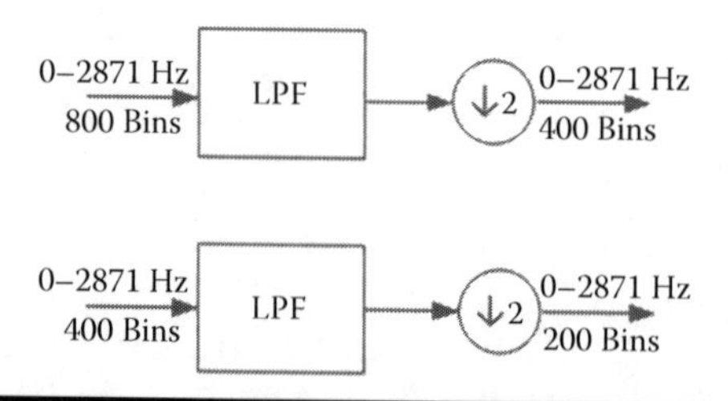

Figure 11.8 Frequency bin decimation to obtain 400- and 200-bin spectrum from 800-bin spectrum.

11.4.2.2.2 Background Noise Correction (Y)

The sound generated by a vessel is propagated in multiple ways through the underwater environment, where it will be mixed with a variety of other sounds generated by biological sources, human intervention, seismic activity, rainfall, etc. Therefore, a passive sonar system will receive this sound spectrum in combination with the background noise. An estimate of the background noise can be used to correct the acquired spectra, aiming at improving the classification efficiency.

This background noise is evaluated by the use of TPSW algorithm [7] to correct received sound spectrum using evaluated background noise, we use following equations. $X_k(n)$ is nth spectrum of kth vessel, $m_k(n) = \text{TPSW}[X_k(n)]$ defines as evaluated background noise from spectrum $X_k(n)$ using the TPSW, and $Y_k(n)$ is the obtained corrected spectrum that can be calculated using one of the equations:

$$y_k(n) = x_k(n) - m_k(n) \tag{11.5}$$

$$y_k(n) = \frac{x_k(n) - m_k(n)}{m_k(n)} \tag{11.6}$$

$$y_k(n) = \frac{x_k(n)}{m_k(n)}. \tag{11.7}$$

11.4.2.2.3 Spectrum Normalization (Z)

After signal received by a passive sonar sensor, a normalization scheme is essential to calibrate signal amplitude. Moreover, to coordinate with the dynamic range of the neural classifier, normalization is typically required. Different spectral normalization schemes, based on information extracted from each incoming spectrum $Y_k(n)$, were submitted. The normalized spectrum ($z_k(n)$) can be calculated using one of these equations:

$$z_k(n) = \frac{y_k(n)}{\max(y_k(n))} \tag{11.8}$$

$$z_k(n) = \frac{y_k(n)}{\text{rms}(y_k(n))} \tag{11.9}$$

$$z_k(n) = \frac{y_k(n)}{\text{mean}(y_k(n))}, \tag{11.10}$$

where we have two values as follows:

$$\text{rms}(y_k(n)) = \sqrt{\frac{1}{p}\sum_{i=1}^{P}(y_{k,i}(n))^2}$$

$$\text{mean}\left(y_k(n)\right)=\frac{1}{P}\sum_{i=1}^{P}y_{k,i}(n). \tag{11.11}$$

11.4.2.2.4 Frequency Normalization (W)

The energy density of the signal generated by the machinery inside the ship is highly concentrated in the first bins of the spectrum [17]. Moreover, the generated signals propagating underwater have a tendency to higher attenuation at higher frequencies [18]. Therefore, some frequency bin normalization may help in revealing characteristics of the relevant signals (Table 11.2).

Table 11.2 Summary of the Methods for STFT Signal Configuration Block

Block Name	*Method*	*Option*	*Description*
F	No. of bins/frequency	F1	200 bin spectra/$\Delta f = 3.59$ Hz
		F2	400 bin spectra/$\Delta f = 7.18$ Hz
		F3	800 bin spectra/$\Delta f = 14.36$ Hz
Y	Background noise correction	Y1	$Y = x$ (no correction)
		Y2	$Y = x - m$
		Y3	$Y = \frac{x-m}{m}$
		Y4	$Y = \frac{x}{m}$
		Y5	$Y = \log 10(x) - \log 10(m)$
		Y6	$Y = \log 10(x)$
Z	Individual spectrum normalization	Z1	$Z = Y$(without normalization)
		Z2	$Z = \frac{Y}{\max(Y)}$
		Z3	$Z = \frac{Y}{\text{rms}(Y)}$
		Z4	$Z = \frac{Y}{\text{mean}(Y)}$
W		W1	$W_i = Z_i$ (without normalization)

(Continued)

Table 11.2 (*Continued*) Summary of the Methods for STFT Signal Configuration Block

W	Frequency bin normalization	W2	$W_i = \dfrac{Z_i}{\text{mean}(Z_i)}$
		W3	$W_i = \dfrac{Z_i - \text{mean}(Z_i)}{\text{mean}(Z)}$
		W4	$W_i = \dfrac{Z_i - m\text{boxmean}(Z_i)}{\text{rmbs}(Z_i)}$

$$w_{k,i}(n) = \frac{z_{k,i}}{\text{mean}\left(z_{(trn),i}\right)} \tag{11.12}$$

$$w_{k,i}(n) = \frac{z_{k,i}(n) - \text{mean}\left(z_{(trn),i}\right)}{\text{mean}\left(z_{(trn),i}\right)} \tag{11.13}$$

$$w_{k,i}(n) = \frac{z_{k,i}(n) - \text{mean}\left(z_{(trn),i}\right)}{\text{rms}\left(z_{(trn),i}\right)}, \tag{11.14}$$

where $z_{k,i}(n)$ shows the ith frequency bin of nth spectrum related to kth class;

$$\text{mean}\left(Z_{(trn),i}\right) = \frac{1}{trn}\sum_{n=1}^{trn} z_{(trn),i} \tag{11.15}$$

and

$$\text{rms}\left(z_{(trn),i}(n)\right) = \sqrt{\frac{1}{trn}\sum_{n=1}^{trn}\left(z_{(trn),i} - \text{mean}\left(z_{(trn),i}\right)\right)^2} \tag{11.16}$$

Since STFT processing blocks are dependently chained together, 288 distinct configurations of preprocessing chain can be implemented. Surely, different configurations will result in different classification efficiencies. Here we use RBF neural network as the classifier with descriptions offered in section. Experiments that have been simulated in [1] denote 10 configurations with the best classification efficiencies. In this work, these 10 configurations have been simulated and the results of classifier efficiencies have been presented.

11.5 Proposed Method: STFT-FE

The signal analyzing in time–frequency domain provides the ability to process some discriminating characteristics of the signal such as frequency variation, time variation, recognizing the number of components, and separating the frequency components (by filtering in time–frequency domain and for processing specific components separately). Several techniques are available to compute the time–frequency distributions. The simplest and practical way to present a time-dependency in Fourier transform is STFT. STFT consists of windowing the signal with short time duration windows and calculating its Fourier transform separately and continuing this procedure for the rest length of the original signal. STFT is a simple and powerful time–frequency analysis to describe the instant spectrum of the signal at a time, which uses the spectrum of a signal segment centered at that time (Table 11.3).

In this section, we are searching for a new features vector which helps the classifier to improve the sonar signal classification efficiency as well as reduce the decision time. The classification efficiency of intelligent systems used to detect and classify the transient signals depends on signal processing methods, FE, or selection techniques and also classifier unit. The systems implemented based on STFT are appropriate to detect the narrowband signals for which their time lengths are according to or more than the window length in STFT method. STFT is capable of analyzing certain characteristics of the signal in the time domain as well as the frequency domain. However, it is not proper for analyzing most of the transient

Table 11.3 Maximum Classification Efficiencies for Each Preprocessing Alternative

Configuration	*Classification Efficiency (%)*
W1Z1Y5F2	90.38
W1Z2Y2F1	88.02
W1Z2Y3F2	89.07
W1Z3Y3F1	88.9
W1Z3Y4F2	90.02
W2Z2Y1F3	90.11
W2Z4Y1F3	89.3
W3Z1Y4F2	89.9
W4Z1Y3F2	89.5
W4Z2Y6F2	87.9

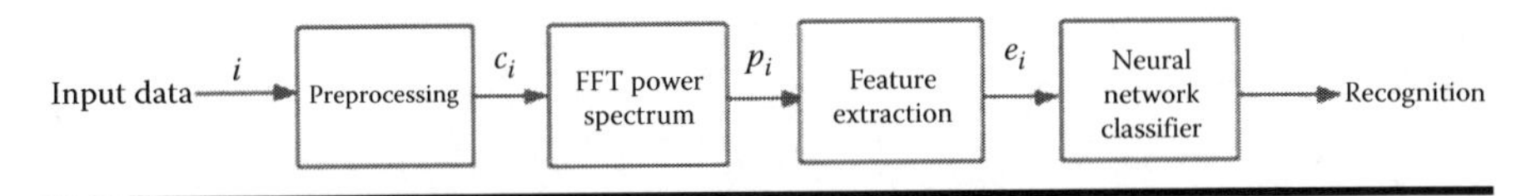

Figure 11.9 Classification of passive sonar signals using STFFT-FE method.

signals, especially when it is used to generate features to detect and classify the interested signals. Regarding the different portion of the signal, each frequency bin contains useful and different information about the signal spectrum. In STFT, features-weighted distribution is a uniform distribution, and therefore, classifier contributes them the same in the weighting of neurons and does not consider the level of each feature capability. Moreover, the high number of featuring as discriminant attributes leads to the computation loads in classifier and so is the increase of decision time and overfitting probability. It seems using feature selection and FE technique can improve detection and classification accuracy of sonar systems. FE is usually associated with two main purposes. It can be used in classification procedure to produce new features, which would preserve class separability as much as possible. It can also be utilized to operate on massive data to form a new lower-dimensional problem representation space. FE techniques, applied either on datasets with high dimensionality or on datasets including indirectly relevant features, can improve the performance of a classifier.

In STFT-FE block diagram (Figure 11.9), the output of FFT power spectrum was a massive matrix with 2048 attributes for each windowed signal. Using a proper FE technique to reduce data dimension and produce features with high capability of class separability could positively affect classifier performance. In this work, we evaluate the performance of two FE techniques, linear discriminant analysis (LDA)-based FE [19] in the context of supervised learning and principal component analysis (PCA)-based FE [20], which is perhaps the most commonly used FE technique. Where the PCA was used as FE technique, we named the procedure the STFT-PCA technique and where the LDA technique was used we named it STFT-LDA. Finally, the extracted feature vectors in each method are used to train an RBF neural network as the ultimate maker.

11.5.1 Feature Extraction

We explain two conventional methods of FE techniques, LDA and PCA which are used to extract the signal features as follows.

11.5.1.1 Linear Discriminant Analysis (LDA) Algorithm

The main idea of LDA is to find a linear transformation from signal original space to new feature space so that feature clusters have the highest separability after transformation. This can be achieved in the analysis of the scattering matrix [21]. For an

M-class problem ($C_1, C_2, \ldots, C_M$, the between and within class scatter matrices S_b and S_w are defined as follows:

$$S_b = S_b = \sum_{i=1}^{M} P\tau(C_i)(\mu_i - \mu)^T = \phi_b \phi_b^T, \tag{11.17}$$

$$Sw = \sum_{i=1}^{M} P\tau(C) \sum_i = \phi_w \phi_w^T, \tag{11.18}$$

where $P\tau(C_i)$ is the prior probability of class (C_i) and usually assigned to $1/M$ with the assumption of equal priors, μ is an overall mean vector; Σ_i is the average scatter of the sample vectors of different classes (C_i) around their reprehensive mean vector μ_i:

$$\sum_i = E\left[(x - \mu_i) - (x - \mu_i)^T \mid C - C_i\right] \tag{11.19}$$

Each class can be measured by a certain criterion separately. A common criterion is the ratio of the determinant of between-class scatters matrix of the projected samples to the within-class scatter matrix of the projected samples:

$$(A) = \arg\max \frac{\left|AS_b A^T\right|}{AS_w A^T}, \tag{11.20}$$

where A is a $m \times n$ matrix with $m \leq n$. For the optimization problem of equation, one solution is to solve the generalized eigenvalue problem:

$$\begin{aligned} S_b A^* &= \lambda S_w A^* \\ D_i(X) &= A^{*T}(X - \mu i), \quad i = 1, 2, \ldots, m. \end{aligned} \tag{11.21}$$

One solution for the equation is to compute the inverse of S_w and solve eigenvalue problem for $S_w^{-1} S_b$ matrix. However, this method is numerically unstable because it involves the direct inversion of a high-dimensional matrix. LDA algorithm is based on simultaneously diagonalizable S_w and S_b, it means:

$$AS_w A^T = I, \quad AS_b A^T = \Lambda \tag{11.22}$$

where Λ is a diagonal matrix with diagonal elements stored in decreasing order. If we want to reduce the dimension of the matrix from n to m, we use first m rows of A as the transformation matrix. Selection of m as $m = M - 1$ is common in LDA algorithm, which corresponds to the largest m eigenvalues of Λ. Most of the algorithms require that the within-class scatter matrix S_w be nonsingular, since the algorithms are diagonalizable as S_w first. As the within-class scatter matrix S_w becomes singular, this procedure breaks down. This can occur when the number of training samples is smaller than the dimensions of sample vectors.

11.5.1.2 Principal Component Analysis

PCA is a statistical technique that extracts a space with lower dimensions by analyzing covariance structure of multivariate statistical observations. The basic idea of PCA is to identify the features that express as much as variations as possible in the data with the few features. PCA transformation matrix computation is based on eigenvalue decomposition of covariance matrix S of input data. Therefore, it is computationally expensive.

$$w \leftarrow \text{eig\_decomposition}\left(S = \sum_{i=1}^{n} (x_i - \mu) - (x_i - \mu)^T \right),$$

where n is the number of observations (samples) x_i is the ith observation and m is the mean vector of input data:

$$\mu = \frac{1}{n} \sum_{i=1}^{n} x_i. \quad (11.23)$$

Principal components are calculated by the following algorithm:

1. Calculate the covariance matrix S of the input data.
2. Compute the eigenvalues and eigenvector of S and sort them in a descending order with respect to the eigenvalues.
3. Form the actual transition matrix by taking the predefined number of components (eigenvectors).
4. Finally, multiply the original feature space with the obtained transition matrix, which yields a lower-dimensional representation.

Although PCA is one of the well-known FE methods, it has some serious problems. Conventional PCA assigned the high weight to features with higher variability, without considering whether these features are useful for classification or not. This may cause the selection of principal components with more variability criteria while these features do not have high separability power. On the other hand, there is no certain way of determining the number of extracted features (m) in this method.

11.5.2 STFT-FE Performance Evaluation

In this work, we used STFT to process our real sonar data and the features extracted from this method were applied to an RBF neural network classifier. The best classification accuracy obtained was 90.38%. Features dimension obtained from STFT block is high, so the STFT-FE method proposed. In this section, we evaluate the performance accuracy of the proposed method by comparing it with a classification accuracy of conventional STFT. We simulate and evaluate proposed method and compare the results with the best classification efficiency achieved in an ordinary STFT method. We use real acoustic data set presented in Section 11.4.2.1, which consists of underwater acoustic records of five marine vessels.

It has been discussed that using the information of classes is very important in FE for supervised learning. Thus, in [22] and [23] it has been shown that PCA gives high weights to features with higher variability regardless of whether they are useful for classification or not. LDA-based FE uses class information but has a serious deficiency because of its parametric nature, so that the number of extracted components cannot be more than the number of classes minus one. Since in this work we use five classes of surface vessels, we had to extract four features in the implementation of STFT-LDA method. For STFT-PCA analysis, to determine the number of extracted features, we used the method presented in [24], which is based on selecting those eigenvectors with eigenvalues larger than the average input variance (average eigenvalues).

$$\lambda_m \geq \frac{1}{n}\sum_{i=1}^{n}\lambda_i, \tag{11.24}$$

where n is the number of primary data and m is the number of extracted features. LDA and PCA techniques were applied to the signal spectrum obtained from W1Z1Y5F2 STFT configuration with 400 frequency bins. After running the proposed method, the following results are achieved (Table 11.4).

Both techniques could reduce the consuming time in the convergence of the neural classifier-learning algorithm. The comparison of the results showed better processing performance of STFT-LDA (96.94%) than STFT-PCA (85.18%). Results obviously show that with the same preprocessing algorithm and classifier, features extracted from LDA technique play a separating role more effective than features extracted from PCA technique. The poor performance of STFT-PCA is due to this reason that PCA gives high weights to features with higher variability

Table 11.4 Evaluation Results

Methods	*Number of Features*	*Classification Efficiency (%)*
STFT	400	90.38
STFT-PCA	50	85.18
STFT-LDA	4	96.94
STFT-LPA	54	98.10

Table 11.5 Confusion Matrix

	STFT					*STFT-LDA*					*STFT-LPA*				
	V1	**V2**	**V3**	**V4**	**V5**	**V1**	**V2**	**V3**	**V4**	**V5**	**V1**	**V2**	**V3**	**V4**	**V5**
V1	2205	50	210	319	6	2416	69	58	247	0	2728	8	2	52	0
V2	69	2671	5	28	8	55	2671	17	38	0	6	2674	1	100	0
V3	70	7	2497	99	57	49	15	2630	35	0	2	2	2715	8	3
V4	0	12	58	2594	24	220	26	25	2407	0	14	21	13	2640	0
V5	123	0	9	36	1215	4	0	0	2	1377	0	0	0	3	1380

regardless of whether they are useful for classification or not. On the other hand, neural network classifiers that have been trained with low features are not reliable to face with the underwater complex environment as the practical classifier. Therefore, the STFT-LDA FE procedure cannot be a good choice, and we need a new feature vector that also conveys reliability and stability of classifier through the unstable situations under water. STFT-LPA uses a feature vector that includes four features extracted from STFT-LDA as well as 50 features extracted from STFT-PCA procedure altogether. Therefore, the feature vectors with 54 attributes would train and test the RBF neural network classifier. The confusion matrix corresponding to the evaluated methods are presented in Table 11.5. As its STFT-LPA could improve classification efficiency and to test the stability and reliability of STFT-LPA feature vectors, a practical environment or a massive dataset of recorded sounds from underwater radiated noise of various targets is required that needs future efforts.

References

1. W. S. Filho, J. M. de Seixas, and N. N. de Moura, "Preprocessing passive sonar signals for neural classification," *IET Radar, Sonar Navig.*, vol. 5, no. 6, p. 605, 2011.
2. G. L. Ogden, L. M. Zurk, M. E. Jones, and M. E. Peterson, "Extraction of small boat harmonic signatures from passive sonar," *J. Acoust. Soc. Am.*, vol. 129, no. 6, pp. 3768–3776, June 2011.
3. C. Chin-Hsing, L. Jiann-Der, and L. Ming-Chi, "Classification of underwater signals using wavelet transforms and neural networks," *Math. Comput. Model.*, vol. 27, no. 2, pp. 47–60, Jan. 1998.
4. J. Ribeiro-Fonseca and L. Correia, "Identification of underwater acoustic noise," *Proc. Ocean.*, vol. 2, p. II/597–II/602, 1994.
5. K. S. Thyagarajan, T. Nguyen, and C. E. Persons, "An image processing approach to underwater acoustic signal classification," *1997 IEEE Int. Conf. Syst. Man, Cybern. Comput. Cybern. Simul.*, vol. 5, pp. 4198–4203, 1997.

6. W. Soares-Filho, J. M. Seixas, and L. P. Caloba, "Enlarging neural class detection capacity in passive sonar systems," *2002 IEEE Int. Symp. Circuits Syst. Proc. (Cat. No.02CH37353)*, vol. 3, pp. 105–108, 2002.
7. C. Kang, X. Zhang, A. Zhang, and H. Lin, "Underwater acoustic targets classification using Welch spectrum estimation and neural networks," *Advances in Neural Networks – ISNN 2004*, vol. 3173. pp. 930–935, 2004.
8. W. Soares-Filho, J. Manoel de Seixas, and L. Pereira Caloba, "Averaging spectra to improve the classification of the noise radiated by ships using neural networks," in *Proceedings. Sixth Brazilian Symposium on Neural Networks*, vol. 1, pp. 156–161, 2000.
9. Q. L. Q. Li, J. W. J. Wang, and W. W. W. Wei, "An application of expert system in recognition of radiated noise of underwater target," *'Challenges of Our Changing Global Environment. Conference Proceedings Ocean. '95 MTS/IEEE*, vol. 1, 1995.
10. C. Chen, J. Lee, and M. Lin, "Classification of underwater signals using neural networks," *Tamkang J. of Science and Engineering*, vol. 3, no. 1, pp. 31–48, 2000.
11. L. Atallah and P. J. Probert Smith, "Using wavelet analysis to classify and segment sonar signals scattered from underwater sea beds," *Int. J. Remote Sens.*, vol. 24, pp. 4113–4128, 2003.
12. J. M. Seixas, D. O. Damazio, P. S. R. Diniz, and W. Soares-Fillho, "Wavelet transform as a preprocessing method for neural classification of passive sonar signals," *ICECS 2001. 8th IEEE Int. Conf. Electron. Circuits Syst. (Cat. No.01EX483)*, vol. 1, pp. 83–86, 2001.
13. S. Mallat, *A Wavelet Tour of Signal Processing*, Burlington, MA: Academic Press, 1999, pp. 20–41. https://www.elsevier.com/books/a-wavelet-tour-of-signal-processing/mallat/978-0-12-374370-1.
14. C. Turner, A. Joseph, M. Aksu, and H. Langdond, "The wavelet and fourier transforms in feature extraction for text-dependent, filterbank-based speaker recognition," *Procedia Comput. Sci.*, vol. 6, pp. 124–129, Jan. 2011.
15. M. W. M. Seah, C. K. Tham, V. Srinivasan, and A. Xin, "Achieving coverage through distributed reinforcement learning in wireless sensor networks," in *Proceedings of the 2007 International Conference on Intelligent Sensors, Sensor Networks and Information Processing, ISSNIP*, vol. 13, pp. 425–430, 2007.
16. M. K. Ward and M. Stevenson, "Sonar signal detection and classification using artificial neural networks," *Can. Conf. Electr. Comput. Eng.*, vol. 2, pp. 717–721, 2000.
17. D. Ross, *Mechanics of Underwater Noise*. Los Altos, CA: Peninsula Publishing, 1976.
18. R. J. Urick, *Principles of Underwater Sound*, 3rd ed. New York: McGraw-Hill, 1983.
19. R. Fisher, "The use of multiple measurements in taxonomic problems," *Ann. Eugen.*, vol. 7, pp. 179–188, 1936.
20. K. Pearson, "LIII. On lines and planes of closest fit to systems of points in space," *London, Edinburgh, Dublin Philos. Mag. J. Sci.*, vol. 2, pp. 559–572, 1901.
21. K. Fukunaga, *Introduction to Statistical Pattern Recognition*, vol. 22. Academic Press, 1990.
22. S. S. Wilks, *Mathematical Statistics*. New York: John Wiley, 1962.
23. N. Oza and K. Tumer, "Input decimation ensembles: Decorrelation through dimensionality reduction," in *Multiple Classifier Systems*, 2001, pp. 238–247. https://link.springer.com/chapter/10.1007/3-540-48219-9_24.
24. C. M. Bishop, *Pattern Recognition and Machine Learning* (Information Science and Statistics). New York: Springer-Verlag New York, 2006.

Chapter 12

Propagation Delay-Aligned Structure for Acoustic Communications

Feng Liu, Shuchao Jiang, Conggai Li, and Shengming Jiang

Shanghai Maritime University

Contents

12.1 Introduction

Acoustic communications have been used for a long time by animals and human beings [1–5]. When multiple messages are sent from the sources to different destinations, the undesired ones will become interference for each receiver, which reduces the dimensions of available signal space for message decoding. Elementary, simple processing of the interference will greatly deteriorate the channel throughput. Thus, mitigating or even eliminating the mutual interference in wireless communication systems has been a challenging and important problem.

Fortunately, the innovative technique of interference alignment (IA) emerged in [6,7] provides an enhanced but still efficient approach to achieve the degrees of freedom (DoF) of single-hop wireless channels, which is an important concept indicating the prefactor of channel capacity, i.e., DoF gives a linear approximation of maximum transmission rate without error. In the multihop scenarios, interference neutralization (IN) based on IA provides a further improved approach to achieve the network DoF, which is also named as aligned IN (AIN) [8]. We should remark that basically full or global channel state information (CSI) is perfectly required at each node for IA and IN. Otherwise, there will be some DoF loss.

The main idea of IA is to overlap the interferences from different transmitters as much as possible to maximize the dimension of signal space for desired messages. An extensive overview on different aspects of IA is given in [9] and [10]. In general, IA can be separately or jointly implemented in the time/frequency/spatial domains. It is simplest to process in the time domain without increasing any cost, while orthogonal frequency division multiplexing modulation can be implemented in the frequency domain [11]. Using multiple microphones or hydrophones, space resource can be further exploited for acoustic communications with a higher cost, just like the multiple-antenna technique for wireless networks with electromagnetic waves.

In this chapter, we focus on the implementation of IA in time domain, where the difference of propagation delay (PD) among links is utilized for IA. This can be referred to as the PD aligned structure (PDAS). We remark that time domain is a proper resource for acoustic communications, since the slow propagation speed leads to relatively large PD suitable for the frame design. When all links share the same propagation speed, PD is proportional to the distance relationship, which can be referred to as distance-aligned structure (DAS). We remark that DAS is a special case of PDAS. Without loss of generality, we will treat them in an exchangeable way.

IA by PD has originally been introduced as a conceptual example in [6] for K-user interference channels (IC), and further provided for a 2×2 X-channel (XC)

[12]. A closer look at IA for IC by PD is taken in [13] and to some extent in [14–16]. In [13,17–23], node placement schemes with DAS are discussed to position transmitters and receivers in the Euclidean space for acoustic communications and networks. Furthermore, the authors of [19] considered IA by PD and node placement for cognitive radio systems. In [24], cyclic IA based on PD is investigated for 2×2 XC and K-user IC. In [25], the performance of IA by PD for randomly placed users is investigated. Possible applications in satellite and underwater networks are studied in [25–27]. We have studied the DAS for IC in [20], DAS/PDAS for $K\times 2$ XC in [21,22], and PDAS for $2\times 3\times 3$ XC in [23]. Based on the similar idea, IA used for grid networks [28] and linear networks [29] were proposed to increase the throughput of underwater acoustic networks. In [30] and [31] the authors apply the IA to multihop network for improving the throughput. In [32], three approaches of imperfect IA are studied for IC without achieving the theoretical DoF.

In this chapter, we mainly introduce the DoF-achieving PDAS schemes for IC and X networks, which are closely related to our recent works [20–23]. They are theoretically suitable for future potential acoustic communications. Other related results will be briefly discussed.

12.2 PDAS with Acoustic Interference Channel

IC is a classical network model with multiple transmitters and multiple receivers in which there are K user pairs interfering each other. As shown in Figure 12.1, for user-pair k, its transmitter $\mathcal{S}_k$ has message W_k to send to its receiver $\mathcal{D}_k$. But for other receivers $\mathcal{D}_j$ $(\forall j \neq k)$, this message will be treated as interference.

We will study the following two aspects of the PDAS for IC.

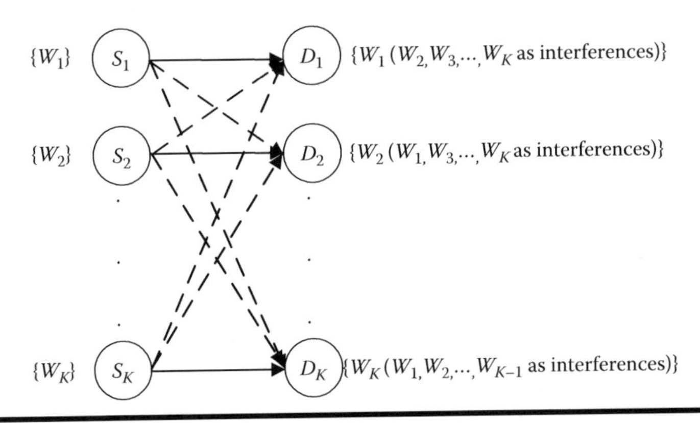

Figure 12.1 Illustration of the K-user IC model, where W_k is the expected signal from $\mathcal{S}_k$ to $\mathcal{D}_k$ interfering all other receivers $\mathcal{D}_j, \forall j \neq k$.

- The feasibility of IA in the two-dimensional (2D) or three-dimensional (3D) Euclidean space is investigated to find outhow many user pairs can be supported for the PDAS with IC model.
- The performance of the related PD-adaptive protocol for medium access control (MAC) is given to show the adaptive switching property among three basic modes.

From the point of view of the physical layer [6], the DoF of K-user IC is $K/2$, among which each user pair obtains a DoF of 1/2. The content of this section is more comprehensive than the simplified version of [20].

12.2.1 System Model

The proposed PDAS base IC model considers two constraints: the signal paths for all user pairs share the same PD (denoted by τ_1), while each interference path has a different PD (τ_2). For the convenience of studying feasibility, we directly transform it into the DAS expression as follows.

Consider a K-user IC, where each user includes a source node and a destination node, or a source–destination pair. Assume that the distance between each user's source and destination is d_1 (signal distance) and that between any interfering source and destination nodes is d_2 (interference distance). Mathematically, we have

$$d\left(S_i D_j\right) = \begin{cases} d_1, & i = j, \forall i, j \in \{1, \cdots, K\} \\ d_2, & \text{Otherwise.} \end{cases} \tag{12.1}$$

Assume that all transmitters send their own independent messages synchronously. With the earlier conditions, if $d_1 \neq d_2$, each destination node of the K users should receive its own message without interference simultaneously, while interferences also arrive at each receiver at the same time. The PDAS model does not work if $d_1 = d_2$. We will discuss cases for $d_2 > d_1$ and $d_2 < d_1$ in the following contexts. We remark that many existing works only focus on the former case and often assume that $d_1 = 1$ and $d_2 = 1$.

For an intuitive demonstration, two example 2D networks with $K = 2$ and $K = 3$ are given in Figure 12.2.

12.2.2 Number of Supported Users

Next, we study the number of user pairs supported by the IC based on PDAS subject to the distance constraints (12.1). We give the 3D feasibility conditions for 2–4 users. However, the numerical computation cannot find any feasible 3D DAS for $K = 5$ with $D_2 > D_1$ and for $K = 4$ with $D_2 < D_1$. This result partially confirms that K-user IC takes K–1 dimensions in Euclidean space for $D_2 > D_1$ case [17]. So we suppose all feasible scenarios have been found.

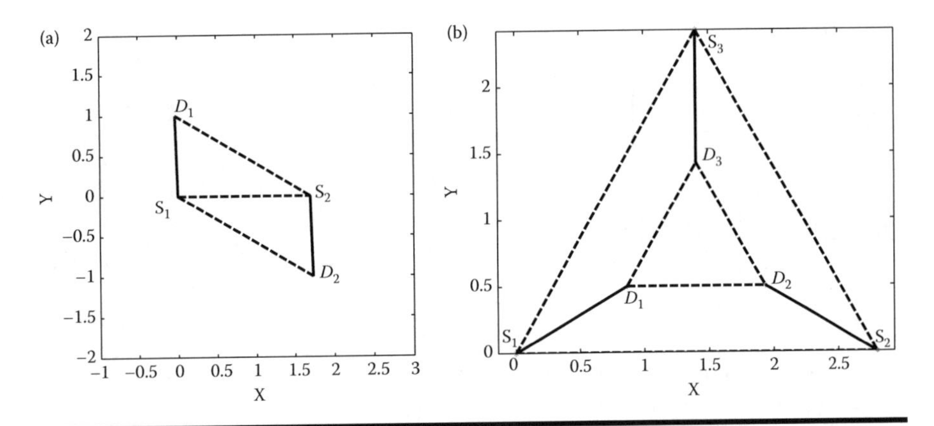

Figure 12.2 Examples of PDAS-based *K*-user IC in 2D Euclidean space with d_1 = 1 and d_2 = 2 for (a) K = 2 and (b) K = 3, respectively.

Case I: $d_2 > d_1$

In this case, we have checked the conditions for a feasible PDAS base IC with the supported number of users K up to 5. Numerical results only show the existence of $K = 2, 3, 4$.

Two Users: Without loss of generality, we assume that the first source node is located at the origin of the 3D Euclidean space, denoted as $\mathcal{S}_1(0,0,0)$, and the second source node at the positive X-axis with coordinate $\mathcal{S}_2(x_1,0,0)$ where $x_1 > 0$.

For the second receiver $\mathcal{D}_2$ with coordinate (x,y,z), from (12.1), the following distance conditions should be satisfied:

$$\begin{cases} x^2 + y^2 + z^2 = d_2^2, \\ (x - x_1)^2 + y^2 + z^2 = d_1^2, \end{cases} \tag{12.2}$$

which can be simplified to

$$\begin{cases} x = \dfrac{d_2^2 - d_1^2 + x_1^2}{2x_1}, \\ y^2 + z^2 = d_1^2 - \dfrac{(d_2^2 - d_1^2 - x_1^2)^2}{4x_1^2}. \end{cases} \tag{12.3}$$

Obviously, we should require that

$$y^2 + z^2 = d_1^2 - \frac{(d_2^2 - d_1^2 - x_1^2)^2}{4x_1^2} > 0 \Rightarrow -2d_1 < \frac{d_2^2 - d_1^2 - x_1^2}{x_1} < 2d_1, \tag{12.4}$$

or equivalently

$$\begin{cases} x_1^2 - 2d_1x_1 - \left(d_2^2 - d_1^2\right) < 0, \\ x_1^2 + 2d_1x_1 - \left(d_2^2 - d_1^2\right) > 0, \end{cases} \tag{12.5}$$

which can be simplified as

$$d_2 - d_1 < x_1 < d_2 + d_1. \tag{12.6}$$

This condition indicates that distance between two source nodes is within the range $d\left(\mathcal{S}_1\mathcal{S}_2\right) \in \left(d_2 - d_1, d_2 + d_1\right)$.

Similar condition can be applied to the first destination node $\mathcal{D}_1$ and the same results are obtained. So (12.6) provides the condition for the feasibility of a 2-user IC with PDAS, which is the basic constraint to be satisfied between any two source nodes if more users are involved, i.e.,

$$d\left(\mathcal{S}_i\mathcal{S}_j\right) \in \left(d_2 - d_1, d_2 + d_1\right), \quad \forall i \neq j. \tag{12.7}$$

With the feasibility condition (12.6), we can easily see that in fact there are infinite feasible instances of the 2-user PDAS-based IC with $d_2 > d_1$. Two examples are depicted by Figure 12.3 for demonstration in 2D/3D Euclidean space.

Three users: In this case, we can determine the feasibility conditions using the symmetry of a triangle connecting the three source nodes.

Assume that the three source nodes have coordinates: $\mathcal{S}_1(0,0,0)$, $\mathcal{S}_2(x_1,0,0)$ and $\mathcal{S}_3(x_2, y_2, z_2)$ with $x_1 > 0$, respectively. We want to get the location information of the third destination node $\mathcal{D}_3$, first.

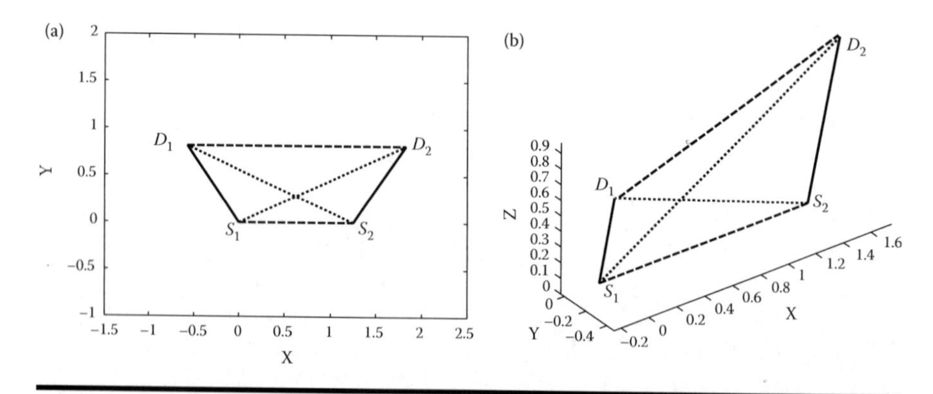

Figure 12.3 Examples of 2-user PDAS-based IC in with $d_1 = 1$ and $d_2 = 2$ in (a) 2D and (b) 3D Euclidean space.

It is easy to find that if $\mathcal{S}_1$–$\mathcal{S}_3$ are on the same line, there is no feasible DAS instance. So we only need to consider the condition with $y_2 \neq 0$ or $z_2 \neq 0 z_2 \neq 0$.

Denote the coordinate of the third destination node as $\mathcal{D}_3(x, y, z)$. Then we have the following constraints:

$$\begin{cases} x^2 + y^2 + z^2 = d_2^2, \\ (x - x_1)^2 + y^2 + z^2 = d_2^2, \\ (x - x_2)^2 + (y - y_2)^2 + (z - z_2)^2 = d_1^2. \end{cases} \tag{12.8}$$

After some manipulation, we obtain

$$\begin{cases} x = x_1/2, \\ y^2 + z^2 = d_2^2 - x_1^2/4, \\ (y - y_2)^2 + (z - z_2)^2 = d_1^2 - (x_1/2 - x_2)^2. \end{cases} \tag{12.9}$$

So we get two circles. The existence of $\mathcal{D}_3$ only needs that the two circles have intersection, i.e.,

$$\left| \sqrt{d_2^2 - \frac{x_1^2}{4}} - \sqrt{d_1^2 - \left(\frac{x_1}{2} - x_2\right)^2} \right| \leq \sqrt{y_2^2 + z_2^2} \leq \sqrt{d_2^2 - \frac{x_1^2}{4}} + \sqrt{d_1^2 - \left(\frac{x_1}{2} - x_2\right)^2}. \tag{12.10}$$

We can see that $\mathcal{D}_3$ must be one of the intersection points.

Similar conditions can be applied to get $\mathcal{D}_1$ and $\mathcal{D}_2$.

Numerical results show that there are infinite feasible instances of the above 3-user PDAS-based IC, among which two examples are shown in Figure 12.4.

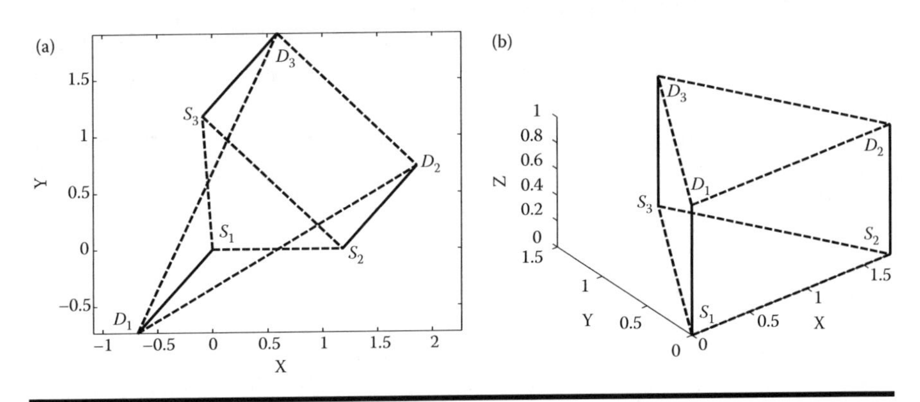

Figure 12.4 Examples of 3-user PDAS-based IC in 3D Euclidean space with $d_1 = 1$ and $d_2 = 2$.

Four users: Assume that the four source nodes have coordinates $\mathcal{S}_1(0,0,0)$, $\mathcal{S}_2(x_1,0,0)$, $\mathcal{S}_3(x_2,y_2,0)$, and $\mathcal{S}_4(x_3,y_3,z_3)$ with $x_1>0$, respectively. Using the same idea for the 3-user case, we want to get the location information of the fourth destination node $\mathcal{D}_4$ and the feasibility conditions for the 4-user PDAS-based IC.

Denote the fourth destination node as $\mathcal{D}_4(x,y,z)$. Following (12.8), we have constraints:

$$\begin{cases} x^2+y^2+z^2=d_2^2, \\ (x-x_1)^2+y^2+z^2=d_2^2, \\ (x-x_2)^2+(y-y_2)^2+z^2=d_2^2, \\ (x-x_3)^2+(y-y_3)^2+(z-z_3)^2=d_1^2, \end{cases} \tag{12.11}$$

which can be further simplified as

$$\begin{cases} x=x_1/2, \\ y^2+z^2=d_2^2-x_1^2/4, \\ 2xx_2+2yy_2=x_2^2+y_2^2, \\ 2x(x_2-x_1)+2yy_2=x_2^2+y_2^2-x_1^2. \end{cases} \tag{12.12}$$

According to (12.12), we get only two intersections that can be chosen as the coordinate of $\mathcal{D}_4$ as follows:

$$\left(\frac{x_1}{2},\frac{x_2^2+y_2^2-x_1x_2}{2y_2},\pm\sqrt{d_2^2-\frac{x_1^2}{4}-\frac{(x_2^2+y_2^2-x_1x_2)^2}{4y_2^2}}\right). \tag{12.13}$$

Similar conditions can be applied to get $\mathcal{D}_1$–$\mathcal{D}_3$.

Therefore, we have found only two feasible instances of the 4-user PDAS-based IC. One of them is shown in Figure 12.5, and the other is its symmetry of the XY plane.

However, we cannot find any feasible instance with five users by numerical searching, which might indicate that such a PDAS-based IC can only support at most four users when $D_2>D_1$.

Case II: $d_2 > d_1$

On the other side, it is possible that the interference distance is shorter than the signal distance, i.e., $d_2<d_1$. We also need to check the feasibility conditions for this case. In the following, we show that such a PDAS-based IC supports users up to three in 3D Euclidean space.

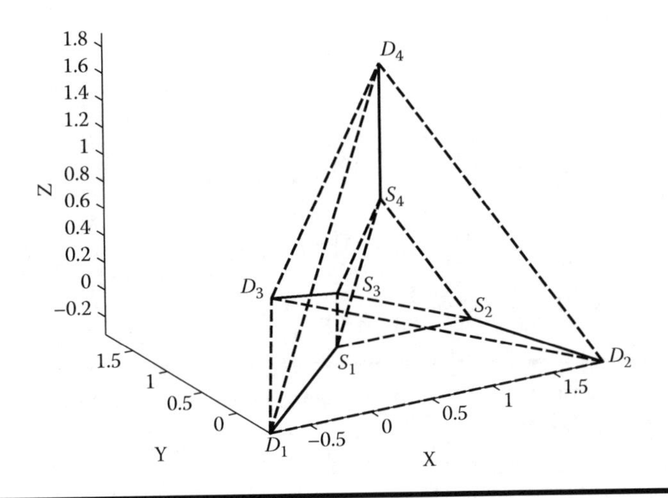

Figure 12.5 Example of the 4-user DAS-based IC in 3D Euclidean space with $d_1 = 1$ and $d_2 = 2$.

Two users: With the same assumptions, the conditions are similar to those of the 2-user scenario in the earlier case. Now the distance between $\mathcal{S}_1$ and $\mathcal{S}_2$ is restricted by

$$d_1 - d_2 < d\left(\mathcal{S}_1\mathcal{S}_2\right) < d_1 + d_2. \tag{12.14}$$

Two examples are depicted in Figure 12.6.

Three users: Similarly, we can get

$$\begin{cases} d\left(\mathcal{S}_i\mathcal{S}_j\right) \leq 2d_2, \\ d_1 - d_2 < d\left(\mathcal{S}_1\mathcal{S}_2\right) < d_1 + d_2, \end{cases} \tag{12.15}$$

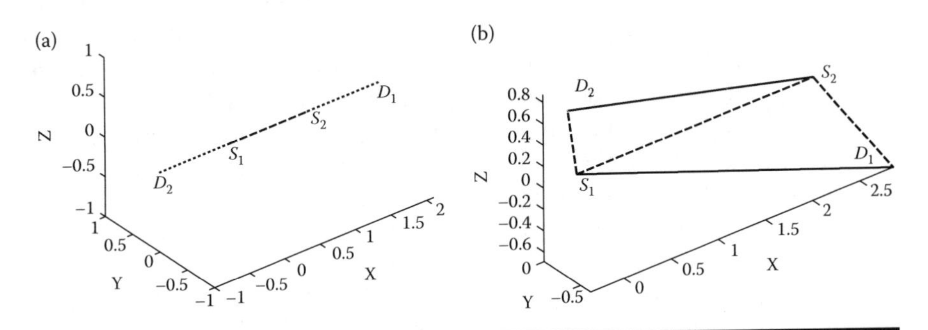

Figure 12.6 Examples of the 2-user PDAS-based IC in 3D Euclidean space with $d_1 = 2$ and $d_2 = 1$.

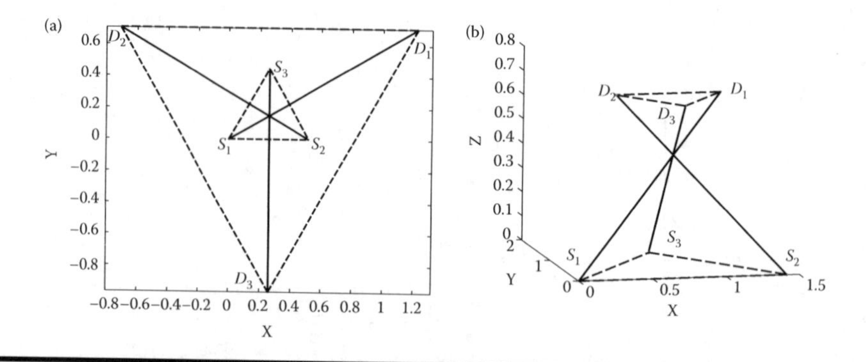

Figure 12.7 Examples of the 3-user PDAS-based IC in Euclidean space with $d_1 = \sqrt{2}$ and $d_2 = 1$.

according to the second equation of (12.9) and some other results. So the range of d_1 changes to $(d_2, 3d_2)$ based on (12.15). Other conditions can be similarly obtained.

Two examples are depicted in Figure 12.7.

However, numerical search has not shown any evidence that such PDAS-based IC can support more users. So, maybe three is the largest number of supported users when $d_2 < d_1$.

12.2.3 An Adaptive MAC Protocol for PDAS-Based IC

Here we propose a new access protocol for the earlier PDAS-base IC, which is adaptive to the distance difference between d_1 and d_2. With IA, this protocol can overlap most unwanted interference during a common time period to improve the spectrum efficiency. According to the relationship between d_1 and d_2, three modes are provided for different scenarios. Then the best one is chosen under that situation. We assume that a perfect synchronization among the transmitters is available. So, all source nodes send their frames at the same time. We can use some efficient approaches reported in [33–35] to obtain the transmission synchronization for all nodes. Meanwhile, it is easy to make sure that all nodes are aware of the network topology by utilizing the information exchanged in the phase of synchronization or adopting some MAC protocols like [36] to locate these nodes. Moreover, a four-step handshaking mechanism is used to control the transmission/reception procedure that is similar to the MAC design method because we consider a distributed protocol for PDAS as illustrated Figure 12.8. There may be contention (hidden terminal problem) when $\mathcal{S}_1$ and $\mathcal{S}'_1$ send messages to $\mathcal{D}_1$ simultaneously. Hence, the request-to-send (RTS)/clear-to-send (CTS) is necessary.

Major requirements of the proposed protocol are listed as follows:

1. Each node works in time-division-duplex mode and cannot transmit/receive simultaneously;

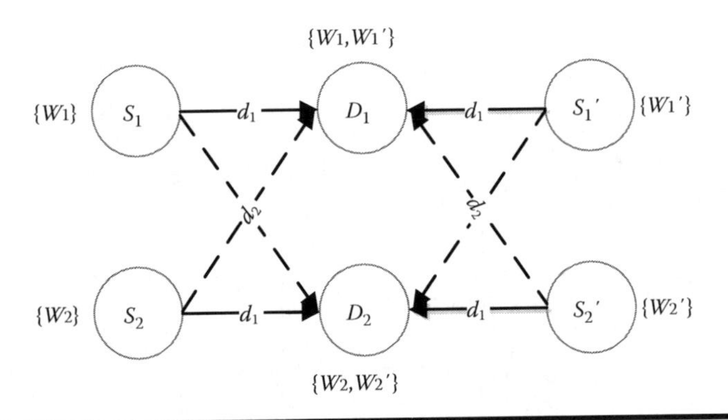

Figure 12.8 Illustration of a distributed protocol for PDAS.

2. Each node transmits only one kind of frames each time;
3. Each type of frame has the same length for all users;
4. There is a guard interval (τ_{gt}) between two useful frames and $\tau_{gt}/2$ between useful frame and unexpected frame(s) to tolerate the variation of transmission delay.

The following notation will be used:

- R_i: RTS frame from transmitter $\mathcal{S}_i$;
- C_i: CTS frame from receiver $\mathcal{D}_i$;
- W_i: data frame from transmitter $\mathcal{S}_i$;
- A_i: acknowledgment (ACK) frame from receiver $\mathcal{D}_i$;
- τ_{R_i}: transmission interval of R_i;
- τ_{C_i}: transmission interval of C_i, approximately equal to τ_{R_i}. For simplification, we assume $\tau_{C_i} = \tau_{R_i}$;
- τ_{A_i}: transmission interval of A_i. Also assume $\tau_{A_i} = \tau_{R_i}$;
- τ_{W_i}: transmission interval of W_i, much longer than τ_{R_i};
- τ_{gt}: guard interval;
- v: acoustic propagation speed, e.g., approximately equal to 1500 m/s in the water.

Now, we will discuss three modes for the proposed MAC protocol and show adaptive switching among them under different conditions to obtain a higher overall performance.

Modes 1 and 2 for Case I: $d_2 > d_1$

In this case, we study two modes as follows.

Mode 1: Here, the data frame is received after the arrival of all RTS frames. Figure 12.9 shows the detail of Mode 1 for the 3-user PDAS-based IC. With the earlier assumptions, the same kind of interference signals is received at the same time.

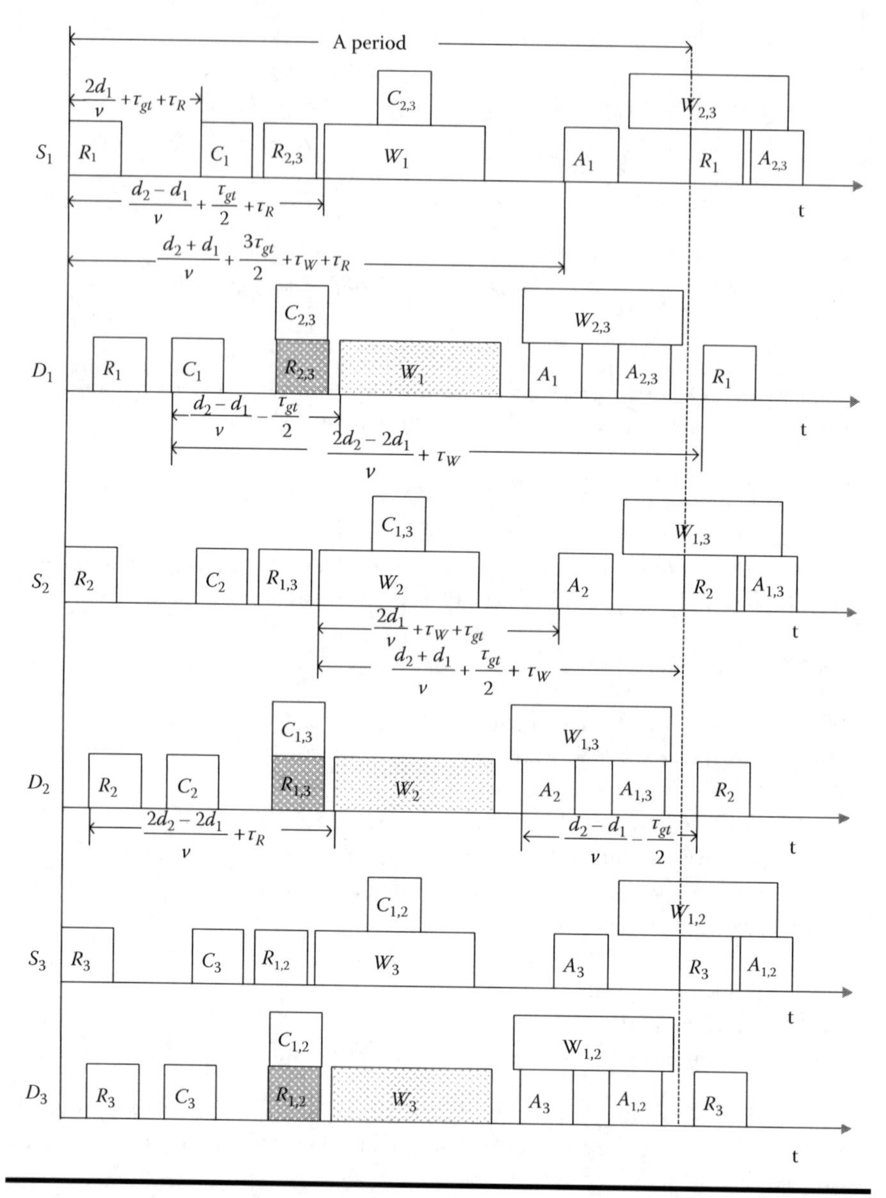

Figure 12.9 Mode 1 demonstration for the 3-user PDAS-based IC with $d_2 > d_1$.

The arrival time of the first interference frame should not overlay the first expected frame for each user. So we have $d_2/v \geq \tau_{R_1} + d_1/v$ and $d_2/v \geq \tau_{W_1} + d_1/v + \frac{\tau_{gt}}{2}$. Thus, the interval difference of arrival times at different receivers should be $\tau_{W_1} + \tau_{gt}/2$. Besides, we can determine the length of the largest data is $(d_2-d_1)/v-\tau_{gt}/2$.

Here we need to consider a special condition that any frame (RTS/DATA) sent by a source node cannot be received by other source nodes when they are receiving their own expected frames such as CTS/ACKs, which brings the following constraints:

$$0 < d\left(\mathcal{S}_i\mathcal{S}_j\right) \leq 2d_1 + \frac{\tau_{gt} v}{2}, \quad \text{or} \tag{12.16}$$

$$2d_1 + \left(\frac{3\tau_{gt}}{2} + 2\tau_R\right)v \leq d\left(\mathcal{S}_i\mathcal{S}_j\right) \leq d_2 + d_1 + \left(\tau_W + \tau_{gt}\right)v, \tag{12.17}$$

for the RTS frame, and

$$0 < d\left(\mathcal{S}_i\mathcal{S}_j\right) \leq 2d_1 + \frac{\tau_{gt} v}{2}, \quad \text{or} \tag{12.18}$$

$$2d_1 + \left(\frac{3\tau_{gt}}{2} + \tau_R + \tau_W\right)v \leq d\left(\mathcal{S}_i\mathcal{S}_j\right) \leq d_2 + d_1 + \left(\tau_{gt} + \tau_R\right)v, \tag{12.19}$$

for the DATA frame, which can be simplified as

$$0 < d\left(\mathcal{S}_i\mathcal{S}_j\right) \leq 2d_1 + \frac{\tau_{gt} v}{2}, \quad \text{or} \tag{12.20}$$

$$2d_1 + \left(\frac{3\tau_{gt}}{2} + \tau_R + \tau_W\right)v \leq d\left(\mathcal{S}_i\mathcal{S}_j\right) \leq d_2 + d_1 + \left(\tau_{gt} + \tau_R\right)v, \tag{12.21}$$

for the distance between source nodes. Similar analysis can be used for the destination nodes, i.e., any frame (CTS/ACK) sent by a destination node cannot be received by other destination nodes when they are receiving their own RTS/DATA. So we have

$$0 < d\left(\mathcal{D}_i\mathcal{D}_j\right) \leq d_2 - d_1 - \left(\tau_R + \tau_{gt}\right)v, \quad \text{or} \tag{12.22}$$

$$d_2 - d_1 + \tau_W v \leq d\left(\mathcal{D}_i\mathcal{D}_j\right) \leq 2d_2 - 2d_1 + \left(\tau_W - \tau_R - \frac{\tau_{gt}}{2}\right)v, \tag{12.23}$$

for the CTS frame, and

$$0 < d\left(\mathcal{D}_i\mathcal{D}_j\right) \leq d_2 - d_1 - \left(\tau_R + \tau_{gt}\right)v, \quad \text{or} \tag{12.24}$$

$$d_2 - d_1 + \tau_W v \leq d\left(\mathcal{D}_i\mathcal{D}_j\right) \leq 2d_2 - 2d_1 - \frac{\tau_{gt} v}{2}, \tag{12.25}$$

for the ACK frame, which can be simplified as

$$0 < d\left(\mathcal{D}_i\mathcal{D}_j\right) \le d_2 - d_1 - \left(\tau_R + \tau_{gt}\right)v, \quad \text{or} \tag{12.26}$$

$$d_2 - d_1 + \tau_W v \le d\left(\mathcal{D}_i\mathcal{D}_j\right) \le 2d_2 - 2d_1 - \frac{\tau_{gt}v}{2}, \tag{12.27}$$

for the distance between destination nodes.

Besides, inspired by [37] we propose an improved version of the earlier scheme to increase the per-user efficiency. Figure 12.10 shows that multiple data frames are controlled by a single RTS/CTS/ACK handshaking. Analyses in next section show that per-user efficiency of this method is higher than the original one.

Mode 2: As shown in Figure 12.11, the data frame is received before the arrival of the interference RTS, which are shown as different relative positions of shaded frames compared with the Mode 1 in Figure 12.9. Unlike Mode 1, in this mode we require that $\tau_W \le (d_2 - 3d_1)/v - 2\tau_R - 5\tau_{gt}/2$, which means that the protocol can only apply to the condition $d_2 > 3d_1 + (5\tau_{gt}/2 + 2\tau_R + \tau_W)v$.

As Mode 1, we need to consider the same condition for the range of $d\left(\mathcal{S}_i\mathcal{S}_j\right)$ and $d\left(\mathcal{D}_i\mathcal{D}_j\right)$. Analyzing the detail of Mode 2, we can get the range of $d\left(\mathcal{S}_i\mathcal{S}_j\right)$ as follows:

$$0 < d\left(\mathcal{S}_i\mathcal{S}_j\right) \le 2d_1 + \frac{\tau_{gt}v}{2}, \quad \text{or} \tag{12.28}$$

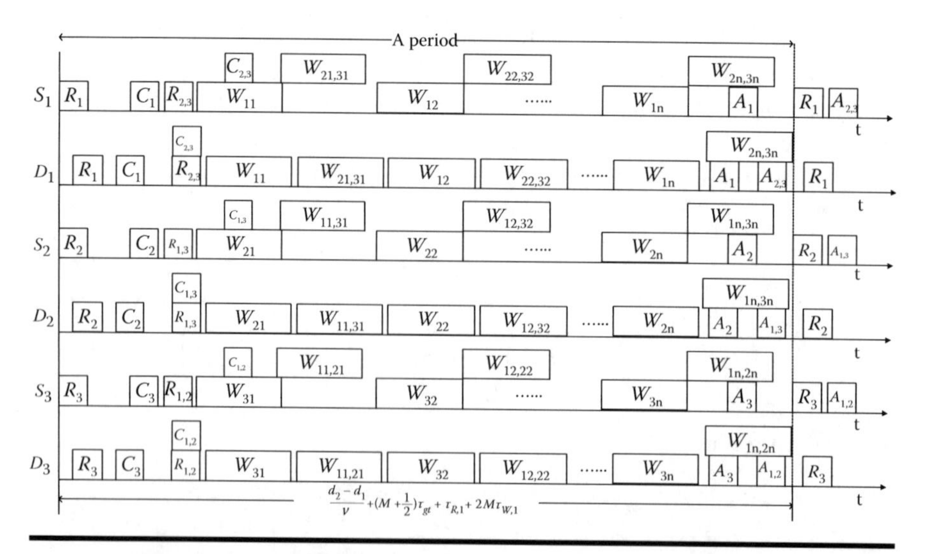

Figure 12.10 Mode 1 demonstration for the 3-user PDAS-based IC with multiple data frames per RTS/CTS/ACK ($d_2 > d_1$).

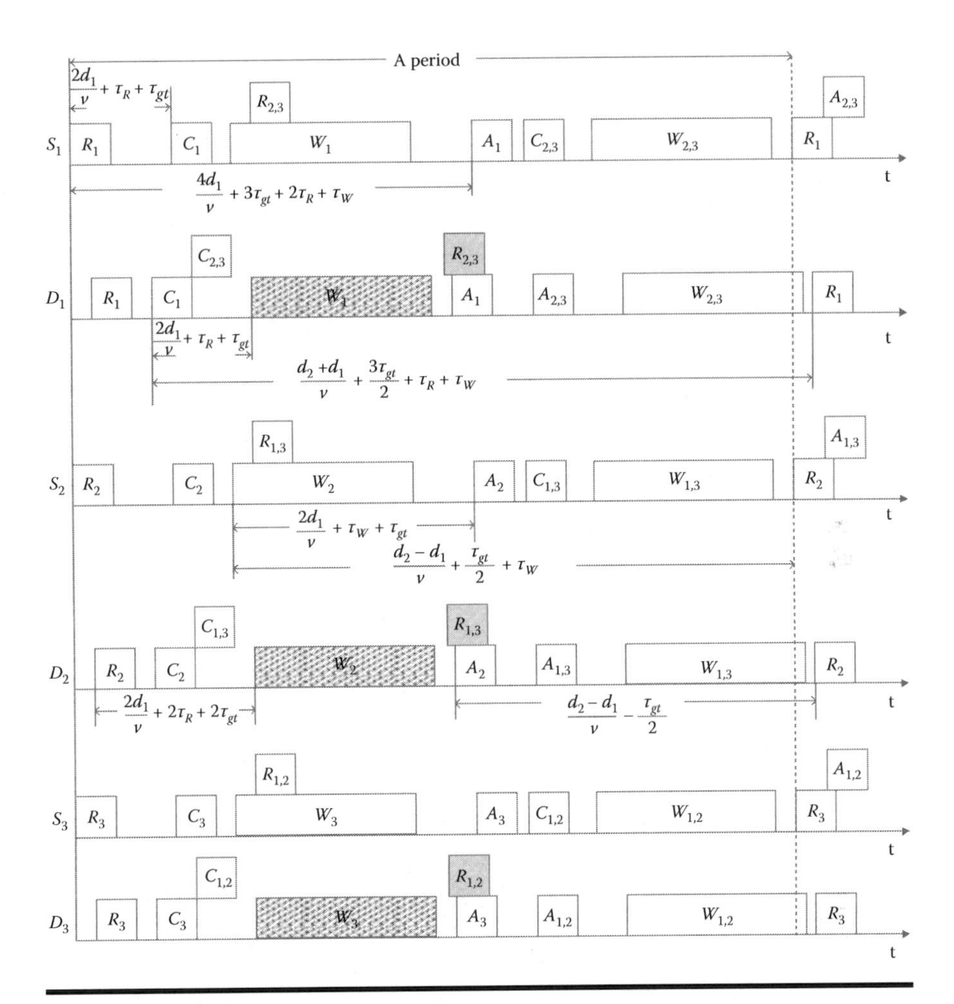

Figure 12.11 Mode 2 demonstration for the 3-user PDAS-based IC with $d_2 > d_1$.

$$2d_1 + \left(2\tau_R + \frac{3\tau_{gt}}{2}\right)v \le d\left(\mathcal{S}_i\mathcal{S}_j\right) \le 4d_1 + \left(\frac{5\tau_{gt}}{2} + \tau_W + \tau_R\right)v, \qquad (12.29)$$

for the RTS frame, and

$$0 < d\left(\mathcal{S}_i\mathcal{S}_j\right) \le 2d_1 + \frac{\tau_{gt}v}{2}, \quad \text{or} \qquad (12.30)$$

$$2d_1 + \left(\tau_R + \tau_W + \frac{3\tau_{gt}}{2}\right)v \le d\left(\mathcal{S}_i\mathcal{S}_j\right) \le d_2 + d_1 + \left(\tau_R + \tau_{gt}\right)v, \qquad (12.31)$$

for the DATA frame, which can be simplified as

$$0 < d\left(\mathcal{S}_i\mathcal{S}_j\right) \le 2d_1 + \frac{\tau_{gt}v}{2}, \quad \text{or} \tag{12.32}$$

$$2d_1 + \left(\frac{3\tau_{gt}}{2} + \tau_R + \tau_W\right)v \le d\left(\mathcal{S}_i\mathcal{S}_j\right) \le 4d_1 + \left(\frac{5\tau_{gt}}{2} + \tau_W + \tau_R\right)v, \tag{12.33}$$

while the range of $d\left(\mathcal{D}_i\mathcal{D}_j\right)$ can be obtained by

$$0 < d\left(\mathcal{D}_i\mathcal{D}_j\right) \le 2d_1 + \frac{\tau_{gt}v}{2}, \quad \text{or} \tag{12.34}$$

$$2d_1 + \left(\frac{3\tau_{gt}}{2} + \tau_R + \tau_W\right)v \le d\left(\mathcal{D}_i\mathcal{D}_j\right) \le d_2 + d_1 + \left(\tau_{gt} + \tau_W\right)v, \tag{12.35}$$

for the CTS frame, and

$$0 < d\left(\mathcal{D}_i\mathcal{D}_j\right) \le d_2 - d_1 - \left(\tau_{gt} + \tau_R\right)v, \quad \text{or} \tag{12.36}$$

$$d_2 - d_1 + \tau_R v \le d\left(\mathcal{D}_i\mathcal{D}_j\right) \le d_2 + d_1 + \left(\tau_{gt} + \tau_R\right)v, \tag{12.37}$$

for the ACK frame, which can be simplified as

$$0 < d\left(\mathcal{D}_i\mathcal{D}_j\right) \le 2d_1 + \frac{\tau_{gt}v}{2}, \quad \text{or} \tag{12.38}$$

$$d_2 - d_1 + \tau_R v \le d\left(\mathcal{D}_i\mathcal{D}_j\right) \le d_2 + d_1 + \left(\tau_{gt} + \tau_R\right)v, \tag{12.39}$$

for the distance between any two destinations.

Like Mode 1, we can also utilize multiple data frames by a single RTS/CTS/ACK to improve the efficiency in Mode 2.

Mode 3 for Case II: $d_2 > d_1$

This mode is designed for case II. With a guard interval, we consider the condition $0 < d_2 < d_1 - \tau_{gt}/2$. We directly apply the multiple data frames to Mode 3, since single data frame per RTS/CTS/ACK is quite inefficient.

Figure 12.12 shows the detail of Mode 3. Similarly, the distance between $\mathcal{S}_i$ and $\mathcal{S}_j$, $\mathcal{D}_i$ and $\mathcal{D}_j$ should be limited by

$$\begin{cases} 0 < d\left(\mathcal{S}_i\mathcal{S}_j\right) \le 2d_1 + \tau_{gt}v/2, \\ 0 < d\left(\mathcal{D}_i\mathcal{D}_j\right) \le 2d_1 + \tau_{gt}v/2. \end{cases} \tag{12.40}$$

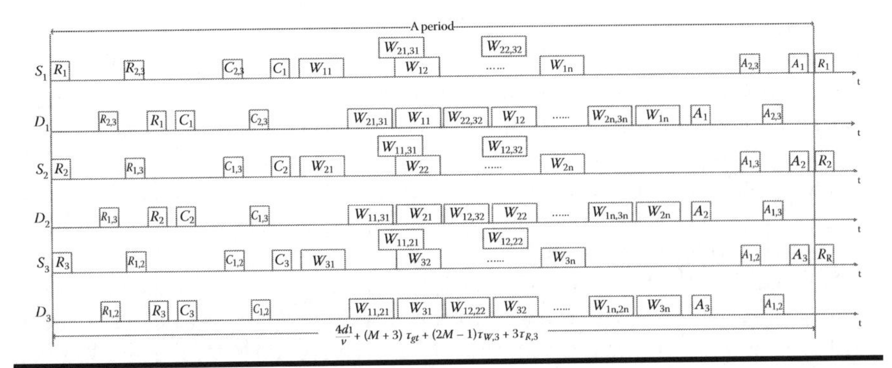

Figure 12.12 Mode 3 demonstration for the 3-user PDAS-based IC with multiple data frames per RTS/CTS/ACK ($d_2 < d_1$).

Switching between Modes 1–3

We have proposed three modes for the MAC protocol. The adaptation of the protocol is based on automatic switching among the earlier three modes with the relative difference between d_1 and d_2, as shown in Figure 12.13. Later, we will discuss the switching detail through performance analysis. Moreover, the proposed protocol is also suitable for any number of users with good extensibility. For the earlier feasible PDAS-based IC with 2/3/4 users, the protocol can be directly applied.

12.2.4 Performance Evaluation

In the earlier protocol, with perfect synchronization, each user has the same opportunity to access the channel. Under the feasibility conditions with the number of

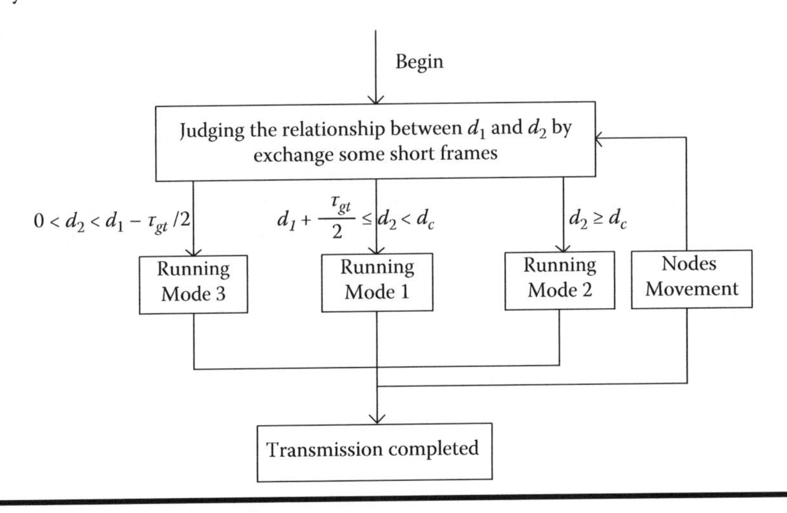

Figure 12.13 Flowchart of the mode switching process.

supported users, fairness is guaranteed for each user to equally share the medium. Now, we focus on the per-user efficiency to show the advantages of the PDAS-based IC. From the practical perspective, we further discuss the effects of node perturbation and show the robustness conditions of our proposal.

12.2.4.1 Per-User Efficiency

Here we evaluate the spectrum efficiency of the proposed MAC protocol in the PDAS-based IC. The per-user efficiency is measured by the ratio of the time occupation of a data frame τ_W to the whole transmission period τ_T, i.e., $\eta = \tau_W/\tau_T$. For convenience, we assume that all the control frames (RTS/CTS/ACK) have the same length τ_R. Then, the minimum length of the data frame can be expressed as N times of that of the control frames, i.e., $\tau_{W_{\min}} = N\tau_R$.

Through the earlier analysis, we found that τ_W and τ_R can be expressed by d_1, d_2 and τ_{gt}. Moreover, we set $d_1 = 3000$ m and $\tau_{gt} = 1/75$ s for initialization. We will use $\tau_{W,k}$ to denote the length of a data frame with Mode k.

$d_2 > d_1$: With Mode 1, the length of data frame is $M\tau_{W,1}$, where M is the number of data frames. The total length of a whole transmit–receive period can be found as

$$\tau_{T,1} = \frac{d_2 - d_1}{v} + (M + 1/2)\tau_{gt} + \tau_{R,1} + 2M\tau_{W,1}. \tag{12.41}$$

With Mode 2, we have the length of data frame: $M\tau_{W,2}$ and a different total length of the whole transmit–receive period:

$$\tau_{T,2} = \frac{Md_2 - (M-2)d_1}{v} + (2 + M/2)\tau_{gt} + 2\tau_{R,2} + M\tau_{W,2}. \tag{12.42}$$

$d_2 < d_1$: In this case, Mode 3 will be activated. The length of data frame is $M\tau_{W,3}$, and the length of the whole period time is

$$\tau_{T,3} = \frac{4d_1}{v} + (M + 3)\tau_{gt} + (2M - 1)\tau_{W,3} + 3\tau_{R,3}. \tag{12.43}$$

In Figures 12.14 and 12.15, we show the protocol performance in terms of per-user efficiency versus d_2 with different numbers of data frames. We can see that there is a cross between the curves Mode 1 and Mode 2. In detail, Mode 1 performs better at the lower d_2 region, while Mode 2 is superior at the higher d_2 region. The cross-point is called the critical point, which is at the position of $d_2 = d_c$. To obtain higher overall performance, we choose Mode 1 when $d_1 + \frac{\tau_{gt}v}{2} \le d_2 < d_c$, Mode 2 when $d_2 \ge d_c$, and Mode 3 when $d_1 - \frac{\tau_{gt}v}{2} \ge d_2$. The adaptation property is shown as the black solid curve in Figures 12.14 and 12.15. For example, with the earlier assumptions, Mode 1 will be active when 3010 m$\le d_2 <$21,746 m, while Mode 2 will be

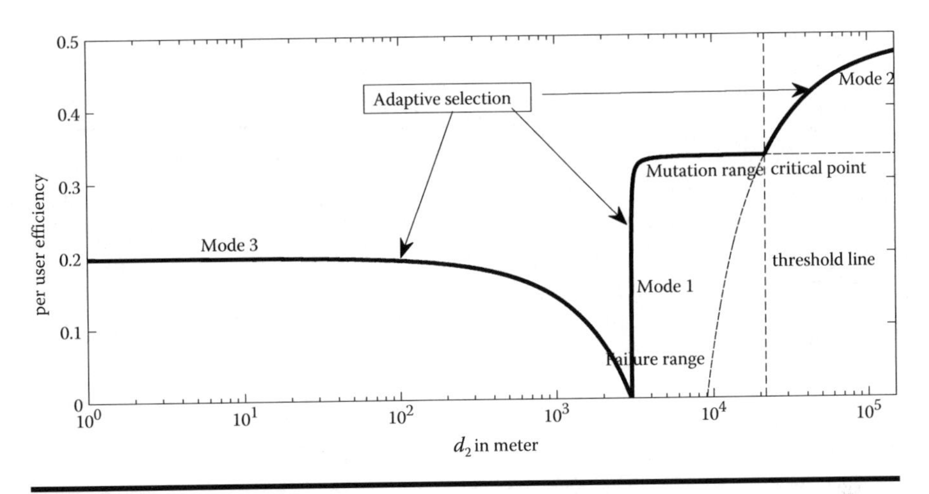

Figure 12.14 Per-user efficiency versus d_2 with $N = 100$, $M = 1$.

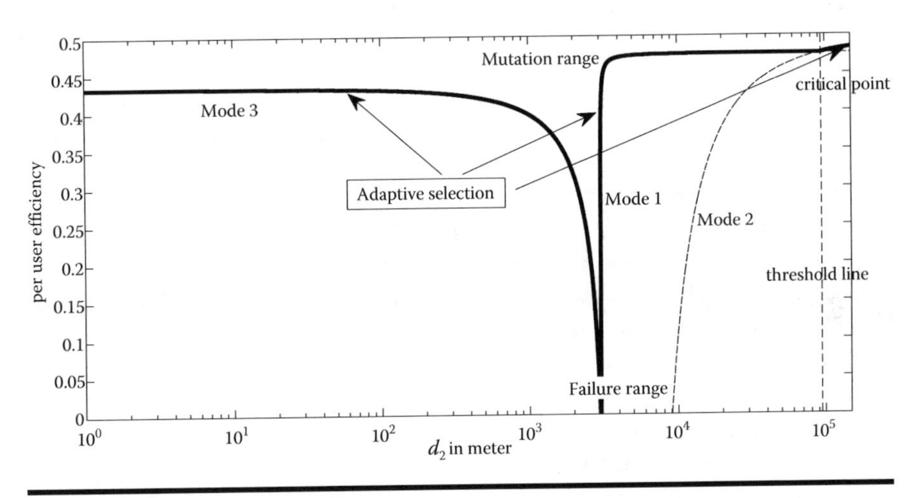

Figure 12.15 Per-user efficiency versus d_2 with $N = 100$, $M = 10$.

active when $d_2 \geq 21{,}746$ m, and Mode 3 will be active when $0\text{ m} < d_2 \leq 2090$ m with $M = 1$. With $M = 10$, the protocol chooses Mode 1 when $3010\text{ m} \leq d_2 < 94{,}879$ m, and Mode 2 when $d_2 \geq 94{,}879$ m, and Mode 3 when $0\text{ m} < d_2 \leq 2090$ m. Obviously, the protocol always chooses the mode with best performance to be adaptive with the interference distance.

Now we discuss the detailed aspects of the performance, including the asymptotic efficiency, the failure range, the mutation range, and the critical position.

Asymptotic behavior: From Figures 12.14 and 12.15, the asymptotic per-user efficiency in Mode 1 is 1/3 with $M = 1$ and 1/2 with $M = 10$, while Mode 2 can further achieve half of the spectrum with both $M = 1$ and $M = 10$, i.e., 1/2. In

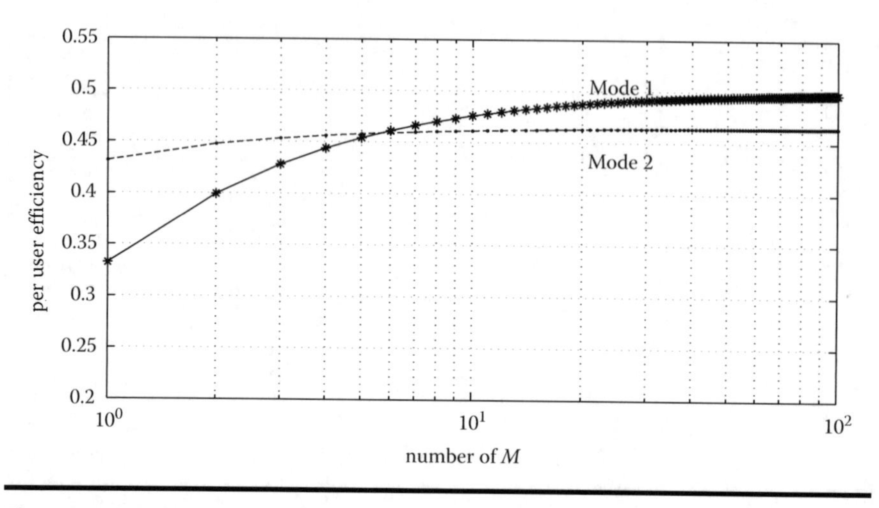

Figure 12.16 Per-user efficiency versus M with N = 100, d_2 = 50,000.

Mode 3 the asymptotic per-user efficiency is about 1/5 with $M = 1$ and 0.45 with $M = 10$. Obviously, a higher M will increase the spectrum efficiency closer to the theoretical value 1/2.

Multiple data frames: According to Figures 12.15 and 12.16, with increasing M, the per-user efficiency of Mode 1 increases to 1/2 rapidly, while in Mode 2 it improves much slower than that in Mode 1. This leads to a very large d_c for the critical point, which might make the network impractical. Hence, Mode 1 might be utilized more widely than Mode 2.

Failure range: The failure range indicates that the proposed protocol for DAS will not be available if d_1 and d_2 are quite close. In detail, the failure range of d_2 happens between Modes 1 and 3 in Figures 12.14 and 12.15. In this range, the efficiency will drop to zero quickly.

Mutation range: The mutation range indicates that with Mode 1 in Figures 12.14 and 12.15, the performance is very sensitive to d_2 when d_2 is a bit longer than d_1 and reaches its asymptotic performance rapidly.

Critical point: We can get the abscissa of the critical point d_c by setting the efficiency value of Modes 1 and 2 equal to each other, i.e.,

$$\begin{aligned}&\frac{\tau_{W,1}}{\frac{d_2-d_1}{v}+(M+1/2)\tau_{gt}+\tau_{R,1}+2M\tau_{W,1}}\\&=\frac{\tau_{W,2}}{\frac{Md_2-(M-2)d_1}{v}+(2+M/2)\tau_{gt}+2\tau_{R,2}+M\tau_{W,2}},\end{aligned} \tag{12.44}$$

where

$$\begin{cases} \tau_{W,1} = \dfrac{d_2 - d_1}{v} - \dfrac{\tau_{gt}}{2}, \\ \tau_{R,1} = \tau_{W,1_{\min}} / N, \\ \tau_{W,2} = \dfrac{d_2 - 3d_1}{v} - 2\tau_{R,2} - \dfrac{5\tau_{gt}}{2}, \\ \tau_{R,2} = \tau_{W,2_{\min}} / N. \end{cases} \tag{12.45}$$

From the earlier equation, we can obtain that d_2 is equal to d_c.

12.2.4.2 *Impact of Node Perturbation and Guard Time*

To make the protocol robust, we add a guard time or half guard time between every two frames for the proposed protocol, as discussed previously. The change of transmission delay can also be modeled as perturbation of nodes.

Let the perturbation amplitude of each node is Δd and denote t_a as the arrival time of all messages if there are no perturbation and PD jitter. Since all expected/interference messages are received in the range $\left[t_a - \dfrac{\tau_{gt}}{2}, t_a + \dfrac{\tau_{gt}}{2}\right]$, we get constraints between Δd and τ_{gt} with all modes as follows

$$\begin{cases} \dfrac{d_1}{v} - \dfrac{\tau_{gt}}{2} \le \dfrac{d_1 \pm \Delta d}{v} \le \dfrac{d_1}{v} + \dfrac{\tau_{gt}}{2}, \\ \dfrac{d_2}{v} - \dfrac{\tau_{gt}}{2} \le \dfrac{d_2 \pm \Delta d}{v} \le \dfrac{d_2}{v} + \dfrac{\tau_{gt}}{2}, \end{cases} \tag{12.46}$$

which indicates

$$\tau_{gt} \ge \frac{2\Delta d}{v}. \tag{12.47}$$

If Δd and τ_{gt} cannot follow the earlier inequalities, there will be collisions at the receivers.

12.3 PDAS with Acoustic Single-Hop X Network

12.3.1 Introduction

The new type of 2×2 XC was first studied in [38,39], which has a higher DoF than the 2-user IC by allowing all potential single-hop unicast messages from the transmitters to the receivers. Later, when there are M transmitters and N receivers (often called X network) equipped with single antenna, it is proven that its DoF is $\frac{MN}{M+N-1}$ [12].

This section introduces our recent works [21,22] on the implementation of perfect IA based on PD for K×2 XC, where each of the K source nodes sends an independent message to each of the two destination nodes. The DoF is $2K/(K+1)$, implicating that each receiver should obtain K interference-free messages in $K+1$ time slots. Thus, the key problem is to align all interferences at one time slot and arrange the required K messages in a proper order.

12.3.2 System Model

The typical $K\times 2$ single-hop X network with two receivers is shown in Figure 12.17. There are K ($K\geq 2$) source nodes denoted by $\mathcal{S}_k, \forall k \in \{1,2,\ldots,K\}$ and two destination nodes denoted by $\mathcal{D}_j, \forall j \in \{1,2\}$. Each source node $\mathcal{S}_k$ wants to send an independent message W_{kj} to the destination node $\mathcal{D}_j$. The links between $\mathcal{S}_k$ and $\mathcal{D}_j$ are fully connected.

Denote the PD between $\mathcal{S}_k$ and $\mathcal{D}_j$ as $\tau_{k,j}$ and organize the PD relationships into a $K\times 2$ matrix T. For simplicity, the shortest PD is set to be the link between $\mathcal{S}_1$ and $\mathcal{D}_1$ as reference and its PD is set as $\tau_{1,1} = 0$. We further normalize the PD increase step among $\tau_{k,j}$ as $\Delta\tau = 1$.

Mathematically, the original message vector from $\mathcal{S}_k$ is

$$\mathbf{x}_k = \begin{bmatrix} x_{k1} & x_{k2} \end{bmatrix}. \tag{12.48}$$

After proper scheduling, the corresponding message vector to be transmitted is given by

$$\tilde{\mathbf{x}}_k = \mathbf{x}_k \mathbf{P}_k, \tag{12.49}$$

where $\mathbf{P}_k$ denotes the binary scheduling/precoding matrix.

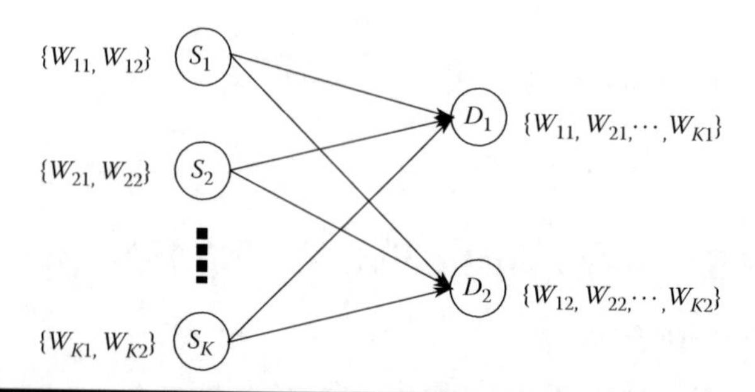

Figure 12.17 System model of the $K\times 2$ X channels.

Let the total number of required messages be m. In the PDAS-based model, the DoF can be defined as

$$\text{DoF} = m/n. \tag{12.50}$$

To achieve maximum DoF of $2K/(K+1)$ with $m = 2K$ for the $K \times 2$ X network, the length of time slots at the receiver side should be equal to $n = K + 1$. Here, we know that each destination node needs K time slots for the required messages. So, only one time slot is available for IA.

Our goal is to achieve the DoF of $2K/(K+1)$ through providing an artificial PDAS matrix T.

12.3.3 DoF-Achieving PDAS Scheme

The problem can be solved by the following theorem.

Theorem 1.

The DoF of $2K/(K+1)$ for $K \times 2$ X channels is achievable by specific PDAS and proper scheduling.

Next we show the achievability from $K = 2$, 3 to general K.

$K = 2$: This is the simplest scenario. Due to symmetry, we can set $\tau_{1,1} = \tau_{2,2} = 0$, and $\tau_{2,1} = \tau_{1,2} = 1$. Thus, we have the following PDAS matrix

$$T = \begin{bmatrix} \tau_{1,1} & \tau_{1,2} \\ \tau_{2,1} & \tau_{2,2} \end{bmatrix} = \begin{bmatrix} 0 & 1 \\ 1 & 0 \end{bmatrix}. \tag{12.51}$$

The scheduling and PDAS is demonstrated by Figure 12.18. We can see that the required messages $x_{1,1}$ and $x_{2,1}$ can be respectively found at time slots 1 and 3. Similarly, at $\mathcal{D}_2$ the required messages $x_{1,2}$ and $x_{2,2}$ can be decoded at time slots 3 and 1, respectively. Here, time slot 2 is used for IA. So, a total of four messages are obtained without interference in three time slots, which shows that a DoF of 4/3 is achieved.

$K = 3$: Let $\tau_{1,1} = \tau_{3,2} = 0$, $\tau_{2,1} = \tau_{1,2} = 1$, and $\tau_{3,1} = 2$. The PDAS matrix is

$$T = \begin{bmatrix} \tau_{1,1} & \tau_{1,2} \\ \tau_{2,1} & \tau_{2,2} \\ \tau_{3,1} & \tau_{3,2} \end{bmatrix} = \begin{bmatrix} 0 & 1 \\ 1 & 1 \\ 2 & 0 \end{bmatrix}. \tag{12.52}$$

The scheduling and PDAS is demonstrated by Figure 12.19. At $\mathcal{D}_1$, the required messages $x_{1,1}$, $x_{2,1}$, and $x_{3,1}$ can be, respectively, found at time slots 1, 2, and 4,

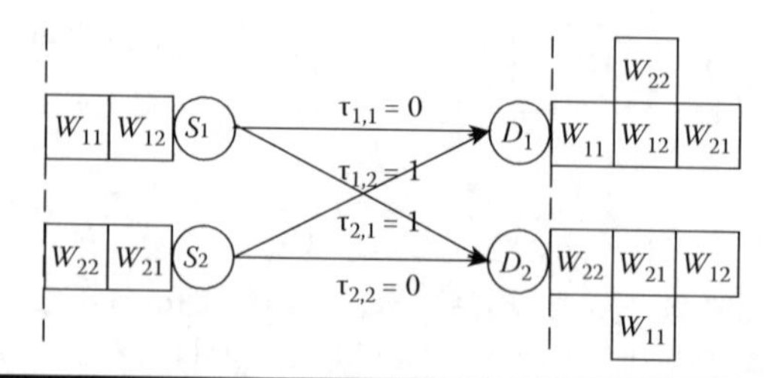

Figure 12.18 Scheduling and PDAS for 2×2 XC.

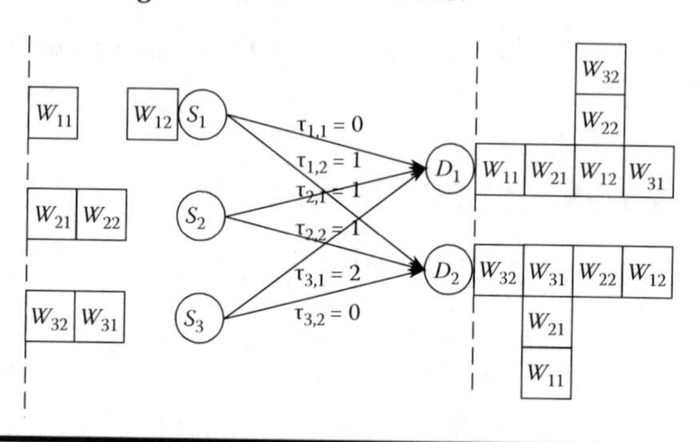

Figure 12.19 Scheduling and PDAS for 3×2 XC.

while time slot 3 is used for IA. Similarly, at $\mathcal{D}_2$, messages $x_{1,2}$, $x_{2,2}$, and $x_{3,2}$ can be decoded at time slots 4, 3, and 1, respectively, while time slot 2 is used for IA.

So, total six messages are obtained without interference in four time slots, which shows that a DoF of 3/2 is achieved.

Case I: *K* is Even

Let $K = 2N$. We generalize the earlier approach by setting $\tau_{1,1} = \tau_{N+1,2} = 0$, $\tau_{2,1} = \tau_{N+2,2} = 1, \ldots, \tau_{N,1} = \tau_{2N,2} = \mathrm{N} - 1$, and $\tau_{N+1,1} = \tau_{N+2,1} = \ldots = \tau_{2N,1} = \tau_{1,2} = \tau_{2,2} = \ldots = \tau_{N,2} = N$. Thus, we have the following PDA matrix

$$T = \begin{bmatrix} \tau_{1,1} & \tau_{1,2} \\ \tau_{2,1} & \tau_{2,2} \\ \vdots & \vdots \\ \tau_{N,1} & \tau_{N,2} \\ \tau_{N+1,1} & \tau_{N+1,2} \\ \tau_{N+2,1} & \tau_{N+2,2} \\ \vdots & \vdots \\ \tau_{2N,1} & \tau_{2N,2} \end{bmatrix} = \begin{bmatrix} 0 & N \\ 1 & N \\ \vdots & \vdots \\ N-1 & N \\ N & 0 \\ N & 1 \\ \vdots & \vdots \\ N & N-1 \end{bmatrix}. \tag{12.53}$$

The scheduling matrices can be correspondingly designed to obtain the following $K \times L$ scheduled message matrix

$$\begin{bmatrix} \tilde{\mathbf{x}}_1 \\ \tilde{\mathbf{x}}_2 \\ \vdots \\ \tilde{\mathbf{x}}_N \\ \tilde{\mathbf{x}}_{N+1} \\ \tilde{\mathbf{x}}_{N+2} \\ \vdots \\ \tilde{\mathbf{x}}_K \end{bmatrix} = \begin{bmatrix} x_{1,1} & 0 & \cdots & 0 & x_{1,2} \\ x_{2,1} & 0 & \cdots & x_{2,2} & 0 \\ \vdots & \vdots & \vdots & \vdots & \vdots \\ x_{N,1} & x_{N,2} & \cdots & 0 & 0 \\ x_{N+1,2} & 0 & \cdots & 0 & x_{N+1,1} \\ x_{N+2,2} & 0 & \cdots & x_{N+2,1} & 0 \\ \vdots & \vdots & \vdots & \vdots & \vdots \\ x_{2N,2} & x_{2N,1} & \cdots & 0 & 0 \end{bmatrix}. \tag{12.54}$$

The received messages can be obtained as follows:

$$\mathbf{y} \triangleq \begin{bmatrix} \mathbf{y}_1 \\ \mathbf{y}_2 \end{bmatrix} = \begin{bmatrix} x_{1,1} & \cdots & x_{N,1} & \sum_{k=1}^{K} x_{k,2} & x_{2N,1} & \cdots & x_{N+1,1} \\ x_{N+1,2} & \cdots & x_{2N,2} & \sum_{k=1}^{K} x_{k,1} & x_{N,2} & \cdots & x_{1,2} \end{bmatrix}. \tag{12.55}$$

Here we see that all interference messages are aligned at time slot $N+1$ for both receivers. The position index of the required messages can be easily obtained and used for decoding.

So the DoF $2K/(K+1)$ is already achieved, since a total of $2K$ messages are obtained without interference in $K+1$ time slots.

General Case II: *K* is Odd

Let $K = 2N-1$. We propose the following PDA matrix

$$T = \begin{bmatrix} \tau_{1,1} & \tau_{1,2} \\ \tau_{2,1} & \tau_{2,2} \\ \vdots & \vdots \\ \tau_{N,1} & \tau_{N,2} \\ \tau_{N+1,1} & \tau_{N+1,2} \\ \tau_{N+2,1} & \tau_{N+2,2} \\ \vdots & \vdots \\ \tau_{2N-1,1} & \tau_{2N-1,2} \end{bmatrix} = \begin{bmatrix} 0 & N-1 \\ 1 & N-1 \\ \vdots & \vdots \\ N-1 & N-1 \\ N & 0 \\ N & 1 \\ \vdots & \vdots \\ N & N-2 \end{bmatrix}. \tag{12.56}$$

Choosing proper scheduling matrices, we can obtain the following scheduled message matrix.

$$\begin{bmatrix} \tilde{\mathbf{x}}_1 \\ \tilde{\mathbf{x}}_2 \\ \vdots \\ \tilde{\mathbf{x}}_N \\ \tilde{\mathbf{x}}_{N+1} \\ \tilde{\mathbf{x}}_{N+2} \\ \vdots \\ \tilde{\mathbf{x}}_K \end{bmatrix} = \begin{bmatrix} x_{1,1} & 0 & \cdots & 0 & x_{1,2} \\ x_{2,1} & 0 & \cdots & x_{2,2} & 0 \\ \vdots & \vdots & \vdots & \vdots & \vdots \\ x_{N,1} & x_{N,2} & \cdots & 0 & 0 \\ x_{N+1,2} & 0 & \cdots & x_{N+1,1} & 0 \\ \vdots & \vdots & \vdots & \vdots & \vdots \\ x_{2N-1,2} & x_{2N-1,1} & \cdots & 0 & 0 \end{bmatrix}. \tag{12.57}$$

Finally, the received message matrix can be found as

$$\mathbf{y} = \begin{bmatrix} x_{1,1} & \cdots & x_{N,1} & \sum_{k=1}^{K} x_{k,2} & x_{2N-1,1} & \cdots & x_{N+1,1} \\ x_{N+1,2} & \cdots & \sum_{k=1}^{K} x_{k,1} & x_{N,2} & x_{N-1,2} & \cdots & x_{1,2} \end{bmatrix}. \tag{12.58}$$

Here we see time slots $N+1$ and N are used for IA at $\mathcal{D}_1$ and $\mathcal{D}_2$, respectively. It's easy to decode the required messages with the position index. Since $2K$ messages are successfully transferred in $K+1$ time slots, the DoF goal is achieved.

12.3.4 Feasibility in Euclidean Space

The feasibility problem turns out to find a solution of node placement in practical 1D/2D/3D Euclidean space. For simplicity, we consider the feasibility with constant propagation speed (v) over all links. The distance between $\mathcal{S}_k$ and $\mathcal{D}_j$ is $d_{k,j} = v\tau_{k,j}$. Denote a as the distance between $\mathcal{D}_1$ and $\mathcal{D}_2$. The distance between $\mathcal{S}_1$ and $\mathcal{D}_1$ is $d_0 = v\tau_0$. To avoid position overlap, we need $d_{k,j} > 0$, $d_0 > 0$, and $a > 0$. The property of equal PD increase step $\Delta\tau$ in our model indicates the corresponding equal distance increase step $\Delta d = v\Delta\tau > 0$.

We find the following feasibility conditions for $K\times 2$ X network with arbitrary integer $K \geq 2$

$$N\Delta d \leq a \leq 2d_0 + N\Delta d, \tag{12.59}$$

when $K = 2N$ is even and

$$N\Delta d \leq a \leq 2d_0 + (N-1)\Delta d, \tag{12.60}$$

when $K = 2N - 1$ is odd. Since $d_0 > 0$, (12.59) always holds, and (12.60) holds when $0 < \Delta d \leq 2d_0$.

Because d_0 can be arbitrarily chosen, by flexibly adjusting d_0 and Δd (equivalently the frame length $\Delta\tau$), the earlier feasibility conditions (12.59) and (12.60) can be widely practical.

For an example of UANs, typically, $v = 1500$ m/s. Assume data rate is 2400 bps and $K = 10$. Fixed $d_0 = 1000$ m, if the range of Δd is [10, 500] m (accordingly with bits in each frame as [16, 800]), we can obtain a wide range of a as [50, 4500] m.

Here we give two examples to illustrate that the proposed PDAS is feasible in 3D Euclidean space.

Figures 12.20 and 12.21 show feasible network examples for $K = 3$ and $K = 4$, respectively. The coordinates of related nodes in the 3D Euclidean space are listed as follows:

$$\begin{aligned} \mathcal{S}_2 &: \left(5/4, 1, -\sqrt{23}/4\right), \\ \mathcal{S}_3 &: \left(57/20, 0.5, \sqrt{251}/20\right), \\ \mathcal{D}_1 &: (0,0,0), \\ \mathcal{D}_2 &: (2.5,0,0), \end{aligned} \tag{12.61}$$

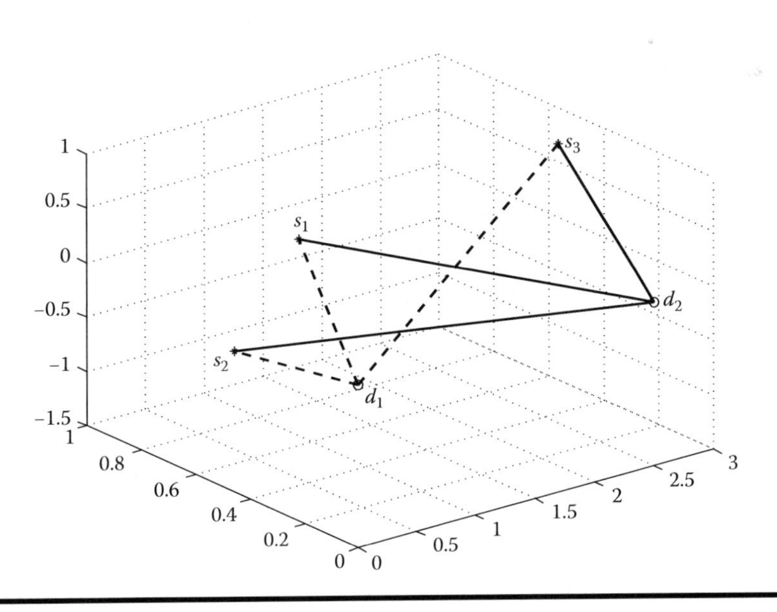

Figure 12.20 Feasible example of 3×2 PDAS-based XC in the 3D Euclidean space.

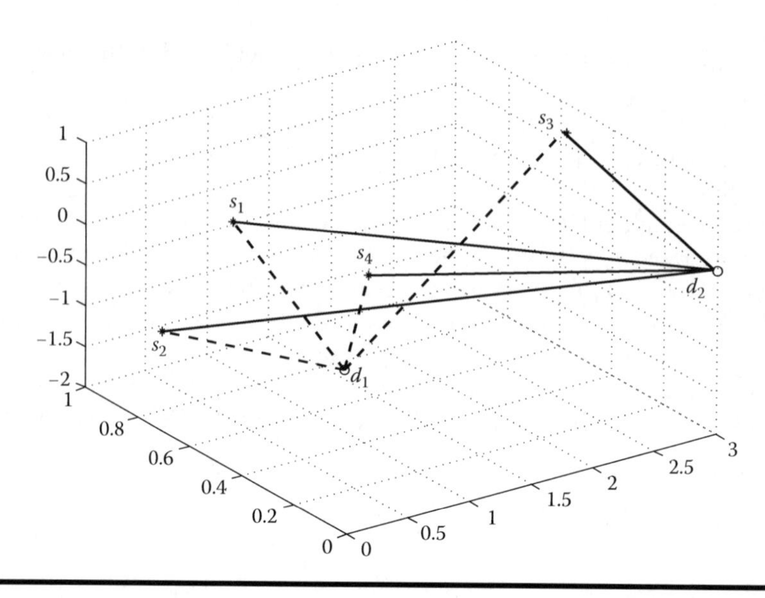

Figure 12.21 Feasible example of 4×2 PDAS-based XC in the 3D Euclidean space.

for $K = 3$, and

$$\begin{aligned}
\mathcal{S}_1 &: \left(1/6, 0.5, \sqrt{26}/6\right), \\
\mathcal{S}_2 &: \left(2/3, 1, -\sqrt{23}/3\right), \\
\mathcal{S}_3 &: \left(17/6, 0.5, \sqrt{26}/6\right), \\
\mathcal{S}_4 &: \left(7/3, 1, -\sqrt{23}/3\right), \\
\mathcal{D}_1 &: (0,0,0), \\
\mathcal{D}_2 &: (3,0,0),
\end{aligned} \tag{12.62}$$

for $K = 4$.

12.4 PDAS with Acoustic Multi-Hop X Network

12.4.1 Introduction

In the multihop scenario, IN/AIN approach can be further used to eliminate interference at each destination, such as the basic 2×2×2 interference networks [8]. However, PD-based IA for multihop has not been widely studied.

In this section, we investigate the implementation of AIN to achieve the theoretical DoF of acoustic 2×3×3 PDAS-based X network. The DoF upper bound is given and shown to be achievable by a specific PDAS. Here, message scheduling at the source and relay layers plays an important role. Moreover, the feasibility of the proposed scheme is demonstrated by providing the coordinates of all nodes in 3D Euclidean space.

12.4.2 System Model

The 2×3×3 layered X network is shown by Figure 12.22.

At the first source layer, there are two nodes $\mathcal{S}_1$ and $\mathcal{S}_2$. At the middle relay layer, there are three nodes $\mathcal{R}_1$, $\mathcal{R}_2$, and $\mathcal{R}_3$. At the last destination layer, there are three nodes $\mathcal{D}_1$, $\mathcal{D}_2$, and $\mathcal{D}_3$. Each source node $\mathcal{S}_i$ wants to send an independent message W_{ji} to destination $\mathcal{D}_j$. We assume there is no direct link between the source and destination layers. So the transmission/reception only occurs between neighboring layers. Moreover, we assume nodes in the same layer work in the same transmission/reception modes in synchronous way. So there is no interference coming from other nodes in the same layer. For relay nodes, full duplex is assumed here. Thus, there are two hops from source to destination.

Denote the distance between any two nodes along the first and second hops as d_{ki}^1 and d_{jk}^2, respectively. So we can obtain two distance matrices as

$$D_1 = \begin{bmatrix} d_{11}^1 & d_{21}^1 & d_{31}^1 \\ d_{12}^1 & d_{22}^1 & d_{32}^1 \end{bmatrix}, \quad D_2 = \begin{bmatrix} d_{11}^2 & d_{21}^2 & d_{31}^2 \\ d_{12}^2 & d_{22}^2 & d_{32}^2 \\ d_{13}^2 & d_{23}^2 & d_{33}^2 \end{bmatrix}, \tag{12.63}$$

for the first and second hops, respectively.

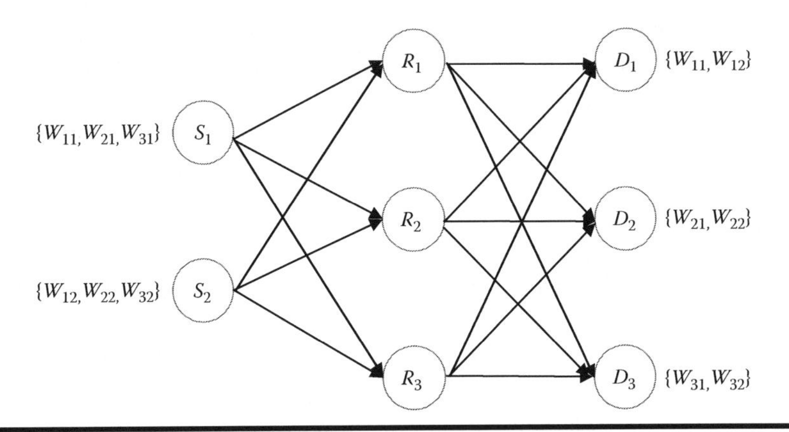

Figure 12.22 System model for the layered 2×3×3 X networks.

Assume the speed of acoustic wave is fixed for all links. Then the earlier distance relationship also presents that of the propagation delay. Denote the PD between any two nodes along the first and second hops as τ_{ki}^1 and τ_{jk}^2, respectively. Then the corresponding two PD matrices are

$$T_1 = \begin{bmatrix} \tau_{11}^1 & \tau_{21}^1 & \tau_{31}^1 \\ \tau_{12}^1 & \tau_{22}^1 & \tau_{32}^1 \end{bmatrix}, \quad T_2 = \begin{bmatrix} \tau_{11}^2 & \tau_{21}^2 & \tau_{31}^2 \\ \tau_{12}^2 & \tau_{22}^2 & \tau_{32}^2 \\ \tau_{13}^2 & \tau_{23}^2 & \tau_{33}^2 \end{bmatrix}.$$

Without loss of generality, we assume each message has the same length, which is equal to the period of one time slot. After normalization, we can simply set the earlier PD matrices with positive integer elements.

12.4.3 DoF-Achieving Scheme

12.4.3.1 DoF Result

First, we give the following lemma to show the DoF upper bound of the discussing $2\times3\times3$ X networks.

Lemma 1.

The upper bound on DoF of $2\times3\times3$ X networks is 3/2.

Proof. We know that for $M\times N$ X channel with single hop, its DoF is $\frac{MN}{M+N-1}$. Based on this result, the DoF of 2×3 X network is 3/2. Since adding relaying layer between the source and destination does not increase the capacity/DoF, the DoF of single-hop network provides an upper bound on that of two-hop layered network with the same number of transmitters and receivers. Thus, the DoF of 2×3 X network is an upper bound on the DoF of the discussing $2\times3\times3$ X network, which is 3/2.

Second, we summarize our main contribution as the following theorem.

Theorem 2.

The upper bound 3/2 on DoF of 2×3×3 X networks is achievable by specific PDAS

$$T_1 = \begin{bmatrix} \tau_1 & \tau_2 & \tau_2 \\ \tau_2 & \tau_1 & \tau_1 \end{bmatrix}, \quad T_2 = \begin{bmatrix} \tau_3 & \tau_4 & \tau_3 \\ \tau_4 & \tau_4 & \tau_3 \\ \tau_4 & \tau_3 & \tau_4 \end{bmatrix} \tag{12.64}$$

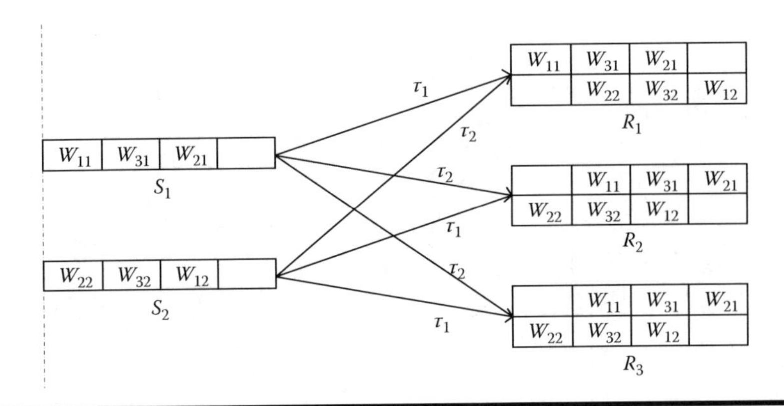

Figure 12.23 Message processing demonstration of the first hop for the proposed PDAS.

for the first and second hops, respectively.

Next, we provide the achievability proof as follows.

To achieve the DoF of 3/2, we further set that $\tau_1+1=\tau_2$ and $\tau_3+1=\tau_4$. With these settings, we can perform IA over the first hop and IN over the second hop, as shown later.

IA over the First Hop:

The main purpose of the relay layer is to perform IA and then choose some messages to forward. The IA over the first hop is demonstrated in Figure 12.23.

Source nodes S_1 and S_2 separately send three independent messages in the three successive time slots. Note both nodes begin to send simultaneously. The fourth time slot is kept silent to avoid further interference. With the earlier PD pattern, we can easily obtain the received message sequence as shown at each relay node.

With the receiving sequence, the relay nodes can choose proper messages to send over the second hop. From Figure 12.23, we can easily see that messages in time slots 1 and 4 can be directly decoded since there is no interference. On the other hand, although messages in time slots 2 and 3 are mixed together and cannot be separated, the mixed messages can be forwarded as a whole part over the second hop.

IN over the Second Hop:

For the second hop, relay nodes choose some messages to send in some time slots. Here we choose time slots 1 and 3 to be active and keep time slots 2 and 4 silent. The goal of this arrangement is to perform IN over the second hop, as shown in Figure 12.24. Desired messages can be correctly decoded without interference.

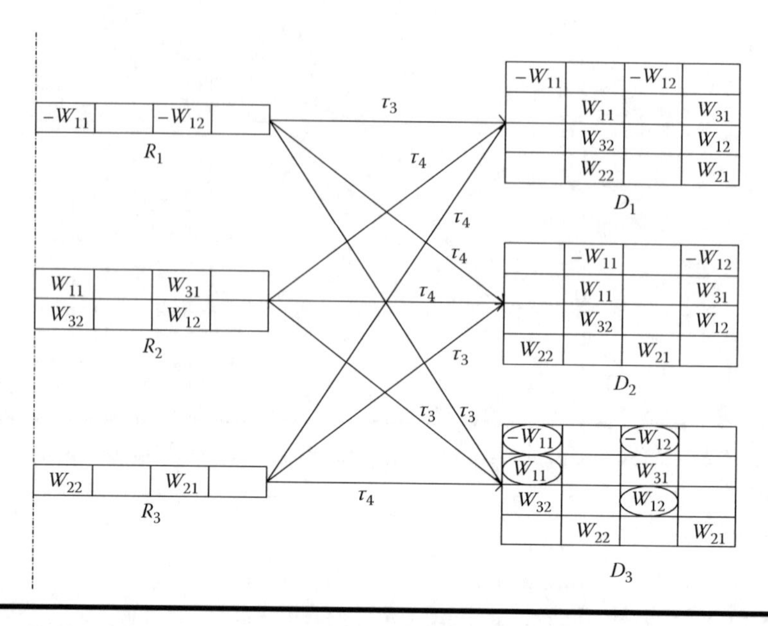

Figure 12.24 Message processing demonstration of the second hop for the proposed PDAS.

In summary, each destination node ($\mathcal{D}_1$, $\mathcal{D}_2$, and $\mathcal{D}_3$) obtains their desired two messages over four successive time slots, which indicates that the upper bound on DoF of 3/2 is achieved perfectly.

12.4.4 Feasible PDAS Solution

For simplification, we normalize the acoustic wave speed as 1; then, the distance matrix will be equal to the PDAS matrix.

According the PDAS matrix, we can obtain the corresponding distance matrix

$$D_1 = \begin{bmatrix} C_1 & C_2 & C_2 \\ C_2 & C_1 & C_1 \end{bmatrix}, \quad D_2 = \begin{bmatrix} C_3 & C_4 & C_3 \\ C_4 & C_4 & C_3 \\ C_4 & C_3 & C_4 \end{bmatrix}, \tag{12.65}$$

for the first and second hops, respectively. Here C_1–C_4 are a constant.

With the given pattern, we have relationships: $C_1+1 = C_2$ and $C_3+1= C_4$. To get the actual location coordinate, we make sure that $C_2 \neq C_3$. Assume the coordinates of the three relay nodes $\mathcal{R}_1$, $\mathcal{R}_2$, and $\mathcal{R}_3$ are $[a,0,0]$, $[0,0,0]$, and $[b,c,0]$, respectively. According to the distance relationship, we can find the position coordinates of all source, relay, and destination nodes.

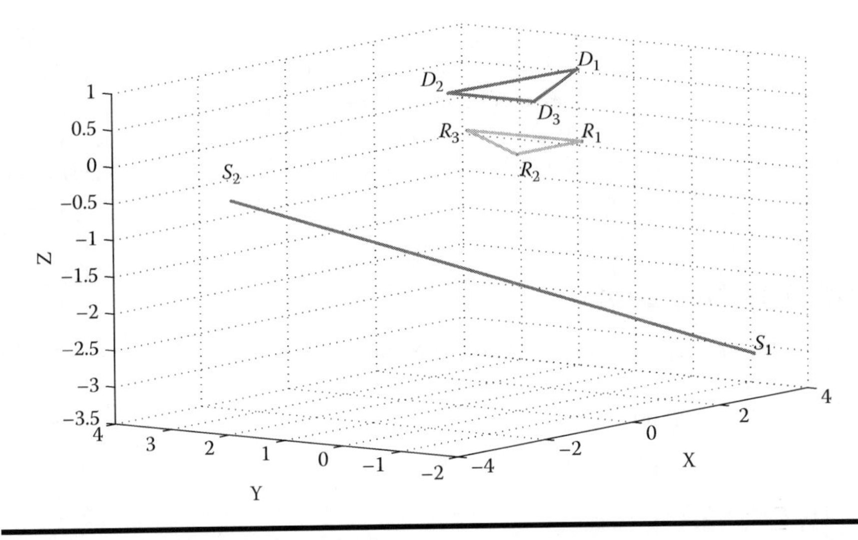

Figure 12.25 A feasible $2\times3\times3$ PDAS-based X network in the 3D Euclidean space.

The numerical computation can be turned over to computer. Here is a feasible example. If $C_1=4$, $C_2=5$, $C_3=1$, $C_4=2$, $a=1.5$, $b=1.5$, $c=2$, the coordinates of related nodes are given as follows:

$$\begin{aligned}
\mathcal{S}_1 &: [3.75,-1.25,-3.062],\\
\mathcal{S}_2 &: [-2.25,3.25,-0.612],\\
\mathcal{R}_1 &: [1.5,0,0],\\
\mathcal{R}_2 &: [0,0,0],\\
\mathcal{R}_3 &: [1.5,2,0],\\
\mathcal{D}_1 &: [1.75,0.25,0.9354],\\
\mathcal{D}_2 &: [0.75,1.75,0.6124],\\
\mathcal{D}_3 &: [0.75,0.25,0.6124].
\end{aligned} \tag{12.66}$$

Such an example is shown in Figure 12.25.

12.5 Conclusions

From the earlier introduction, we can see that the DoF-achieving PDAS/DAS schemes for IC/XC or X network can be feasible in the Euclidean space for acoustic communications. In fact, we can use the mathematical notation of cyclic codes as the cyclic IA in [24] for all related communication models, which provides a powerful framework for the analysis and design of PDAS schemes. The feasibility problem turns out to be the problem of node placement in the 3D Euclidean space. Interestingly, here algebra meets geometrics, which indicates the problem might become quite complicated.

Acknowledgments

This work was partially supported by the National Natural Science Foundation of China (61271283, 61472237) and the Innovation Program of Shanghai Municipal Education Commission (14YZ113).

References

1. R. H. Wiley and D. G. Richards, "Physical constraints on acoustic communication in the atmosphere: Implications for the evolution of animal vocalizations," *Behavioral Ecology & Sociobiology*, vol. 3, no. 1, pp. 69–94, 1978.
2. R. H. Wiley and D. G. Richards, "5 adaptations for acoustic communication in birds: Sound transmission and signal detection," *Acoustic Communication in Birds*, vol. 33, no. 5, pp. 131–181, 1982.
3. D. B. Kilfoyle and A. B. Baggeroer, "The state of the art in underwater acoustic telemetry: Special issue papers on underwater acoustic communication," *IEEE Journal of Oceanic Engineering*, vol. 25, no. 1, pp. 4–27, 2000.
4. H. C. Gerhardt and F. Huber, *Acoustic Communication in Insects and Anurans: Common Problems and Diverse Solutions*, Chicago: University of Chicago Press, 2002.
5. H. Brumm and H. Slabbekoorn, "Acoustic communication in noise," *Advances in the Study of Behavior*, vol. 35, pp. 151–209, 2005.
6. V. R. Cadambe and S. A. Jafar, "Interference alignment and degrees of freedom of the *K*-User interference channel," *IEEE Transactions on Information Theory*, vol. 54, no. 54, pp. 3425–3441, 2008.
7. M. A. Maddah-Ali, A. S. Motahari, and A. K. Khandani, "Signaling over MIMO multi-base systems: Combination of multi-access and broadcast schemes," in *IEEE International Symposium on Information Theory*, 2006, pp. 2104–2108.
8. T. Gou, S. A. Jafar, C. Wang, S. W. Jeon, and S. Y. Chung, "Aligned interference neutralization and the degrees of freedom of the 2×2×2 interference channel," *IEEE Transactions on Information Theory*, vol. 58, no. 7, pp. 4381–4395, 2010.
9. S. A. Jafar, "Interference alignment: A new look at signal dimensions in a communication network," *Foundations & Trends® in Communications & Information Theory*, vol. 7, no. 1, pp. 1–136, 2010.
10. S. W. Jeon and M. Gastpar, "A survey on interference networks: Interference alignment and neutralization," *Entropy*, vol. 14, no. 10, pp. 1842–1863, 2012.
11. S. Zhou and Z. Wang, *OFDM for Underwater Acoustic Communications*. Chichester: Wiley Publishing, 2014.

12. V. R. Cadambe and S. A. Jafar, "Interference alignment and the degrees of freedom of wireless *X* networks," *IEEE Transactions on Information Theory*, vol. 55, no. 9, pp. 3893–3908, 2009.
13. V. R. Cadambe and S. A. Jafar, "Degrees of freedom of wireless networks—What a difference delay makes," in *Asilomar Conference on Signals, Systems and Computers*, 2007, pp. 133–137.
14. V. R. Cadambe and S. A. Jafar, "Can 100 speakers talk for 30-minutes each in one room within one hour and with zero interference to each other's audience?" in *45th Annual Allerton Conference on Communications, Control and Computing*, 2007, p. 1141–1148.
15. V. R. Cadambe, S. A. Jafar, and S. Shamai, "Interference alignment on the deterministic channel and application to fully connected gaussian interference networks," *IEEE Transactions on Information Theory*, vol. 55, no. 1, pp. 269–274, 2009.
16. H. L. Grokop, N. C. David, and R. D. Yates, "Interference alignment for line-of-sight channels," *IEEE Transactions on Information Theory*, vol. 57, no. 9, pp. 5820–5839, 2011.
17. R. Mathar and M. Zivkovic, "How to position n transmitter-receiver pairs in n–1 dimensions such that each can use half of the channel with zero interference from the others," in *Global Telecommunications Conference, 2009. GLOBECOM*, 2009, pp. 1–4.
18. R. Mathar and G. Bocherer, "On spatial patterns of transmitter-receiver pairs that allow for interference alignment by delay," in *International Conference on Signal Processing and Communication Systems*, 2009, pp. 1–5.
19. H. Zhou and T. Ratnarajah, "A novel interference draining scheme for cognitive radio based on interference alignment," in *New Frontiers in Dynamic Spectrum, 2010 IEEE Symposium on*, 2010, pp. 1–6.
20. S. Jiang, F. Liu, and S. Jiang, "Distance-alignment based adaptive MAC protocol for underwater acoustic networks," in *the IEEE Wireless Communications and Networking Conference (WCNC2016)*, 2016, pp. 761–766.
21. S. Jiang, F. Liu, S. Jiang, and X. Geng, "A feasible distance aligned structure for underwater acoustic X networks with two receivers," *IEICE Transactions on Fundamentals of Electronics Communications & Computer Sciences*, vol. E100.A, no. 1, pp. 332–334, 2017.
22. F. Liu, S. Jiang, S. Jiang, and C. Li, "DoF achieving propagation delay aligned structure for $K \times 2$ X channels," *IEEE Communications Letters*, vol. 15, no. 4, p. 897–900, 2017.
23. Z. Bao, F. Liu, S. Jiang, and S. Jiang, "DoF-achieving distance-aligned structure for layered underwater acoustic $2 \times 3 \times 3$ X networks," in *ACM International Conference on Underwater Networks & Systems*, 2016, p. 39.
24. H. Maier, J. Schmitz, and R. Mathar, "Cyclic interference alignment by propagation delay," in *Communication, Control, and Computing*, 2012, pp. 1761–1768.
25. F. L. Blasco, F. Rossetto, and G. Bauch, "Time interference alignment via delay offset for long delay networks," *IEEE Transactions on Communications*, vol. 62, no. 2, pp. 590–599, 2011.
26. M. Chitre, M. Motani, and S. Shahabudeen, "A scheduling algorithm for wireless networks with large propagation delays," in *IEEE Oceans*, 2010, pp. 1–5.
27. M. Chitre, M. Motani, and S. Shahabudeen, "Throughput of networks with large propagation delays," *IEEE Journal of Oceanic Engineering*, vol. 37, no. 4, pp. 645–658, 2012.

28. S. Lmai, M. Chitre, C. Laot, and S. Houcke, "Throughput-maximizing transmission schedules for underwater acoustic multihop grid networks," *IEEE Journal of Oceanic Engineering*, vol. 40, no. 4, pp. 853–863, 2015.
29. S. Lmai, M. Chitre, C. Laot, and S. Houcke, "Throughput-efficient super-TDMA MAC transmission schedules in ad hoc linear underwater acoustic networks," *IEEE Journal of Oceanic Engineering*, vol. PP, no. 99, pp. 1–19, 2016.
30. H. Zeng, Y. T. Hou, Y. Shi, W. Lou, S. Kompella, and S. F. Midkiff, "Shark-IA: An interference alignment algorithm for multi-hop underwater acoustic networks with large propagation delays," *Proceedings of the International Conference on Underwater Networks & Systems*, 6, 2014, pp. 1–8.
31. H. Zeng, Y. Shi, Y. T. Hou, and W. Lou, "An analytical model for interference alignment in multi-hop MIMO networks," *IEEE Transactions on Mobile Computing*, vol. 15, no. 1, pp. 17–31, 2015.
32. F. Blasco, F. Rossetto, and G. Bauch, "Time interference alignment via delay offset for long delay networks," *IEEE Transactions on Communications*, vol. 62, no. 2, pp. 590–599, 2014.
33. N. Chirdchoo, W. S. Soh, and K. C. Chua, "MU-Sync: A time synchronization protocol for underwater mobile networks," in *The Workshop on Underwater Networks*, 2008, pp. 35–42.
34. F. Yuan, S. Chen, X. Lin, and E. Cheng, "Time-frequency synchronization method for multi-point underwater acoustic communication system," in *Control Conference*, 2010, pp. 2054–2061.
35. D. Zennaro, B. Tomasi, L. Vangelista, and M. Zorzi, "Light-sync: A low overhead synchronization algorithm for underwater acoustic networks," in *Oceans*, 2012, pp. 1–7.
36. H. Ramezani and G. Leus, "Localization packet scheduling for underwater acoustic sensor networks," *IEEE Journal on Selected Areas in Communications*, vol. 33, no. 7, pp. 1345–1356, 2015.
37. M. Chitre and W. S. Soh, "Reliable point-to-point underwater acoustic data transfer: To juggle or not to juggle?" *IEEE Journal of Oceanic Engineering*, vol. 40, no. 1, pp. 93–103, 2015.
38. S. A. Jafar and S. Shamai, "Degrees of freedom region of the MIMO *X* channel," *IEEE Transactions on Information Theory*, vol. 54, no. 1, pp. 151–170, 2008.
39. M. A. Maddah-Ali, A. S. Motahari, and A. K. Khandani, "Communication over MIMO X channels: Interference alignment, decomposition, and performance analysis," *IEEE Transactions on Information Theory*, vol. 54, no. 8, pp. 3457–3470, 2008.

Chapter 13

Propagation over Earth

Mohammad N. Abdallah and Tapan K. Sarkar
Syracuse University

Magdalena Salazar-Palma
Universidad Carlos III de Madrid

Contents

13.1 Introduction

Analysis of propagation over Earth is introduced for horizontal and vertical loop antennas. To simulate the electromagnetic propagation path loss in a cellular environment using the physical parameters related to the environment, a method of moments code based on the Sommerfeld Green's function to treat the imperfectly conducting planar ground has been used. The computer code used to perform this analysis is commercially available and is called *Analysis of Wire Antennas and Scatterers* (AWAS). In this code, the effect of the real imperfect and infinite ground plane is simulated using the Sommerfeld formulation. It is shown that the propagation path loss inside a cellular communication cell is first −30 dB per decade of distance and, later on, usually outside the cell, it is −40 dB per decade of distance between the transmitter and the receiver. This implies that the magnetic field decays first at a rate of $\rho^{-1.5}$ inside the cell and, later on, usually outside the cell, as ρ^{-2}, where ρ stands for the horizontal distance between the transmitter and the receiver. It is seen that near the transmitting antenna there is interference between the direct space wave and the field from the image produced by the imperfect ground, providing variation of the total field strength. This is often labeled as large-scale fading. This interference pattern stops at an approximate distance of $4H_{TR}H_{RX}/\lambda$ (H_{TR} represents the height of the transmitting antenna over the ground, H_{RX} represents the height of the receiving antenna over the ground, and λ is the wavelength of operation), then a monotonic decay of the field occurs with a slope of −30 dB/decade and continues approximately to a distance of $8H_{TR}H_{RX}/\lambda$ from where the far field of the antenna starts and the slope becomes roughly −40 dB/decade. Also, it is shown that the electrical properties of the ground has very little effect since, for a fixed height of the transmitting antenna radiating over an imperfect ground, the ground parameters do not change the nature of the distant magnetic fields, whereas near the antenna the shape of the interference pattern can be slightly different.

13.2 Point Charge Located on Top of a Perfect Electric Conductor

Let us start with a simple electrostatic problem. Assume that we have a point charge q in free space at a distance d meters above a grounded perfect electric conductor (PEC) plane located at $z = 0$, we can formulate the potential as follows:

$$v(x,y,z) = \frac{1}{4\pi\varepsilon_o}\left[\frac{q}{\sqrt{x^2+y^2+(z-d)^2}} + \iint_s \frac{\rho_s}{r}\,ds\right], \tag{13.1}$$

where r is the distance between ds and the point under consideration (x, y, z).

The problem is that we do not know ρ_s till we solve for $v(x, y, z)$. Even if we know ρ_s it is difficult to perform the integral on each ds on the infinite plane. For these reasons, we apply the image theory and try to solve for $v(x, y, z)$ from which we can calculate ρ_s.

If we regenerate an equivalent problem in the free space using image theory, which means adding charges in the region of no interest and keeping the region of interest (ROI) unchanged and assuming that all the space is filled of the same medium as the ROI, the added charges in the region of no interest along with the charges in the ROI have to satisfy the boundary conditions, which is in this case $v(x, y, z = 0) = 0$.

It turns out that the potential can be determined in the free space above the PEC ground plane if we assume that there is a charge $-q$, which is located at d meters below the $z = 0$ plane and assuming that there is no PEC ground so there is free space everywhere. In this case

$$v(x,y,z)=\frac{1}{4\pi\varepsilon_o}\left[\frac{q}{\sqrt{x^2+y^2+(z-d)^2}}-\frac{q}{\sqrt{x^2+y^2+(z+d)^2}}\right]. \tag{13.2}$$

This equivalent problem satisfies the boundary condition, namely $v(x, y, z = 0) = 0$, which means that this represents the potential in any point $v(x, y, z)$, where $z > 0$; i.e., ROI. Obviously there is no need to determine the potential in the PEC region of the original problem since the PEC is an equipotential surface, where $v = 0$ for all the points where $z < 0$.

Now we can easily find ρ_s,

$$\rho_s=\varepsilon_o E_{1z}(z=0)=-\varepsilon_o\frac{\partial v}{\partial z}(z=0)=-\frac{2qd}{4\pi(x^2+y^2+d^2)^{3/2}}$$

It is interesting to observe that $\iint_s \rho_s\, ds = -q$

where s covers the entire ground plane.

13.3 Point Charge on Top of a Dielectric Substrate

The previous problem was easy to analyze using the image theory since we knew the specific boundary condition namely $v(x, y, z = 0) = 0$, if we let the second medium to be dielectric with relative permittivity ε_r, then the new boundary conditions are the continuity of both the tangential component of the electric field **E** and the normal component of the electric flux **D**. We do not know the value of the tangential component of the electric field neither in the upper half space (free space) nor in the lower half space (the dielectric). However, using the image theory, we can construct two equivalent problems, one for the upper half space and another one for the lower half space and then equate the tangential component of the electric field in the upper half space at $z = 0$ to the tangential component of the electric field in the lower half space at $z = 0$. The normal component of the electric flux in the upper half space at $z = 0$ is equal to the normal component of the electric flux in the lower half space at $z = 0$. The equivalent problem for $z > 0$ can be generated by assuming that there is a charge $-q_2$ at a distance d meters below the $z = 0$ plane and that the whole space is a free space (no dielectric). The equivalent problem for $z < 0$ can be generated by assuming that there is another point charge $+q_1$ at d meters above the $z = 0$ plane along with the original point charge q at the same location and that the whole space is filled with the dielectric. Now, we can formulate the potential in both regions as follows:

$$v_1(x, y, z > 0) = \frac{1}{4\pi\varepsilon_o}\left[\frac{q}{\sqrt{x^2 + y^2 + (z-d)^2}} - \frac{q_2}{\sqrt{x^2 + y^2 + (z+d)^2}}\right] \tag{13.3}$$

$$v_2(x, y, z < 0) = \frac{1}{4\pi\varepsilon_o\varepsilon_r}\left[\frac{q + q_1}{\sqrt{x^2 + y^2 + (z-d)^2}}\right] \tag{13.4}$$

$$v_1(x, y, z = 0) = \frac{1}{4\pi\varepsilon_o}\left[\frac{q}{\sqrt{x^2 + y^2 + d^2}} - \frac{q_2}{\sqrt{x^2 + y^2 + d^2}}\right]$$

$$v_2(x, y, z = 0) = \frac{1}{4\pi\varepsilon_o\varepsilon_r}\left[\frac{q + q_1}{\sqrt{x^2 + y^2 + d^2}}\right].$$

Applying the appropriate boundary conditions we have:

$$v_1(x, y, z = 0) = v_2(x, y, z = 0) \Rightarrow q - q_2 = \frac{q + q_1}{\varepsilon_r}$$

$$\varepsilon_o \frac{\partial v_1}{\partial z}(z = 0) = \varepsilon_o\varepsilon_r \frac{\partial v_2}{\partial z}(z = 0) \Rightarrow q + q_2 = q + q_1 \Rightarrow q_2 = q_1$$

So

$$q_2 = q_1 = \frac{\varepsilon_r - 1}{\varepsilon_r + 1} q. \tag{13.5}$$

This completes the solutions for the potential for $z > 0$ and for $z < 0$.

The conclusion is that even in electrostatic problems things became very different when we changed the PEC into dielectric. This is also the case for antennas radiating over PEC ground plane versus antennas radiating over imperfect ground plane. The nature of added complexity is similar to the problem just addressed and provides a completely different physics and conclusions for both the problems.

13.4 A Vertical Hertzian Dipole on Top of a PEC

Next, consider a z-directed Hertzian dipole located at d meters on top of a PEC ground plane. The tangential component of the electric field at $z = 0$ has to be zero. We can construct an equivalent problem using image theory in which the new equivalent problem satisfies the same boundary conditions. This results in the introduction of another z-directed Hertzian dipole located at d meters below the $z = 0$ plane and assuming that the whole space is a free space and there is no PEC ground plane. The geometry of the equivalent problem is shown in Figure 13.1. This equivalent problem gives the correct fields for $z > 0$.

$$\mathbf{E} = j\omega\mu \frac{e^{-jkr_1}\sin\theta_1 Il}{4\pi r_1}\left(1+\frac{1}{jkr_1}+\frac{1}{(jkr_1)^2}\right)\hat{\mathbf{a}}_{\theta_1} + \frac{e^{-jkr_1}\cos\theta_1 Il\eta}{2\pi r_1}\left(\frac{1}{r_1} - j\frac{1}{kr_1^2}\right)\hat{\mathbf{a}}_{r_1}$$
$$+ j\omega\mu \frac{e^{-jkr_2}\sin\theta_2 Il}{4\pi r_2}\left(1+\frac{1}{jkr_2}+\frac{1}{(jkr_2)^2}\right)\hat{\mathbf{a}}_{\theta_2} + \frac{e^{-jkr_2}\cos\theta_2 Il\eta}{2\pi r_2}\left(\frac{1}{r_2} - j\frac{1}{kr_2^2}\right)\hat{\mathbf{a}}_{r_2} \tag{13.6}$$

$$r_1 = \sqrt{\rho^2 + (d - z)^2} \tag{13.7}$$

$$r_2 = \sqrt{\rho^2 + (d + z)^2} \tag{13.8}$$

$$\sin\theta_1 = \frac{\rho}{r_1}, \quad \cos\theta_1 = -\frac{(d - z)}{r_1}, \quad \sin\theta_2 = \frac{\rho}{r_2}, \quad \cos\theta_2 = \frac{(d + z)}{r_2}, \tag{13.9}$$

where ρ and z are the cylindrical coordinates of the field point.

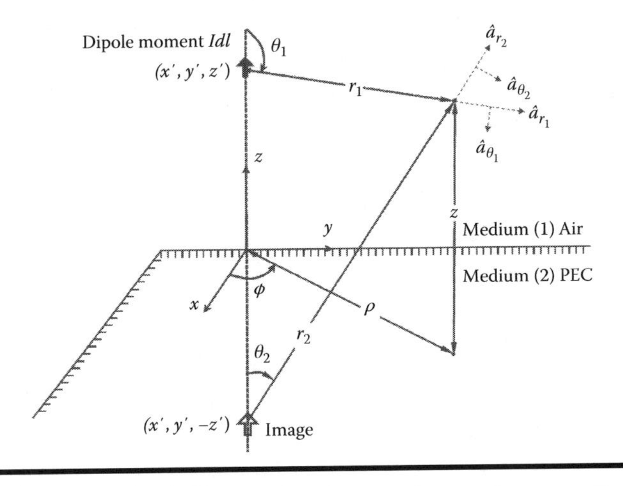

Figure 13.1 Equivalent problem using image theory for a vertical Hertzian dipole on top of the PEC ground plane.

We need to recognize that $\hat{a}_{\theta_1}$ and $\hat{a}_{\theta_2}$ are two different directions. Also, $\hat{a}_{r_1}$ and $\hat{a}_{r_2}$ are two different directions. We can simplify the mathematics by converting these into cylindrical coordinates.

$$\hat{a}_\theta = \hat{a}_\rho \cos\theta - \hat{a}_z \sin\theta$$

$$\hat{a}_r = \hat{a}_\rho \sin\theta + \hat{a}_z \cos\theta$$

If we apply all the previous equations, we will see clearly that, for the new equivalent problem, we have:

$$\begin{aligned} E_\rho &= j\omega\mu \frac{e^{-jkr_1}\sin\theta_1\cos\theta_1\, Il}{4\pi r_1}\left(1+\frac{1}{jkr_1}+\frac{1}{(jkr_1)^2}\right) \\ &+ j\omega\mu \frac{e^{-jkr_2}\sin\theta_2\cos\theta_2\, Il}{4\pi r_2}\left(1+\frac{1}{jkr_2}+\frac{1}{(jkr_2)^2}\right) \\ &+ \frac{e^{-jkr_1}\cos\theta_1\sin\theta_1\, Il\eta}{2\pi r_1}\left(\frac{1}{r_1}-j\frac{1}{kr_1^2}\right)+\frac{e^{-jkr_2}\cos\theta_2\sin\theta_2\, Il\eta}{2\pi r_2}\left(\frac{1}{r_2}-j\frac{1}{kr_2^2}\right) \end{aligned} \tag{13.10}$$

at $z = 0$, we have $r_1 = r_2$, $\sin\theta_1 = \sin\theta_2$, $\cos\theta_1 = -\cos\theta_2$. So we have zero tangential component of the electric field at $z = 0$, i.e., $E_\rho = 0$ at $z = 0$. This means that this equivalent problem has a closed form solution and gives the correct result for $z > 0$.

Also, we can formulate the z component of the electric field as follows:

$$\begin{aligned} E_z &= -j\omega\mu \frac{e^{-jkr_1}\sin^2\theta_1\, Il}{4\pi r_1}\left(1+\frac{1}{jkr_1}+\frac{1}{(jkr_1)^2}\right)+\frac{e^{-jkr_1}\cos^2\theta_1\, Il\eta}{2\pi r_1}\left(\frac{1}{r_1}-j\frac{1}{kr_1^2}\right) \\ &- j\omega\mu \frac{e^{-jkr_2}\sin^2\theta_2\, Il}{4\pi r_2}\left(1+\frac{1}{jkr_2}+\frac{1}{(jkr_2)^2}\right)+\frac{e^{-jkr_2}\cos^2\theta_2\, Il\eta}{2\pi r_2}\left(\frac{1}{r_2}-j\frac{1}{kr_2^2}\right) \end{aligned} \tag{13.11}$$

When $d + z \ll \rho$ and $kr \ll 1$ then,

$$\theta_1 = \theta_2 = \pi/2$$

$$r_1 = \rho\sqrt{1+\left(\frac{d-z}{\rho}\right)^2} \approx \rho\left(1+\frac{1}{2}\left(\frac{d-z}{\rho}\right)^2\right) = \rho+\frac{(d-z)^2}{2\rho}$$

$$r_2 = \rho\sqrt{1+\left(\frac{d+z}{\rho}\right)^2} \approx \rho\left(1+\frac{1}{2}\left(\frac{d+z}{\rho}\right)^2\right) = \rho+\frac{(d+z)^2}{2\rho}$$

$$E_z = -j\omega\mu \frac{Il}{4\pi\rho}\left(e^{-jk\left(\rho+\frac{(d-z)^2}{2\rho}\right)} + e^{-jk\left(\rho+\frac{(d+z)^2}{2\rho}\right)} \right)$$

$$= -j\omega\mu \frac{Il}{4\pi\rho}\left(e^{-jk\left(\rho+\frac{d^2-2dz+z^2}{2\rho}\right)} + e^{-jk\left(\rho+\frac{d^2+2dz+z^2}{2\rho}\right)} \right) \tag{13.12}$$

$$= -j\omega\mu \frac{Il}{4\pi\rho}\left(e^{-jk\left(\rho+\frac{d^2-2dz+z^2}{2\rho}\right)} + e^{-jk\left(\rho+\frac{d^2+2dz+z^2}{2\rho}\right)} \right)$$

$$= -j\omega\mu \frac{Il}{2\pi\rho} e^{-jk\left(\rho+\frac{d^2+z^2}{2\rho}\right)} cos\left(\frac{kdz}{\rho}\right)$$

From (13.12), we conclude that there will be fluctuations in the z component of the electric field. The locations of the nulls are given by

$$\cos\left(\frac{kdz}{\rho}\right) = 0 \Rightarrow \frac{2\pi}{\lambda}\frac{dz}{\rho} = m\frac{\pi}{2}$$

$$\rho = \frac{4dz}{m\lambda}, \quad m = 1, 3, 5, \ldots$$

The last null is given by

$$\rho = \frac{4dz}{\lambda}. \tag{13.13}$$

Also, we conclude that the z component of the electric far field decays as $1/\rho$, which is –20 dB/decade. We were able to get all these closed-form solutions for the problem of a vertical Hertzian dipole radiating on top of a PEC ground plane using the image theory.

13.5 A Horizontal Hertzian Dipole on Top of a PEC

Now, let us discuss an x-oriented Hertzian dipole radiating on top of a PEC ground plane ($z = 0$ plane) located at a height of d meters. We can use the image theory to generate an equivalent problem by assuming another x-oriented Hertzian dipole located at a height of $-d$ meters, which carries an opposite directed current with respect to the original x-oriented Hertzian dipole located at d meters above the ground plane. In this new equivalent problem, we remove the PEC ground plane and assume that the whole space is a free space. The geometry of the equivalent problem is shown in Figure 13.2. This equivalent problem gives the correct fields for $z > 0$. The tangential component of the electric field has to be zero at $z = 0$ for the new equivalent problem.

The fields at (ρ,ϕ,$z > 0$) is given by

$$E_\phi = \eta Il \frac{e^{-jkr_1}}{4\pi r_1}\left(jk + \frac{1}{r_1} + \frac{1}{jkr_1^2}\right)\sin\phi - \eta Il \frac{e^{-jkr_2}}{4\pi r_2}\left(jk + \frac{1}{r_2} + \frac{1}{jkr_2^2}\right)\sin\phi, \tag{13.14}$$

where r_1 and r_2 are given by (13.7) and (13.8), respectively.

We see that, at $z = 0$, we have $r_1 = r_2$. This means that the tangential component of the electric field at $z = 0$ is zero, i.e., $E_\phi = 0$ at $z = 0$.

When $d + z << \rho$ and $kr << 1$ then, as illustrated in the previous section, we get:

$$E_\phi = -\frac{jk\eta Il \sin\phi}{4\pi\rho} e^{-jk\left(\rho + \frac{d^2+z^2}{2\rho}\right)} \sin\left(\frac{kdz}{\rho}\right). \tag{13.15}$$

From (13.15), we conclude that there will be fluctuations in the phi component of the electric field. The locations of the nulls are given by

$$\sin\left(\frac{kdz}{\rho}\right) = 0 \Rightarrow \frac{2\pi}{\lambda}\frac{dz}{\rho} = m\pi$$

$$\rho = \frac{2dz}{m\lambda}, \quad m = 1,2,3,\ldots$$

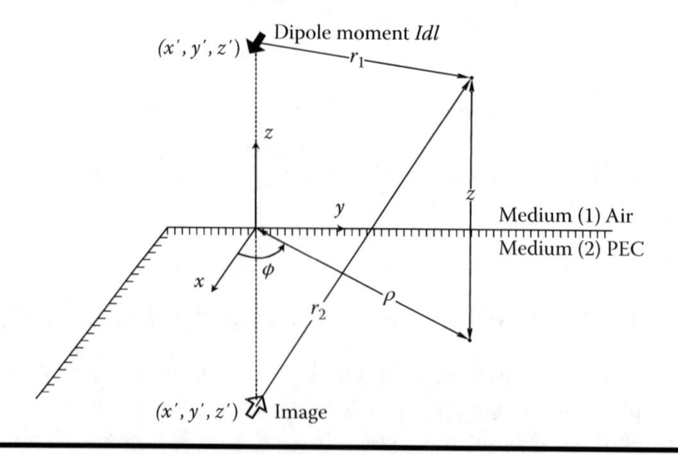

Figure 13.2 Equivalent problem using image theory for a horizontal Hertzian dipole on top of the PEC ground plane.

The last null is given by

$$\rho = \frac{2dz}{\lambda} \tag{13.16}$$

Also, we conclude that the ϕ component of the electric far field decays as $1/\rho^2$ (since $\sin x \approx x$ for small x) which is −40 dB/decade. We were able to get all these closed-form solutions for the problem of a horizontal Hertzian dipole radiating on top of a PEC ground plane using the image theory.

Solutions become more complicated when we have the Hertzian dipole radiating on top of an imperfect ground plane, since applying the image theory is a little complicated. Remember that even for the electrostatic problem it was not easy when we had a dielectric ground instead of a PEC. Sommerfeld was the first one who tackled this problem of a Hertzian dipole radiating on top of an imperfect ground plane.

13.6 Green's Function for a Vertical Hertzian Dipole Operating over an Imperfect Ground Plane

In the original Sommerfeld formulation, one considers an elementary dipole of moment Idz oriented along the z-direction and located at (x', y', z'). The dipole is situated over an imperfect ground plane characterized by a complex relative dielectric constant ε as seen in Figure 13.3. The complex relative dielectric constant is given by $\varepsilon = \varepsilon_r - \dfrac{j\sigma}{\omega\,\varepsilon_0}$, where ε_r represents the relative permittivity of the medium, ε_0 is the permittivity of vacuum, σ is the conductivity of the medium, ω stands

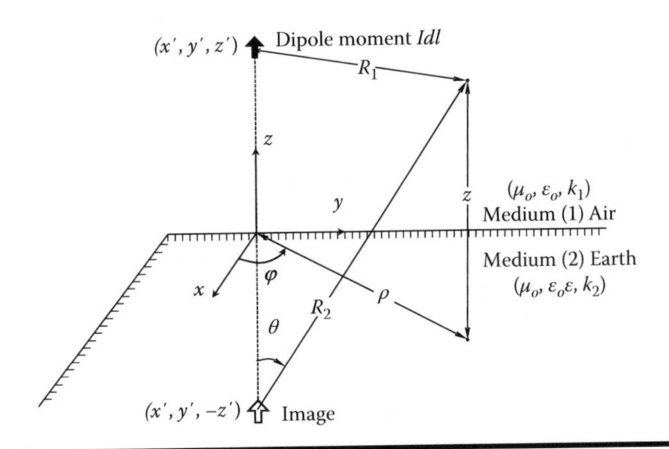

Figure 13.3 Vertical Hertzian dipole on top of an imperfect ground plane.

for the angular frequency, and j is the imaginary unit, i.e., $j=\sqrt{-1}$. It is possible to formulate a solution to the problem of radiation from the dipole operating in the presence of the imperfect ground in terms of a single Hertzian vector potential $\hat{\mathbf{a}}_z\Pi_z$ of the electric type, where $\hat{\mathbf{a}}_z$ stands for the unit vector in the positive z direction. A time variation of $\exp(j\omega t)$ is assumed throughout the analysis, where t is the time variable.

The Hertzian vector potential $\hat{\mathbf{a}}_z\Pi_z$ in this case satisfies the wave equation

$$\left(\nabla^2+k_1^2\right)\Pi_{1z}=\frac{-I\,dz}{j\omega\,\varepsilon_0}\delta(x-x')\delta(y-y')\delta(z-z') \tag{13.17}$$

$$\left(\nabla^2+k_2^2\right)\Pi_{2z}=0 \tag{13.18}$$

where

$$k_1^2=\omega^2\mu_0\varepsilon_0 \tag{13.19}$$

$$k_2^2=\omega^2\mu_0\varepsilon_0\varepsilon \tag{13.20}$$

and δ represents the delta function. The primed and unprimed coordinates are for the source and field points, respectively. The subscript 1 denotes the upper half space which is air and the subscript 2 denotes the lower half space which is the imperfectly conducting earth characterized by a complex relative dielectric constant ε. The electric and the magnetic field vectors are derived from the Hertzian vector potential using

$$\vec{E}_i=\vec{\nabla}\left(\vec{\nabla}\bullet\overline{\Pi_i}\right)+k_i^2\overline{\Pi_i} \tag{13.21}$$

and

$$\overline{\mathrm{H}}_i=j\omega\varepsilon_0\varepsilon_i\left(\vec{\nabla}\times\overline{\Pi_i}\right) \tag{13.22}$$

respectively, with $i=1, 2$.

In medium 1, $\varepsilon_1=1$, and for medium 2, $\varepsilon_2=\varepsilon$. The propagation constants in medium 1 and 2, called k_1 and k_2 respectively, are related by $\frac{k_2}{k_1}=\sqrt{\varepsilon}$. At the interface $z=0$, the tangential electric and magnetic field components must be continuous, conditions which in terms of the Hertzian vector potential components can be written as

$$\frac{\partial\Pi_{1z}}{\partial y}=\varepsilon\frac{\partial\Pi_{2z}}{\partial y} \tag{13.23a}$$

Equation (13.23a) represents the continuity of H_y.

$$\frac{\partial \Pi_{1z}}{\partial x} = \varepsilon \frac{\partial \Pi_{2z}}{\partial x} \tag{13.23b}$$

Equation (13.23b) represents the continuity of H_x.

$$\frac{\partial}{\partial y}\left(\frac{\partial \Pi_{1z}}{\partial z}\right) = \frac{\partial}{\partial y}\left(\frac{\partial \Pi_{2z}}{\partial z}\right) \tag{13.23c}$$

Equation (13.23c) represents the continuity of E_y.

$$\frac{\partial}{\partial x}\left(\frac{\partial \Pi_{1z}}{\partial z}\right) = \frac{\partial}{\partial x}\left(\frac{\partial \Pi_{2z}}{\partial z}\right) \tag{13.23d}$$

Equation (13.23d) represents the continuity of E_x.

Since all the boundary conditions must hold at $z = 0$ for all x and y, the x and y dependence of the fields on either side of the interface must be the same. Therefore,

$$\Pi_{1z} = \varepsilon \Pi_{2z} \tag{13.24a}$$

$$\frac{\partial \Pi_{1z}}{\partial z} = \frac{\partial \Pi_{2z}}{\partial z} \tag{13.24b}$$

The complete solutions for the Hertz vector potentials satisfying the wave equations (13.17) and (13.18) and the boundary conditions (13.24) have been derived by many researchers over the last century. An incomplete partial list [1–5] that will be important to our discussions is provided starting with Sommerfeld [1]. The solutions of the Hertz potentials are

$$\Pi_{1z} = P\left[\frac{\exp(-jk_1R_1)}{R_1} + \int_0^\infty \frac{J_0(\lambda\rho)}{\sqrt{\lambda^2 - k_1^2}} \frac{\varepsilon\sqrt{\lambda^2 - k_1^2} - \sqrt{\lambda^2 - k_2^2}}{\varepsilon\sqrt{\lambda^2 - k_1^2} + \sqrt{\lambda^2 - k_2^2}} \right.$$
$$\left. \times \exp\left(-\sqrt{\lambda^2 - k_1^2}\,(z + z')\right)\lambda\, d\lambda\right] \tag{13.25}$$

and

$$\Pi_{2z} = 2P\int_0^\infty \frac{J_0(\lambda\rho)\exp\left(\sqrt{\lambda^2 - k_2^2}\,z - \sqrt{\lambda^2 - k_1^2}\,z'\right)}{\varepsilon\sqrt{\lambda^2 - k_1^2} + \sqrt{\lambda^2 - k_2^2}}\lambda\, d\lambda \tag{13.26}$$

for Real$\left[\sqrt{\lambda^2 - k_{1,2}^2}\right] > 0$. Here

$$P = \frac{I\,dz}{j\omega 4\pi \varepsilon_0} \tag{13.27}$$

$$\rho = \sqrt{(x - x')^2 + (y - y')^2} \tag{13.28}$$

$$R_1 = \sqrt{\rho^2 + (z - z')^2} \tag{13.29}$$

and λ is the variable of integration. For Π_{1z}, the first term inside the brackets can be interpreted as the particular solution or the direct line-of-sight contribution from the dipole antenna source, i.e., a spherical wave or a direct wave originating from the source and reaching the observation point, and the second term can be interpreted as the complementary solution or a reflection term (reflection from the imperfect ground plane). This second term in the potential, Π_{1z}, is responsible for the fields of the *ground wave*, as per IEEE Standard Definitions of Terms for Radio Wave Propagation [6]. Observe in (25) that the second term of this potential is the strongest one near the surface of the earth and exponentially decays as we go away from the interface.

Similarly, the solution for Π_{2z} can be interpreted as a partial transmission of the wave from medium 1 into medium 2. With these thoughts in mind, the potential Π_{1z} can be split up into two terms, or equivalently, the potentials responsible for the direct and the ground wave as

$$\Pi_{1z} = \Pi_{1z}^{\text{direct}} + \Pi_{1z}^{\text{reflected}} = P(g_0 + g_s) \tag{13.30}$$

where

$$\begin{aligned}\Pi_{1z}^{\text{direct}} &= \frac{P\exp(-jk_1R_1)}{R_1} \\ &= Pg_0\end{aligned} \tag{13.31}$$

$$\begin{aligned}\Pi_{1z}^{\text{reflected}} &= P\int_0^{\infty}\left(\frac{\varepsilon\sqrt{\lambda^2 - k_1^2} - \sqrt{\lambda^2 - k_2^2}}{\varepsilon\sqrt{\lambda^2 - k_1^2} + \sqrt{\lambda^2 - k_2^2}}\right)\frac{J_0(\lambda\rho)\exp\left[-\sqrt{\lambda^2 - k_1^2}\,(z + z')\right]}{\sqrt{\lambda^2 - k_1^2}}\lambda d\lambda \\ &= Pg_s\end{aligned} \tag{13.32}$$

A physical explanation to the two components of the Hertz potential Π_{1z} can now be given. The first one Π_{1z}^{direct} can be explained as a spherical wave originating from the source dipole. This term is easy to deal with. The difficult problem lies in the evaluation of $\Pi_{1z}^{\text{reflected}}$. Therefore, we chose to interpret $\Pi_{1z}^{\text{reflected}}$ as a superposition of plane waves resulting from the reflection of the various plane waves into which a spherical wave from the image point can be expanded. This arises from the identity

$$\frac{\exp(-jk_1R_2)}{R_2} = \int_0^\infty \frac{J_0(\lambda\rho)\exp\left[-\sqrt{\lambda^2-k_1^2}\,(z+z')\right]}{\sqrt{\lambda^2-k_1^2}}\lambda\,d\lambda \tag{13.33}$$

and

$$R_2 = \sqrt{\rho^2+(z+z')^2} \tag{13.34}$$

The term under the integral sign in (13.33) can be recognized as a multiple plane wave decomposition of the spherical wave source. Upon reflection of the plane waves from the dipole source as expressed in $\Pi_{1z}^{\text{reflected}}$, the amplitude of each wave must be multiplied by the reflection coefficient $R(\lambda)$. The complex reflection coefficient $R(\lambda)$ takes into account the phase change as the wave travels from the source (x', y', z') to the boundary and then to the point of observation (x, y, z). The reflection coefficient $R(\lambda)$ is then defined as the term inside the brackets in (13.32) as

$$R(\lambda) = \frac{\varepsilon\sqrt{\lambda^2-k_1^2}-\sqrt{\lambda^2-k_2^2}}{\varepsilon\sqrt{\lambda^2-k_1^2}+\sqrt{\lambda^2-k_2^2}} \tag{13.35}$$

where the semi-infinite integral over λ in $\Pi_{1z}^{\text{reflected}}$ takes into account all the possible plane waves. As $\varepsilon \to \infty$, i.e., a perfect conductor for the earth, then g_s of (13.32) reduces to (13.33) and represents a simple spherical wave originating at the image point. The reflection coefficient takes into account the effects of the ground plane in all the wave decomposition of the spherical wave and sums it up as a ray originating from the image of the source dipole but multiplied by a specular reflection coefficient $\Gamma(\theta)$, where $\Gamma(\theta) = \dfrac{\varepsilon\cos\theta - \sqrt{\varepsilon - \sin^2\theta}}{\varepsilon\cos\theta - \sqrt{\varepsilon-\sin^2\theta}}$.

The crux of the problem lies in the characterization of the various branch points and singularities associated with (13.35). The first point to observe is that the second term of the Hertz potential denoted by a complex integral and particularly $R(\lambda)$ has four branch points located at $\pm k_1$ and $\pm k_2$. Associated with these branch points are four branch cuts, and this give rise to four Riemann sheets. On the four Riemann sheets, the following conditions are satisfied:

Sheet 1: Real $(\sqrt{\lambda^2-k_1^2}) > 0$ and Real $(\sqrt{\lambda^2-k_2^2}) > 0$
Sheet 2: Real $(\sqrt{\lambda^2-k_1^2}) < 0$ and Real $(\sqrt{\lambda^2-k_2^2}) > 0$

Sheet 3: Real $(\sqrt{\lambda^2 - k_1^2}) > 0$ and Real $(\sqrt{\lambda^2 - k_2^2}) < 0$
Sheet 4: Real $(\sqrt{\lambda^2 - k_1^2}) < 0$ and Real $(\sqrt{\lambda^2 - k_2^2}) < 0$

Sheet 1 is the proper Riemann sheet. Now, the function described in (13.35) has two zeros corresponding to the zeros of the numerator, usually called Brewster zeros, and two poles corresponding to the zeros of the denominator, usually called the surface wave poles. The zeros and poles occur exactly at the same location $\lambda_S = \pm \frac{k_1 k_2}{\sqrt{k_1^2 + k_2^2}}$. However, on some Riemann sheets they appear as poles and on other Riemann sheets as zeros. So, the four poles and zeros are distributed on the four Riemann sheets. The most confusing stuff is that on the proper Riemann sheet whether it will be a pole or zero depends on the value of the dielectric constant ε. For example, examining the denominator of $R(\lambda)$, one can observe that for real values of ε on the proper Riemann sheet, there is a zero at λ_S, whereas for a complex value of the dielectric constant ε, that zero becomes a pole on the proper Riemann sheet. Therefore, unless one specifies the values for the dielectric constant and chooses the proper Riemann sheet, it is difficult to know whether the pole will occur or not in the presentation of (13.35).

The zero of the reflection coefficient in (13.35) illustrates the Brewster's phenomenon (i.e., the wave goes into the second medium for a particular angle of incidence without reflecting any energy) and a pole for the reflection coefficient illustrates the presence of a surface wave (i.e., a wave propagating close to the interface). Also, in general, it is difficult to distinguish between a Zenneck wave and a surface wave as both decay exponentially as one moves away from the planar interface and the wave propagates with a low loss along the radial direction [5]. In addition, it is well known that the Brewster's angle, which illustrates that a wave will penetrate into the second medium without reflection, is independent of frequency, whereas the surface wave phenomenon is highly dependent on frequency. As the frequency increases, the fields of the wave are more confined to the planar boundary. In addition a surface wave does not radiate, whereas a Zenneck wave does. These points have been illustrated in [5].

In the original Sommerfeld formulation [1], there was no error in the sign, but the presentation by Sommerfeld was not complete. First of all Sommerfeld demonstrated in his expression that there was a pole and he wrote the partial solution—the contribution from the pole and associated with that some physical properties. However, as has been pointed out in [4,5,7–9] that, when $\varepsilon_r > 0$, the pole should lie within the circle defined by the locus whose center is at $\lambda = 0$ and of radius k_1. In addition, when the integration along the branch cut is evaluated using the saddle point method, part of this branch cut integral actually cancels the contribution from the pole, and so in the final solution, the effect of the pole is not seen! This has been explicitly demonstrated in [10], and so the conjecture that there was an error in sign in Sommerfeld's formulation as initially intimated by Norton is a myth! The fact that the pole does not contribute to the total solution makes sense, as in some cases the poles may not exist on the proper Riemann sheet, as we have seen particularly for real values of the dielectric constant. Also, when the hyperbolic branch cuts in the Sommerfeld's solution are replaced by vertical branch cuts, the poles migrate into a different Riemann sheet and are not relevant [11,12].

13.7 Accurate Numerical Evaluation of the Fields near an Earth–Air Interface

We will use a purely numerical methodology, which evaluates the integrals presented in (13.32) in an essentially very-accurate way. There have been several methodologies and user-friendly codes that have been published in the literature to numerically compute the fields from a transmitting to a receiving antenna using the exact formulation, and commercial codes are available to perform these computations. To find the total solution, one needs to solve for the correct current distribution on the transmitting antenna as it is radiating in the presence of the ground and then uses this current distribution to compute the radiated fields in various ROIs.

The electric field $\vec{E}_A$ in medium 1 due to a $\vec{z}$ directed current element I_A of length ℓ_A radiating over Earth is given by

$$\vec{E}_A = \left[\vec{z}\, k^2 + \vec{\nabla} \frac{\partial}{\partial z} \right]$$

$$\Pi_{1z} = \frac{-j\omega\,\mu_0}{4\pi} \int_{\ell A} \vec{z}\, I_A \left(g_o + g_s \right) dz'_A + \frac{\vec{\nabla}}{j\omega\; 4\pi\varepsilon_0} \frac{\partial}{\partial z} \int_{\ell A} I_A \left(g_o + g_s \right) dz'_A \, , \quad (13.36)$$

where g_0 and g_1 are defined in (13.31) and (13.32), respectively and $\vec{\nabla}$ is the gradient operator. Since $\frac{\partial g_0}{\partial z} = -\frac{\partial g_0}{\partial z'_A}$ and $\frac{\partial g_s}{\partial z} = \frac{\partial g_s}{\partial z'_A}$, then assuming that the current goes to zero at the ends of the open wires, the derivative $\frac{\partial}{\partial z}$ operation on $(g_o + g_s)$ can be transformed to $\frac{d}{dz'_A}$ now operating on I_A instead, by applying integration by parts. It is important to note that only one of the derivatives can be interchanged with the integral in the second term in (13.36). Hence,

$$\vec{E}_A = \frac{-j\omega\,\mu_0}{4\pi} \int_{\ell A} \vec{z}\, I_A \left(g_o + g_s \right) dz'_A + \frac{\vec{\nabla}}{j\omega\, 4\pi\varepsilon_0} \int_{\ell A} \frac{dI_A}{dz'_A} \left(g_o - g_s \right) dz'_A \quad (13.37)$$

At this point, it is important to remember because of the nature of the singularity of the Green's function, the gradient operator on the second term cannot be interchanged with the integral sign. Because, if the gradient operator is interchanged with the integral sign and it operates on the Green's functions, then it would result in a divergent integral.

The mutual impedance Z_{BA} between two $\vec{z}$ directed current elements I_A and I_B of lengths ℓ_A and ℓ_B is expressed as

$$\begin{aligned} Z_{BA} &= -\int_{\ell_B} \vec{E}_A \bullet \vec{I}_B \, dz'_B \\ &= \frac{j\omega\mu_0}{4\pi} \int_{\ell_B} I_B \, dz'_B \int_{\ell_A} I_A \left(g_o + g_s\right) dz'_A \\ &\quad - \frac{1}{j\omega 4\pi\varepsilon_0} \int_{\ell_B} I_B \frac{\partial}{\partial z'_B} \left[\int_{\ell_A} \frac{dI_A}{dz'_A} \left(g_o - g_s\right) dz'_A \right] dz'_B . \end{aligned} \tag{13.38}$$

Transferring the derivative operation on I_B in the second term of the earlier expression and assuming I_B to go to zero at the open ends of the wires, one obtains through integration by parts,

$$\begin{aligned} Z_{BA} &= \frac{j\omega\mu_0}{4\pi} \int_{\ell_B} I_B \, dz'_B \int_{\ell_A} I_A \left(g_o + g_s\right) dz'_A \\ &\quad + \frac{1}{j\omega 4\pi\varepsilon_0} \int_{\ell_B} \frac{dI_B}{dz'_B} \left[\int_{\ell_A} \frac{dI_A}{dz'_A} \left(g_o - g_s\right) dz'_A \right] dz'_B \end{aligned} \tag{13.39}$$

Observe that the Green's function in (13.39) is never differentiated, and therefore, one can obtain a stable accurate solution using this methodology.

13.8 Green's Function for a Horizontal Hertzian Dipole on Top of an Imperfect Ground Plane

Here we consider an elementary dipole of moment Idx oriented along the x-direction and located at (x', y', z'). The dipole is situated over an imperfect ground plane characterized by a complex relative dielectric constant ε as seen in Figure 13.4. The complex relative dielectric constant is given by $\varepsilon = \varepsilon_r - \frac{j\sigma}{\omega\varepsilon_0}$, where ε_r represents the relative permittivity of the medium, ε_o is the permittivity of vacuum, σ is the conductivity of the medium, ω stands for the angular frequency, and j is the imaginary unit, i.e., $j = \sqrt{-1}$.

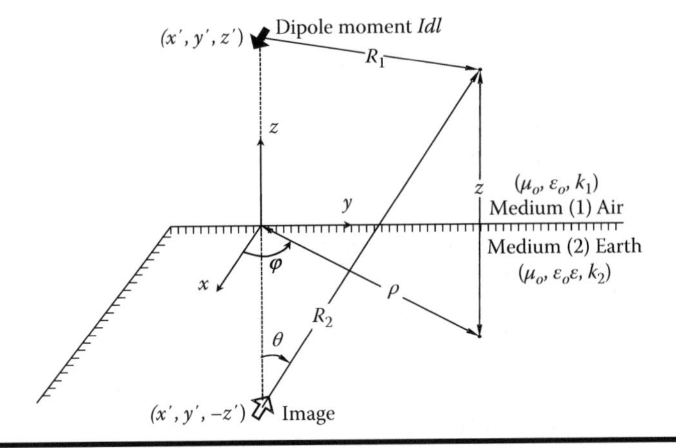

Figure 13.4 Horizontal Hertzian dipole on top of an imperfect ground plane.

It is possible to formulate a solution to the problem of radiation from the dipole operating in the presence of the imperfect ground in terms of a Hertzian vector potential. The Hertzian vector potential in this case satisfies the wave equation

$$\left(\nabla^2 + k_1^2\right)\vec{\Pi}_1 = \frac{-I\,dx}{j\omega\varepsilon_0}\delta\left(x-x'\right)\delta\left(y-y'\right)\delta\left(z-z'\right) \tag{13.40}$$

$$\left(\nabla^2 + k_2^2\right)\vec{\Pi}_2 = 0, \tag{13.41}$$

where

k_1 is given in (13.19) and k_2 is given in (13.20).

and δ represents the delta function. The electric and the magnetic field vectors are derived from the Hertzian vector potential using (13.21) and (13.22), respectively.

At the interface $z = 0$, the tangential electric and magnetic field components must be continuous, conditions which in terms of the Hertzian vector potential components can be written as

$$k_1^2\Pi_{1x} + \frac{\partial}{\partial x}\left(\frac{\partial \Pi_{1x}}{\partial x} + \frac{\partial \Pi_{1z}}{\partial z}\right) = k_2^2\Pi_{2x} + \frac{\partial}{\partial x}\left(\frac{\partial \Pi_{2x}}{\partial x} + \frac{\partial \Pi_{2z}}{\partial z}\right) \tag{13.42a}$$

Equation (13.42a) represents the continuity of E_x.

$$\frac{\partial}{\partial y}\left(\frac{\partial \Pi_{1x}}{\partial x} + \frac{\partial \Pi_{1z}}{\partial z}\right) = \frac{\partial}{\partial y}\left(\frac{\partial \Pi_{2x}}{\partial x} + \frac{\partial \Pi_{2z}}{\partial z}\right) \tag{13.42b}$$

Equation (13.42b) represents the continuity of E_y, this must hold at $z = 0$ for all x and y, the x and y dependence of the fields on either side of the interface must be the same. Therefore,

$$\frac{\partial \Pi_{1x}}{\partial x} + \frac{\partial \Pi_{1z}}{\partial z} = \frac{\partial \Pi_{2x}}{\partial x} + \frac{\partial \Pi_{2z}}{\partial z} \tag{13.43a}$$

$$k_1^2 \Pi_{1x} = k_2^2 \Pi_{2x} \tag{13.43b}$$

$$\frac{\partial \Pi_{1z}}{\partial y} = \varepsilon \frac{\partial \Pi_{2z}}{\partial y} \tag{13.44a}$$

Equation (13.44a) represents the continuity of H_y.

$$\left(\frac{\partial \Pi_{1x}}{\partial z} - \frac{\partial \Pi_{1z}}{\partial x} \right) = \varepsilon \left(\frac{\partial \Pi_{2x}}{\partial z} - \frac{\partial \Pi_{2z}}{\partial x} \right) \tag{13.44b}$$

Equation (13.44b) represents the continuity of H_x.

Since all the boundary conditions must hold at $z = 0$ for all x and y, the x and y dependence of the fields on either side of the interface must be the same. Therefore,

$$\Pi_{1z} = \varepsilon \Pi_{2z} \tag{13.45a}$$

$$\frac{\partial \Pi_{1x}}{\partial z} = \varepsilon \frac{\partial \Pi_{2x}}{\partial z} \tag{13.45b}$$

The complete solutions for the Hertz vector potentials satisfying the wave equations (13.40) and (13.41) and the boundary conditions (13.43) and (13.45) are

$$\Pi_{1x} = P' \left[\frac{\exp(-jk_1 R_1)}{R_1} + \int_0^\infty \frac{J_0(\lambda \rho)}{\sqrt{\lambda^2 - k_1^2}} \frac{\sqrt{\lambda^2 - k_1^2} - \sqrt{\lambda^2 - k_2^2}}{\sqrt{\lambda^2 - k_1^2} + \sqrt{\lambda^2 - k_2^2}} \times \exp\left(-\sqrt{\lambda^2 - k_1^2}\,(z + z') \right) \lambda d\lambda \right] \tag{13.46}$$

and

$$\Pi_{2x} = 2P' \frac{1}{\varepsilon} \int_0^\infty \frac{J_0(\lambda \rho) \exp\left(\sqrt{\lambda^2 - k_2^2}\, z - \sqrt{\lambda^2 - k_1^2}\, z' \right)}{\sqrt{\lambda^2 - k_1^2} + \sqrt{\lambda^2 - k_2^2}} \lambda d\lambda \tag{13.47}$$

$$\Pi_{1z} = 2P' \frac{\partial}{\partial x} \int_0^\infty J_0(\lambda\rho) \frac{\sqrt{\lambda^2 - k_1^2} - \sqrt{\lambda^2 - k_2^2}}{k_2^2\sqrt{\lambda^2 - k_1^2} + k_1^2\sqrt{\lambda^2 - k_2^2}} \exp\left(-\sqrt{\lambda^2 - k_1^2}\,(z + z')\right)\lambda d\lambda$$

$$\triangleq P' \frac{\partial}{\partial x}(g_2) \tag{13.48}$$

$$\Pi_{2z} = 2P' \frac{1}{\varepsilon} \frac{\partial}{\partial x} \int_0^\infty J_0(\lambda\rho) \exp\left(\sqrt{\lambda^2 - k_2^2}\,z - \sqrt{\lambda^2 - k_1^2}\,z'\right)$$

$$\times \frac{\sqrt{\lambda^2 - k_1^2} - \sqrt{\lambda^2 - k_2^2}}{k_2^2\sqrt{\lambda^2 - k_1^2} + k_1^2\sqrt{\lambda^2 - k_2^2}} \lambda d\lambda \tag{13.49}$$

for Real$\left[\sqrt{\lambda^2 - k_{1,2}^2}\right] > 0$. Here

$$P' = \frac{Idx}{j\omega 4\pi\varepsilon_0} \tag{13.50}$$

ρ is given in (13.28) and R_1 is given in (13.29) and λ is the variable of integration.

The potential Π_{1x} can be split up into two terms, or equivalently, the potentials responsible for the direct and the ground wave as

$$\Pi_{1x} = \Pi_{1x}^{\text{direct}} + \Pi_{1x}^{\text{reflected}} = P'(g_0 - g_1 + g_s) \tag{13.51}$$

where

$$\Pi_{1x}^{\text{direct}} = P' \exp(-jk_1R_1)/R_1 = P'\, g_0 \tag{13.52}$$

$$\Pi_{1x}^{\text{reflected}} = P'\left[-\exp(-jk_1R_2)/R_2 + 2\int_0^\infty \frac{J_0(\lambda\rho)\exp\left[-\sqrt{\lambda^2 - k_1^2}\,(z + z')\right]\lambda d\lambda}{\sqrt{\lambda^2 - k_1^2} + \sqrt{\lambda^2 - k_2^2}}\right]$$

$$= P'(-g_1 + g_s) \tag{13.53}$$

13.9 Accurate Numerical Evaluation of the Fields near an Earth–Air Interface

The electric field $\vec{E}_A$ in medium 1 due to a $\vec{x}$ directed current element I_A of length ℓ_A radiating over Earth is given by

$$\vec{E}_A = \left[k_1^2 + \vec{\nabla}\vec{\nabla}\bullet\right]\vec{\Pi}_1 = k_1^2\left(\Pi_{1x}\hat{\mathbf{a}}_x + \Pi_{1z}\hat{\mathbf{a}}_z\right) + \vec{\nabla}\left(\frac{\partial \Pi_{1x}}{\partial x} + \frac{\partial \Pi_{1z}}{\partial z}\right) \tag{13.54}$$

$$\begin{aligned}
\frac{\partial \Pi_{1x}}{\partial x} + \frac{\partial \Pi_{1z}}{\partial z} &= P'\frac{\partial}{\partial x}\left[g_0 - g_1 + \left(g_s + \frac{\partial}{\partial z}g_2\right)\right] \\
&= P'\frac{\partial}{\partial x}\left[g_0 - g_1 + 2k_1^2\int_0^\infty \frac{J_0(\lambda\rho)\exp\left[-\sqrt{\lambda^2 - k_1^2}\,(z+z')\right]\lambda d\lambda}{k_2^2\sqrt{\lambda^2 - k_1^2} + k_1^2\sqrt{\lambda^2 - k_2^2}}\right], \\
&\triangleq P'\frac{\partial}{\partial x}\left[g_0 - g_1 + g_3\right]
\end{aligned} \tag{13.55}$$

where g_0 is given in (13.52), g_1 and g_s are given in (13.53), and g_2 is given in (13.48).

$$E_{Ax} = \frac{-j\omega\mu_0}{4\pi}\int_{\ell_A} I_A\left(g_o - g_1 + g_s\right)dx'_A + \frac{\partial}{\partial x}\left[\frac{1}{j\omega 4\pi\varepsilon_0}\int_{\ell_A} I_A\frac{\partial}{\partial x}\left(g_o - g_1 + g_3\right)dx'_A\right] \tag{13.56}$$

g_3 is given in (13.55).

Since $\frac{\partial g_0}{\partial x} = -\frac{\partial g_0}{\partial x'_A}$, $\frac{\partial g_1}{\partial x} = -\frac{\partial g_1}{\partial x'_A}$, and $\frac{\partial g_3}{\partial x} = -\frac{\partial g_3}{\partial x'_A}$ then assuming that the current goes to zero at the ends of the open wires, the derivative $\partial/\partial x$ operation can be transformed to d/dx'_A now operating on I_A instead by applying integration by parts. Hence,

$$E_{Ax} = \frac{-j\omega\mu_0}{4\pi}\int_{\ell_A} I_A\left(g_o - g_1 + g_s\right)dx'_A + \frac{\partial}{\partial x}\left[\frac{1}{j\omega 4\pi\varepsilon_0}\int_{\ell_A}\frac{dI_A}{dx'_A}\left(g_o - g_1 + g_3\right)dx'_A\right] \tag{13.57}$$

The mutual impedance Z_{BA} between two x-directed current elements I_A and I_B of lengths ℓ_A and ℓ_B is expressed as

$$\begin{aligned}
Z_{BA} &= -\int_{\ell_B} E_{Ax} I_B dx'_B \\
&= \frac{j\omega\mu_0}{4\pi}\int_{\ell_B} I_B\, dx'_B \int_{\ell_A} I_A\left(g_o - g_1 + g_s\right)dx'_A \\
&\quad - \frac{1}{j\omega 4\pi\varepsilon_0}\int_{\ell_B} I_B\frac{\partial}{\partial x'_B}\left[\int_{\ell_A}\frac{dI_A}{dx'_A}\left(g_o - g_1 + g_3\right)dx'_A\right]dx'_B.
\end{aligned} \tag{13.58}$$

Transferring the derivative operation on I_B in the second term of the earlier expression and assuming I_B to go to zero at the open ends of the wires, one obtains through integration by parts,

$$
\begin{aligned}
Z_{BA} &= \frac{j\omega\mu_0}{4\pi}\int_{\ell_B} I_B\, dx'_B \int_{\ell_A} I_A\left(g_o - g_1 + g_s\right)dx'_A \\
&+ \frac{1}{j\omega 4\pi\varepsilon_0}\int_{\ell_B} \frac{dI_B}{dx'_B}\left[\int_{\ell_A} \frac{dI_A}{dx'_A}\left(g_o - g_1 + g_3\right)dx'_A\right]dx'_B
\end{aligned}
\tag{13.59}
$$

This completes the analysis now, as we can deal with any arbitrary orientation of a dipole radiating on top of an imperfect ground. We can decompose any arbitrary orientation into x-directed, y-directed, and z-directed components and then, using the previous Green's functions, we solve each problem and then sum all the results together to give the final result. This procedure (discussed in Sections 13.6–13.9) has been implemented in a general-purpose computer code to analyze arbitrary shaped and oriented wire antennas over an imperfect ground plane. The methodology chosen has also been implemented in a commercially available software package titled AWAS (V. 2) [13], that computes accurately the fields from an arbitrary shape wire antenna radiating over an imperfectly conducting ground. AWAS is a complete electromagnetic field simulator for wire-like structures, utilizing the accurate Sommerfeld formulation for taking into account the effects of an imperfect ground plane.

13.10 Analysis of Propagation over Earth

We now numerically calculate the exact Green's function in (13.30), (13.46), and (13.48) using a computational electromagnetic code evaluating the semi-infinite integrals in an essentially exact way. This is carried out using the code AWAS [13]. As an example, consider a vertical-oriented half-wave dipole located at different heights over an average ground with parameters for the permittivity $\varepsilon_r = 15$ and $\sigma = 0.005$ mhos/m, where the ground parameters were taken from [14]. The frequency of operation was 900 MHz. The height of the field point is 2 m. Figure 13.5 plots the field strength as a function of the horizontal distance for different heights of the antenna above the ground. In this figure, it is seen that near the transmitting antenna there is an interference between the direct space wave and the field from the image produced by the imperfect ground, providing variation of the total field strength. This is often labeled as fading. This interference pattern stops at an approximate distance of $4H_{TR}H_{RX}/\lambda$ (H_{TR} represents the height of the transmitting antenna over the ground, H_{RX} represents the height of the receiving antenna over the ground, and λ is the wavelength of operation) and a monotonic decay of the field occurs with a slope of −30 dB/decade and continues approximately to a

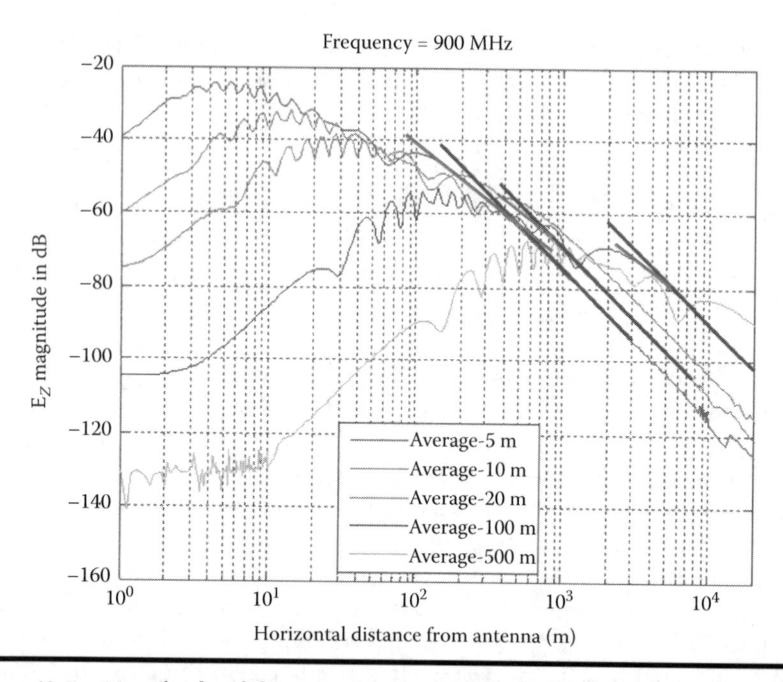

Figure 13.5 Magnitude of the z-component of the electric field in decibels radiated from a vertical-oriented half-wave dipole antenna over an average imperfect ground $\varepsilon_r = 15$ and $\sigma = 0.005$ mhos/m, different curve belongs to different heights above the imperfect ground, and the frequency of operation is 900 MHz. The slope of −30 dB/decade is marked by a thick purple straight line and the slope of −40 dB/decade is marked by a thick black straight line.

distance of $8H_{TR}H_{RX}/\lambda$ from where the far field of the antenna starts and the slope becomes roughly −40 dB/decade [14]. In Figure 13.5, the slope of −30 dB/decade is marked by a thick purple straight line, and the slope of −40 dB/decade is marked by a thick black straight line. Observations seem to verify this prediction carried out by the electromagnetic macromodel [14]. There is a height gain in the far field of the antenna, but in the near field, which is of importance in cellular communication, there is actually a height loss. Hence, it is proposed that a better solution will then be to deploy the transmitting antenna closer to the ground. In that case, the region of the variation in the field strength would be quite small, and the field strength will decay monotonically inside the remainder of the cell minimizing fading. Since there will not be any interference pattern, then it is possible to reduce the transmitting power at least by a factor of 10 (say), providing a better, safe, and cheaper system, as in most cases the tower costs more than the antenna system. However, one could deploy more base stations as the power is reduced. An interesting scenario of this can be seen perhaps in some South American cities, where WiFi is delivered to individual houses by deploying base station antennas on every lamp post. Such a discussion is quite relevant as there is a second channel from the mobile to the base station in which there is no height gain as the mobile is near the ground and the mobile transmits a fraction of the power of the base station. Moreover, the antenna on the mobile can be oriented in any direction, and hence one should question the validity of the state-of-the-art rules of thumb developed for deploying base station antennas, as they do not consider the environment of the second mobile channel that is quite important and does not satisfy any of the rules of thumb associated with the base stations.

Figure 13.6 shows another simulation result for a vertical half-wave dipole located 20 m above different types of imperfect grounds. The frequency of operation was 900 MHz. The ground parameters chosen were for the poor ground with a dielectric permittivity and a conductivity of $\varepsilon_r = 4$ and $\sigma = 0.001$ mhos/m, for the average ground, $\varepsilon_r = 15$ and $\sigma = 0.005$ mhos/m, for the good ground $\varepsilon_r = 25$ and $\sigma = 0.02$ mhos/m, for the sea water $\varepsilon_r = 81$ and $\sigma = 5$ mhos/m, and for the fresh water $\varepsilon_r = 81$ and $\sigma = 0.01$ mhos/m. The ground parameters were taken from [14]. The height of the field point is 2 m. By observing the plots in Figure 13.6, it is seen that within the cell the electrical properties of the ground has little effects as

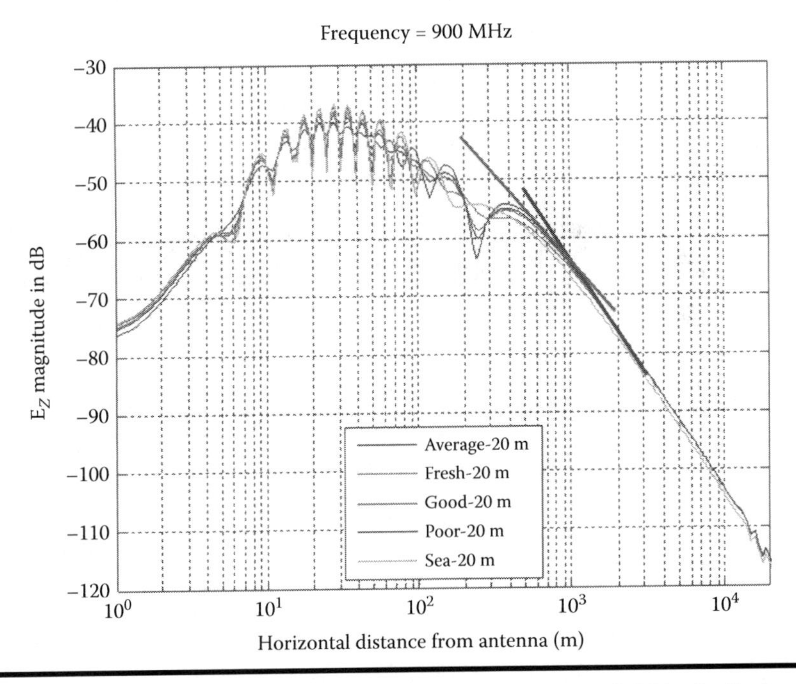

Figure 13.6 Magnitude of the *z*-component of the electric field in decibels radiated from a vertical-oriented half-wave dipole antenna over different imperfect grounds, height of the antenna is 20 m, different curve belongs to different imperfect grounds, and the frequency of operation is 900 MHz. The slope of −30 dB/decade is marked by a thick purple straight line and the slope of −40 dB/decade is marked by a thick black straight line.

for a fixed height of the transmitting antenna radiating over an imperfect ground. The ground parameters do not change the nature of the distant fields, whereas near the antenna, the shape of the interference pattern can be slightly different. In Figure 13.6, the slope of −30 dB/decade is marked by a thick purple straight line and the slope of −40 dB/decade is marked by a thick black straight line. This implies that whether we consider propagation in urban, suburban, industrial, rural, or over water areas, the results should not differ too much from each other and, furthermore, an electromagnetic macromodel can accurately make such predictions!

Now, we can study loop antennas on top of an imperfect ground using the previous Green's functions. As an example, consider a horizontal-oriented square loop antenna with half-wave length circumference located at different heights over an average ground. All other simulation parameters are the same as before. This problem is equivalent for having a vertical-oriented magnetic current. Figure 13.7 plots the electric field strength as a function of the horizontal distance for different heights of the antenna above the ground. Figure 13.8 plots the magnetic field strength as a function of the horizontal distance for different heights of the antenna above the ground. In Figures 13.7 and 13.8, it is seen that near the transmitting antenna there

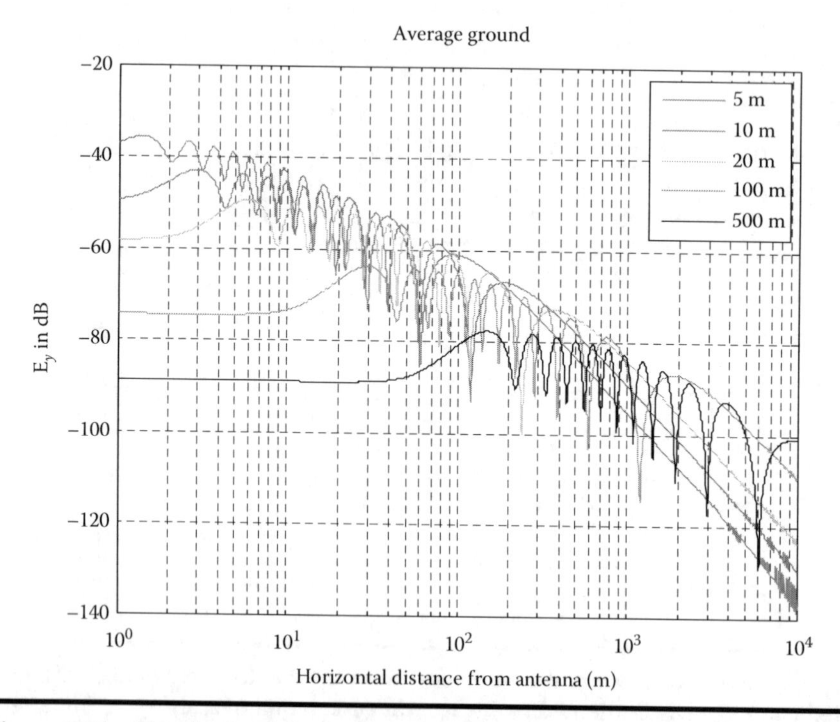

Figure 13.7 Magnitude of the *y*-component of the electric field in decibels radiated from a horizontal-oriented loop antenna over an average imperfect ground $\varepsilon_r = 15$ and $\sigma = 0.005$ mhos/m, different curve belongs to different heights above the imperfect ground, and the frequency of operation is 900 MHz.

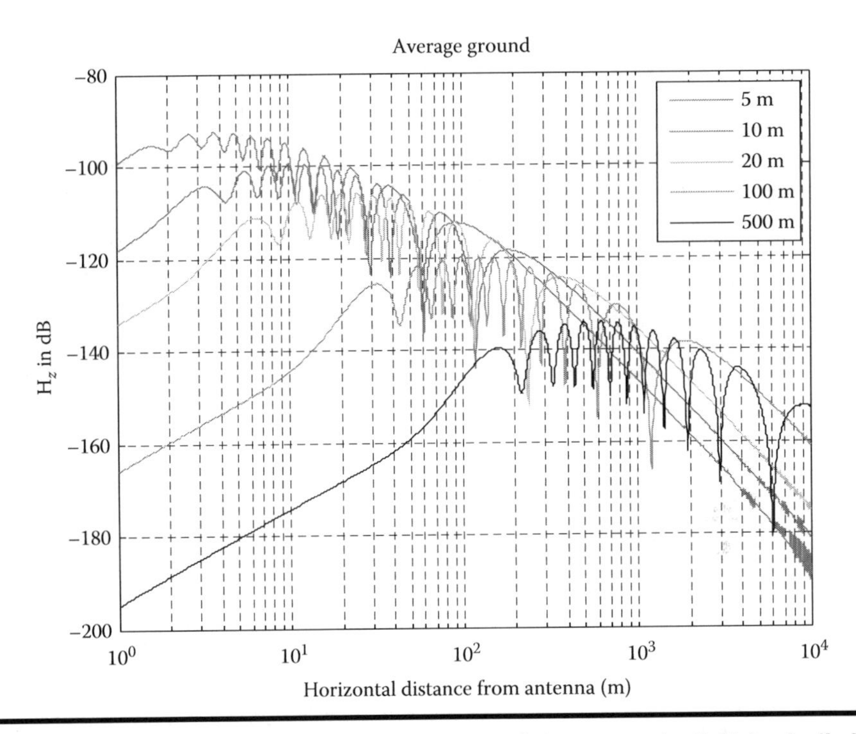

Figure 13.8 Magnitude of the *z*-component of the magnetic field in decibels radiated from a horizontal-oriented loop antenna over an average imperfect ground $\varepsilon_r = 15$ and $\sigma = 0.005$ mhos/m, different curve belongs to different heights above the imperfect ground, and the frequency of operation is 900 MHz.

is interference between the direct space wave and the field from the image produced by an imperfect ground, providing variation of the total field strength. This is often labeled as fading. This interference pattern stops at an approximate distance of $4H_{TR}H_{RX}/\lambda$ and a monotonic decay of the field occurs with a slope of −30 dB/decade and continues approximately to a distance of $8H_{TR}H_{RX}/\lambda$ from where the far field of the antenna starts and the slope becomes roughly −40 dB/decade [14].

Figures 13.9 and 13.10 show another simulation result for a horizontal-oriented square loop antenna with half-wave length circumference located at 20 m above different types of imperfect grounds. All other simulation parameters are the same as before. This problem is equivalent for having a vertical-oriented magnetic current. By observing the plots in Figures 13.9 and 13.10, it is seen that within the cell the electrical properties of the ground has very little effect, as for a fixed height of the transmitting antenna radiating over an imperfect ground, the ground parameters do not change the nature of the distant fields, whereas near the antenna the shape of the interference pattern can be slightly different. As expected, the conclusions are the same as for the vertical half-wave dipole above an imperfect ground plane case.

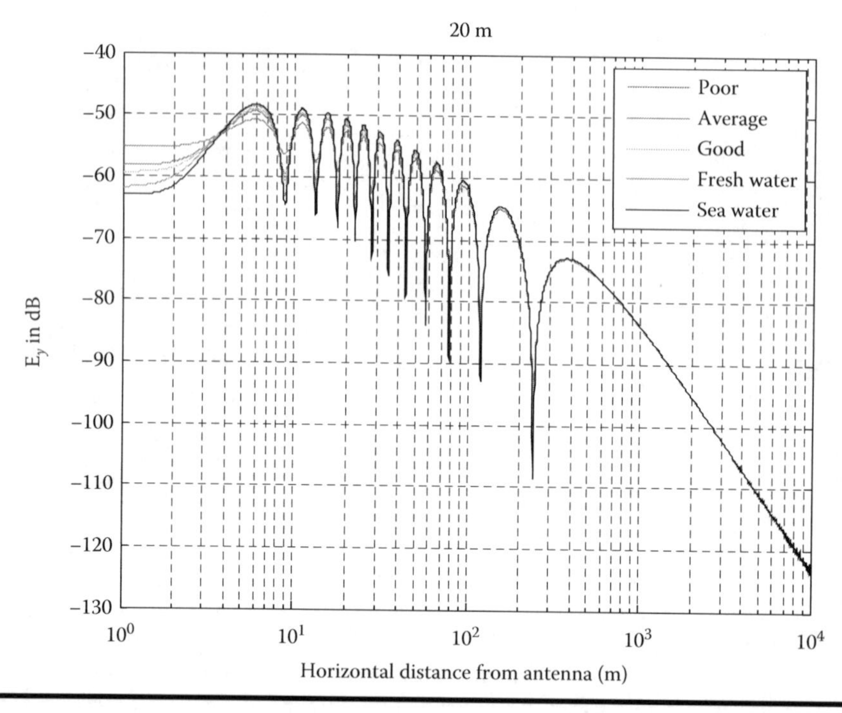

Figure 13.9 Magnitude of the *y*-component of the electric field in decibels radiated from a horizontal-oriented loop antenna over different imperfect grounds, height of the antenna is 20 m, different curve belongs to different imperfect ground, and the frequency of operation is 900 MHz.

Let us consider a vertical-oriented square loop antenna with half-wave length circumference located at different heights over an average ground. All other simulation parameters are the same as before. This problem is equivalent for having a horizontal-oriented magnetic current. Figure 13.11 plots the electric field strength as a function of the horizontal distance for different heights of the antenna above the ground. Figure 13.12 plots the magnetic field strength as a function of the horizontal distance for different heights of the antenna above the ground. In Figures 13.11 and 13.12, it is seen that near the transmitting antenna, there is interference between the direct space wave and the field from the image produced by the imperfect ground, providing variation of the total field strength. This is often labeled as fading. This interference pattern stops at an approximate distance of $4H_{TR}H_{RX}/\lambda$ and a monotonic decay of the field occurs with a slope of −30 dB/decade and continues approximately to a distance of $8H_{TR}H_{RX}/\lambda$ from where the far field of the antenna starts and the slope becomes roughly −40 dB/decade [14].

Figures 13.13 and 13.14 show another simulation results for a vertical-oriented square loop antenna with half-wave length circumference located at 20 m above

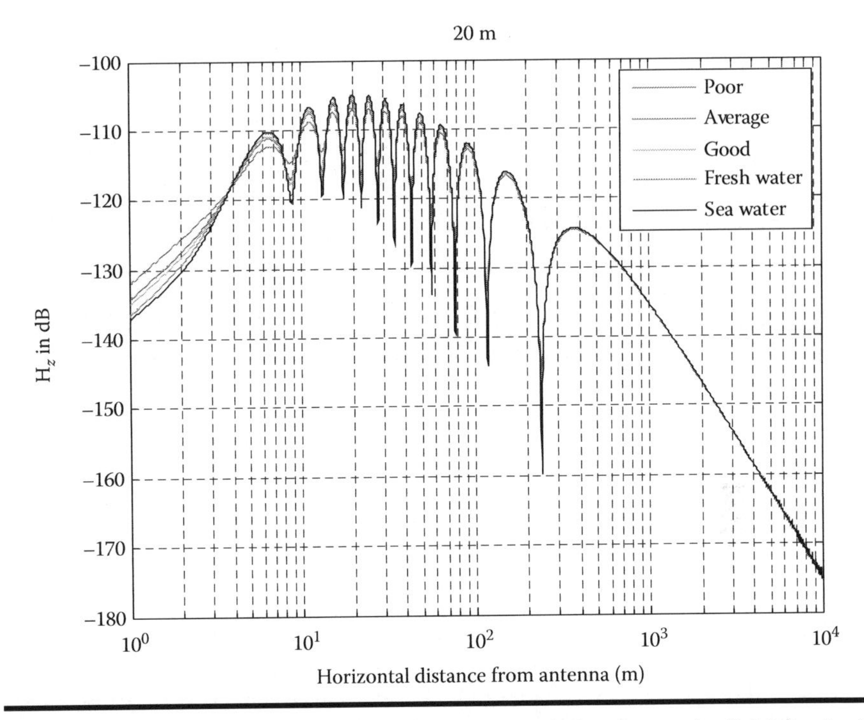

Figure 13.10 Magnitude of the *z*-component of the magnetic field in decibels radiated from a horizontal-oriented loop antenna over different imperfect grounds, height of the antenna is 20 m, different curve belongs to different imperfect ground, and the frequency of operation is 900 MHz.

different types of imperfect grounds. All other simulation parameters are the same as before. This problem is equivalent for having a horizontal-oriented magnetic current. By observing the plots in Figures 13.13 and 13.14, it is seen that within the cell the electrical properties of the ground has little effects, as for a fixed height of the transmitting antenna radiating over an imperfect ground, the ground parameters do not change the nature of the distant fields, whereas near the antenna the shape of the interference pattern can be slightly different.

Furthermore, let us check the effect of varying the frequency on the propagation over earth results. Figure 13.15 and 13.16 plot the results for a vertical-oriented half-wave dipole located 20 m above an average ground with parameters for the permittivity $\varepsilon_r = 15$ and $\sigma = 0.005$ mhos/m. The frequency of operation was varied from 400 MHz, 900 MHz, 1 GHz, and 2 GHz. The height of the field point is 2 m. In these figures, it is seen that near the transmitting antenna, there is an interference between the direct space wave and the field from the image produced by the imperfect ground, providing variation of the total field strength. This is often labeled as fading. This interference pattern stops at an approximate distance

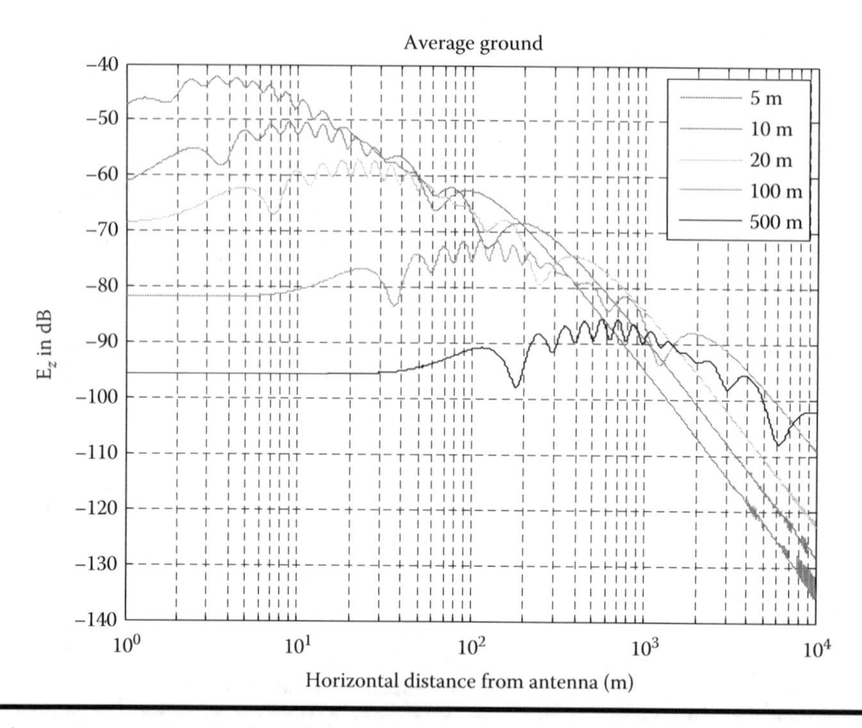

Figure 13.11 Magnitude of the *z*-component of the electric field in decibels radiated from a vertical-oriented loop antenna over an average imperfect ground $\varepsilon_r =$ 15 and $\sigma = 0.005$ mhos/m, different curve belongs to different heights above the imperfect ground, and the frequency of operation is 900 MHz.

of $4H_{TR}H_{RX}/\lambda$ (H_{TR} represents the height of the transmitting antenna over the ground, H_{RX} represents height of the receiving antenna over the ground, and λ is the wavelength of operation) and a monotonic decay of the fields occurs with a slope of −30 dB/decade and continues approximately to a distance of $8H_{TR}H_{RX}/\lambda$ from where the far field of the antenna starts and the slope becomes roughly −40 dB/decade [14].

Figures 13.17 and 13.18 show another simulation result for a horizontal-oriented square loop antenna with half-wave length circumference located at 20 m above an average ground. The frequency of operation was varied from 400 MHz, 900 MHz, 1 GHz, and 2 GHz. The height of the field point is 2 m. This problem is equivalent for having a vertical-oriented magnetic current. By observing the plots in Figures 13.17 and 13.18, as expected, the conclusions are the same as for the vertical half-wave dipole above an imperfect ground plane case.

Now we can discuss how we can benefit from the previous analysis to study propagation over Earth in practical scenarios like the cellular network case. Cellular networks are based on dividing a given area into smaller cells, and each one of these cells is fed by a specific base station. This aims to serve a large number of subscribers

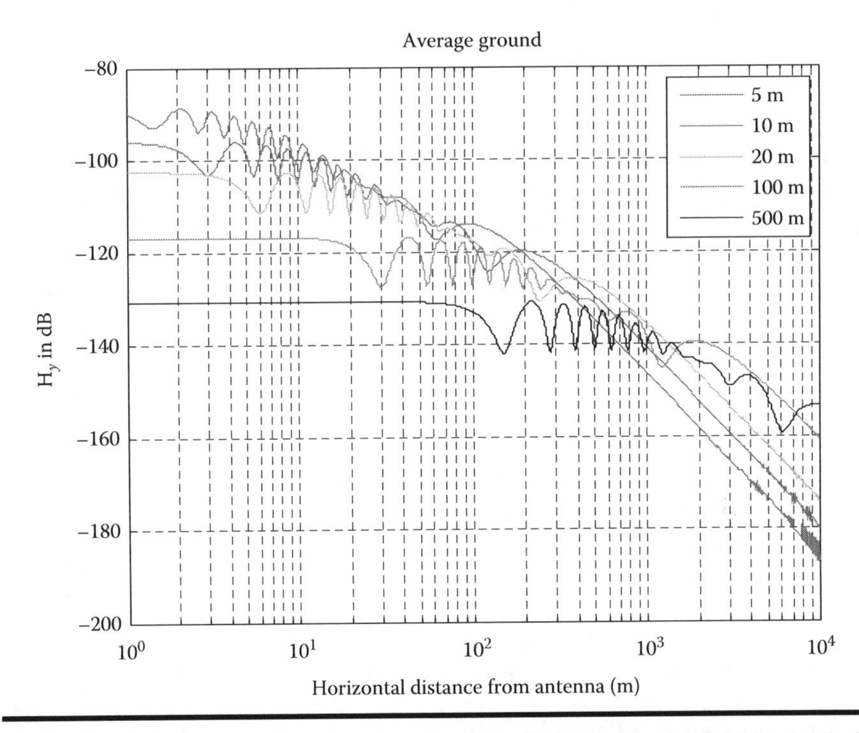

Figure 13.12 Magnitude of the *y*-component of the magnetic field in decibels radiated from a vertical-oriented loop antenna over an average imperfect ground $\varepsilon_r = 15$ and $\sigma = 0.005$ mhos/m, different curve belongs to different heights above the imperfect ground, and the frequency of operation is 900 MHz.

with a small number of available channels. It turns out that the propagation path loss exponent is the essential factor in planning how to divide a given area into cells. The path loss exponent will give the designer an idea how the signal strength will change while moving away from the base station; so it is important to know whether the base station can radiate sufficient power to reach the desired subscribers of that area. At the same time, there will be neighboring base stations that use the same frequency channels to support their subscribers. Thus, the interference level must be kept under some threshold level to guarantee high Quality of Service (QoS) in the cell of interest. The drive test data in cellular networks is crucial for providing information about the real performance of any network, as they illustrate how the signal strength varies while moving away from the base station. The philosophy of channel modeling describes methodologies to predict the performance of a base station antenna inside a given cell. Despite large efforts being made in the field of channel modeling, drive tests data remain irreplaceable, as this checks the accuracy of the model to predict the propagation path loss inside a cell.

In this section, it is shown that a physics-based electromagnetic macromodel can provide accurate data for the path loss exponent in a cellular network using

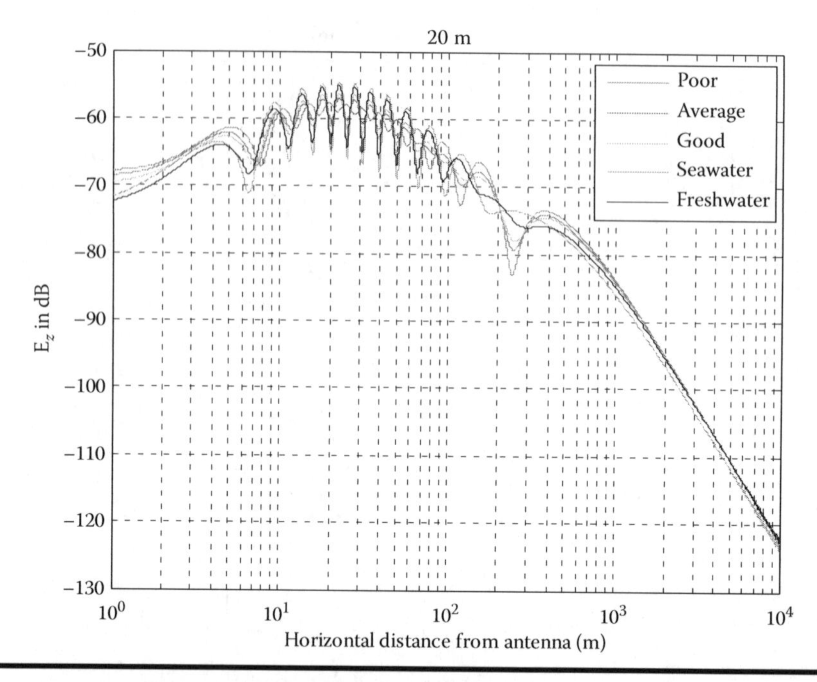

Figure 13.13 Magnitude of the *z*-component of the electric field in decibels radiated from a vertical-oriented loop antenna over different imperfect grounds, height of the antenna is 20 m, different curve belongs to different imperfect ground, and the frequency of operation is 900 MHz.

electromagnetic simulation tools that depend only on some physical parameters of the macromodel. In a macromodel, one needs to include only the electrical parameters of the environment without including the clutter effect factors such as buildings, trees, and so on. As illustrated in [15] and displayed in Figure 13.19a, the reason that the 0.8- to 2.1-GHz cellular band was chosen for mobile broadband is that the reflection from buildings is negligible, and yet the signals can penetrate buildings and terrain and do not significantly bounce inside the rooms. Furthermore in Figure 13.19b, the one-way attenuation through common building materials illustrates that signals at the cellular frequency band can penetrate buildings very easily [16]. Hence, the clutter effects generated by buildings or trees are considered to be second-order effects, as the primary being the effect of the imperfectly conducting ground which is seldom accounted for in any propagation model. Information on Figure 13.19a,b makes it possible for us to use a macro model for the propagation path loss in mobile communication using physical parameters like antenna's height, antenna's gain, antenna's tilt, and the most important factor, namely the effects of an imperfect ground plane. The nonplanar nature of the real ground is deemed not to be a show stopper for the propagation analysis using a macromodel illustrated in [8] and [9]. It is primarily because of

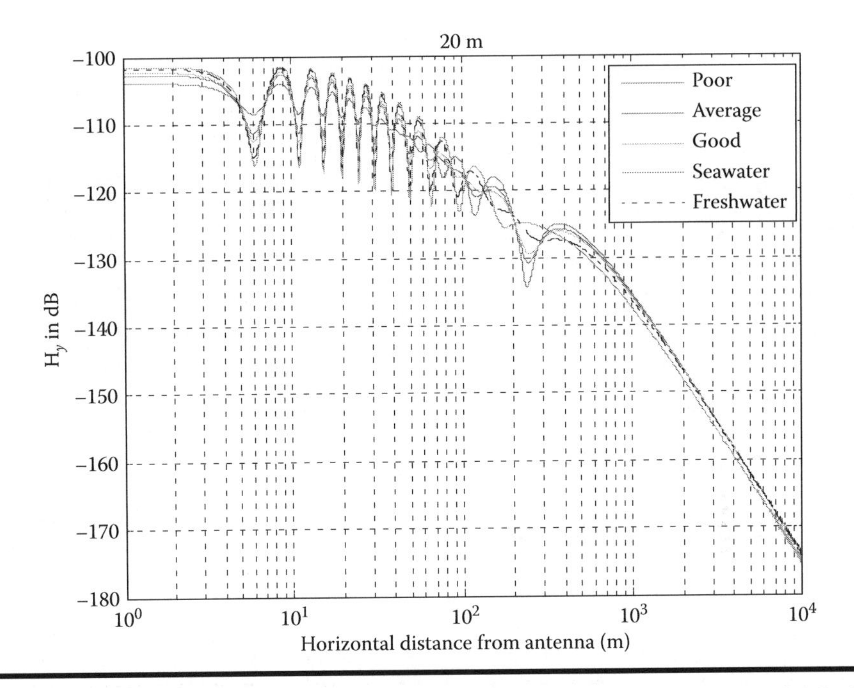

Figure 13.14 Magnitude of the *y*-component of the magnetic field in decibels radiated from a vertical-oriented loop antenna over different imperfect grounds, height of the antenna is 20 m, different curve belongs to different imperfect ground, and the frequency of operation is 900 MHz.

this preponderance of evidence that wireless signals penetrate through buildings and do not get reflected significantly from them, that an electromagnetic macro-model has been proposed based on fundamental physics replacing the statistical models that do not address the underlying electromagnetics, namely the effect of the propagation over Earth.

The main goal is to introduce a physical insight of the propagation in cellular networks instead of blindly applying some empirical or statistical models that do not provide a physics-based view for propagation in cellular networks. It seems that considering buildings, trees, and other obstacles in the propagation model is an overkill, since the clutter effects are considered second-order effects as explained earlier, and also this complicates the analysis so much to the point that researchers start to deviate from physics. In other words, we introduce a pure electromagnetic viewpoint for propagation in cellular networks, and this viewpoint only considers the macroparameters of the environment to model the average path loss. There are variations around the average due to the clutter effects, input power variations, and measurement errors. These variations around the average path loss could be analyzed later on separately. At the end of the day, we will have a physical insight of what is going on and this is the goal.

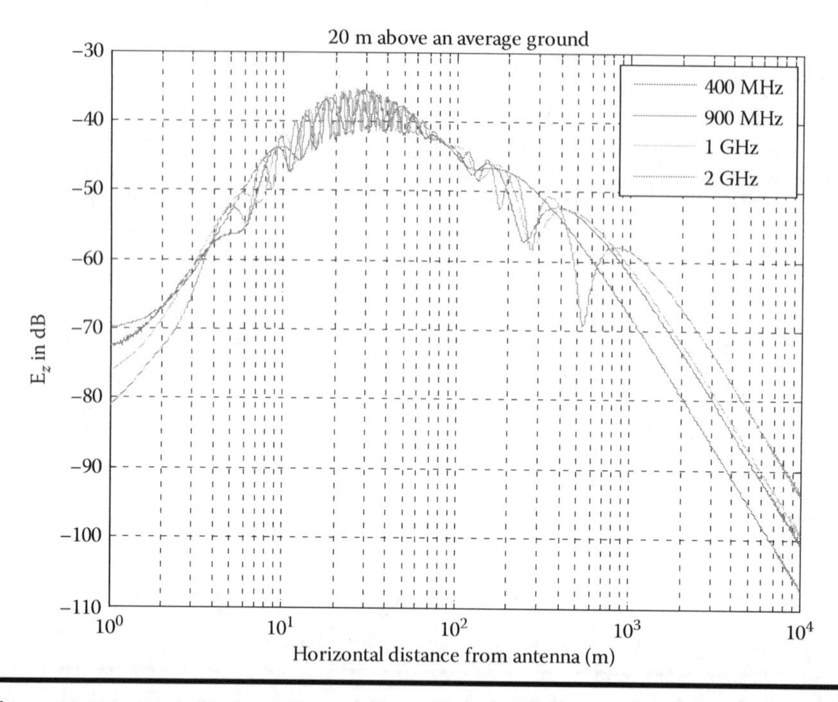

Figure 13.15 Magnitude of the *z*-component of the electric field in decibels radiated from a vertical-oriented half-wave dipole antenna over an average imperfect ground $\varepsilon_r = 15$ and $\sigma = 0.005$ mhos/m, and different curve belongs to different frequencies.

We can use the proposed macromodel based on Sommerfeld formulation to regenerate the propagation data measured by Okumura et al. [17] in their classic propagation measurements in the city of Tokyo. Okumura placed a transmitting antenna with different heights in Tokyo, Japan, and choose a height of 140 m. The signal is received by another vertical polarized antenna located on top of a van 3 m above the ground. The receiving antenna had a gain of 1.5 dB. The transmitting antenna was a 5-element Yagi having a gain of approximately 11 dB and radiating 150 W of power. The van was then driven in the city of Tokyo from 1 to 100 km from the transmitting antenna. Here we consider the measurements done at 453 MHz. Since the 5-element Yagi was an antenna composed of wires, in our simulations we used an optimized 5-element Yagi antenna array that had a gain of 11 dB. In our analysis, the Yagi antenna was synthesized and used in the computations.

Sommerfeld Green's function was used to solve for the current distribution on the transmitting antenna, and then these currents were used to compute the radiating fields for antennas radiating over an imperfect ground. An example for such a code is AWAS [13]. The parameters for the urban ground were relative

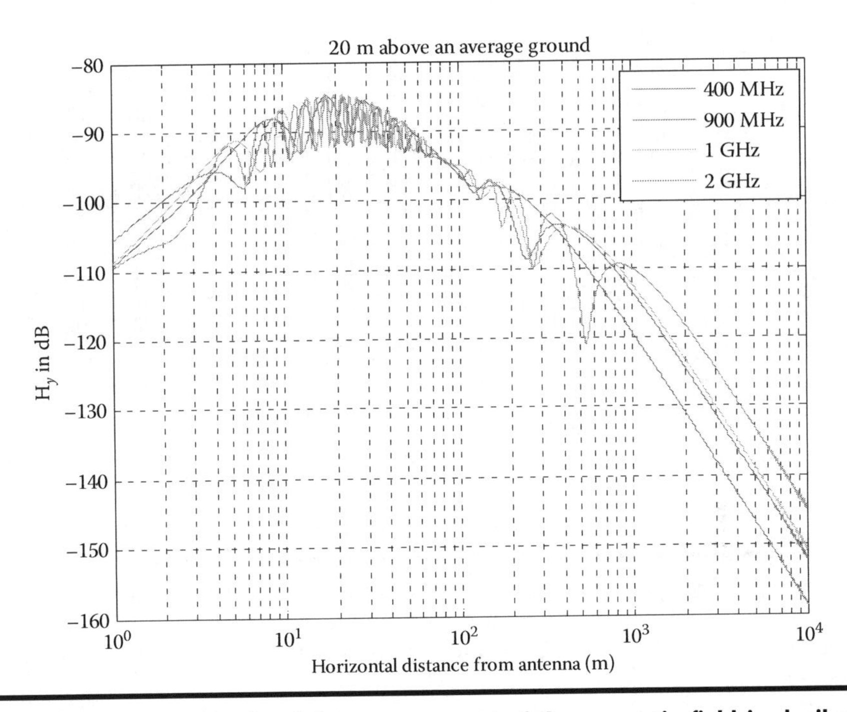

Figure 13.16 Magnitude of the *y*-component of the magnetic field in decibels radiated from a vertical-oriented half-wave dipole antenna over an average imperfect ground $\varepsilon_r = 15$ and $\sigma = 0.005$ mhos/m, and different curve belongs to different frequencies.

permittivity of $\varepsilon_r = 4$ and $\sigma = 2 \times 10^{-4}$ mhos/m [18]. However, since we did not know how Okumura et al. matched their antennas and how it was exactly fed, we used a simple 1 V as an excitation for the Yagi. Then, we shifted all of our computations by a constant value in decibels (98 dB), so that the two plots matched at 7 km as shown in Figure 13.20a. We then overlaid the two plots, theoretical prediction by the Sommerfeld formulation and Okumura et al.'s experimental data. The two plots show remarkable similarity. This is also shown in Figure 13.20. In AWAS, the results for the fields became somewhat unstable when the horizontal distance from the transmitting antenna becomes quite large, say greater than 10 km. This is due to the Sommerfeld integral tails problem. This is solved using the Schelkunoff's formulation. It is also important to point out that, for both the theoretical and experimental data, the slope for the path loss exponent between 1 and 10 km is about 30 dB per decade, which was expected from theoretical analysis. The slope between 10 and 100 km is 40 dB per decade, as predicted for the far field. This illustrates that an accurate electromagnetic macromodeling of the environment is sufficient to predict the path loss, as evidenced by the comparison between theory and experiment. Use of an electromagnetic macromodel makes

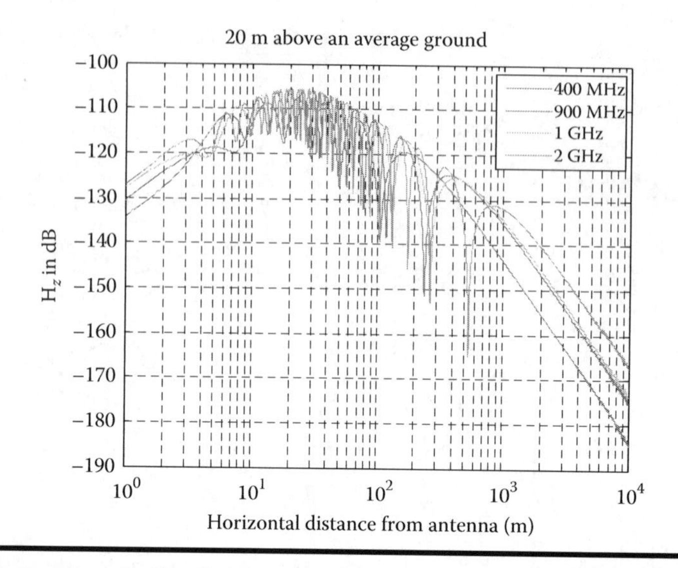

Figure 13.17 Magnitude of the *z*-component of the magnetic field in decibels radiated from a horizontal-oriented loop antenna over an average imperfect ground $\varepsilon_r = 15$ and $\sigma = 0.005$ mhos/m, and different curve belongs to different frequencies.

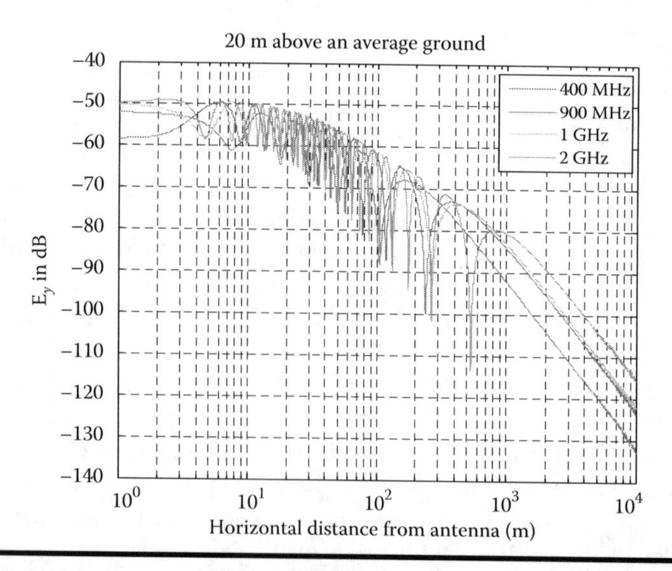

Figure 13.18 Magnitude of the *y*-component of the electric field in decibels radiated from a horizontal-oriented loop antenna over an average imperfect ground $\varepsilon_r = 15$ and $\sigma = 0.005$ mhos/m, and different curve belongs to different frequencies.

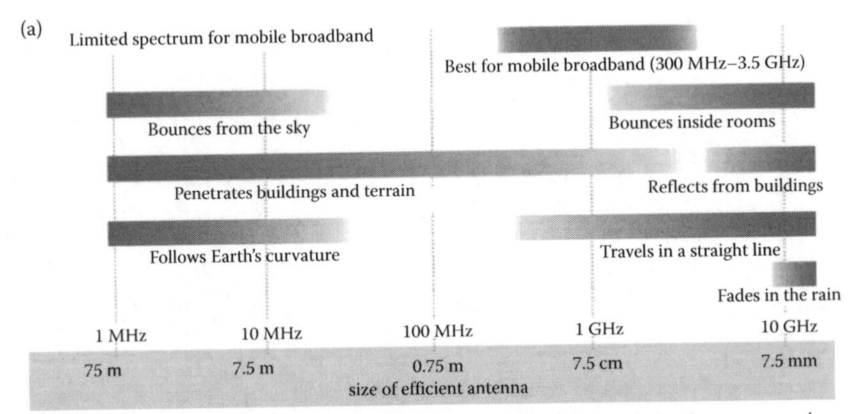

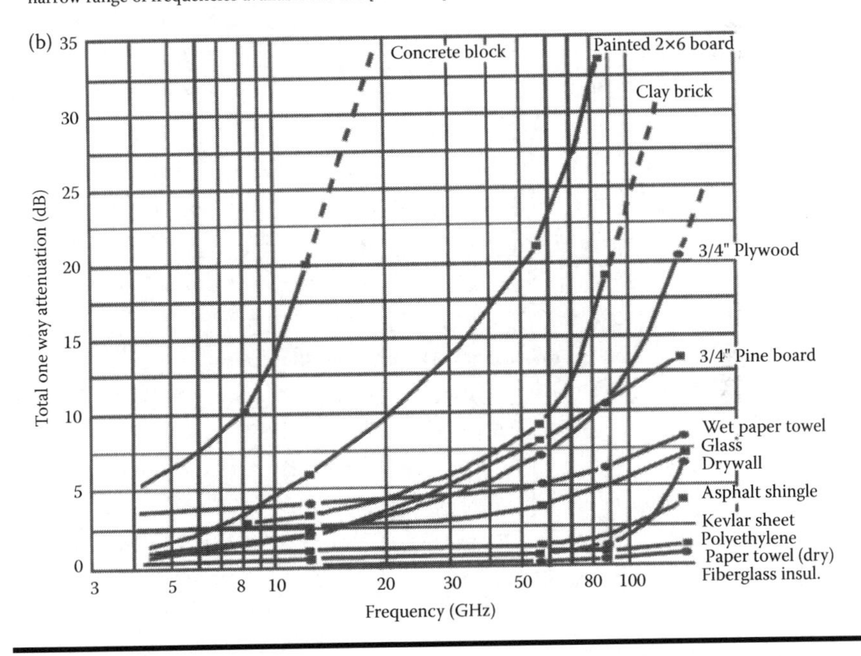

Figure 13.19 (a) Suitable bands for mobile communications in the electromagnetic spectrum [15]. (b) One-way attenuation through common building materials as a function of frequency [16].

sense as the image generated by the transmitting antenna located over an imperfect ground is a semi-infinite line source.

The same numerical analysis of Figure 13.20a is shown in Figure 13.20b for 922 MHz (with 130 dB added to the theoretical data) and in Figure 13.20c for 1920 MHz (with 125 dB added to the theoretical data, note that those numbers

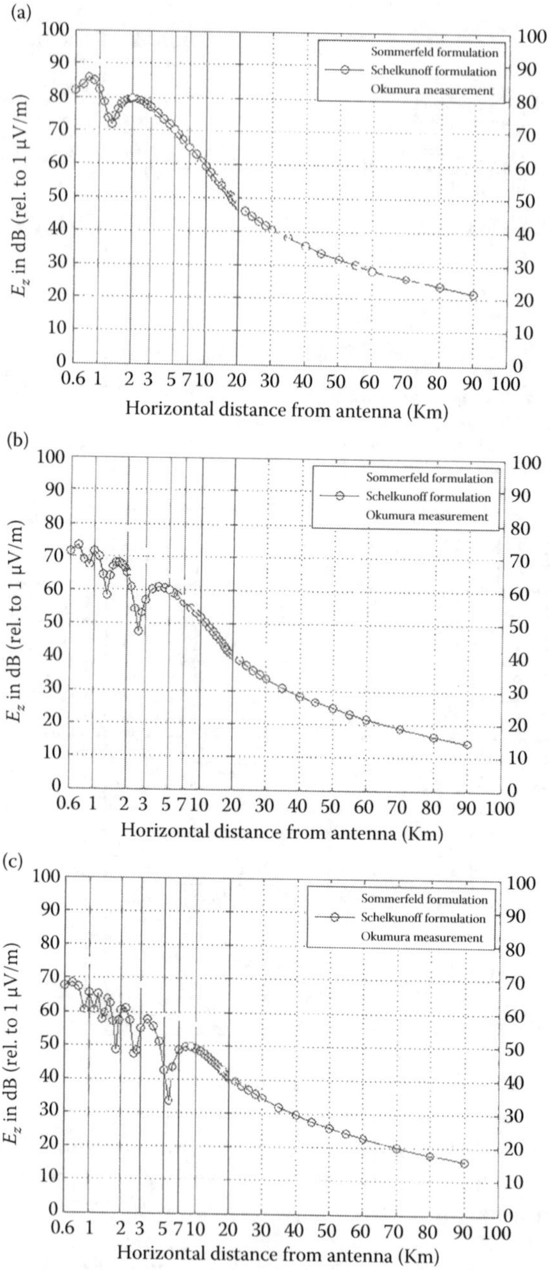

Figure 13.20 Comparison between the experimental (Okumura et al.) and theoretical prediction (Sommerfeld's formulation and Schelkunoff's formulation) computed through a macromodel for predicting propagation path loss in an urban environment at (a) 453 MHz, (b) 922 MHz, and (c) 1920 MHz.

include the adjustment of the reference from 1 V/m to 1 μV/m). For these two frequencies, the plots in Figure 13.20b,c are not with perfect match with the experiment, because in our simulations, at these two frequencies, we used simply a half-wave dipole as the transmitter, whereas Okumura et al. used a highly directive antenna like a parabolic reflector whose dimensions were not reported in their article.

References

1. A. N. Sommerfeld, "Propagation of waves in wireless telegraphy," *Annals of Physics*, Vol. 28, Mar. 1909, pp. 665–736.
2. T. K. Sarkar, "Analysis of arbitrarily oriented thin wire antennas over a plane imperfect ground," *AEÜ*, Band 31, Heft 11, 1977, pp. 449–457.
3. A. R. Djordjevic, M. B. Bazdar, T. K. Sarkar, and R. F. Harrington, *AWAS Version 2.0: Analysis of Wire Antennas and Scatterers, Software and User's Manual*, Artech House, Norwood, MA, 2002.
4. W. M. Dyab, T. K. Sarkar, and M. Salazar-Palma, "A physics-based green's function for analysis of vertical electric dipole radiation over an imperfect ground plane," *IEEE Transactions on Antennas and Propagation*, Vol. 61, No. 8, Aug 2013, pp. 4148–4157.
5. T. K. Sarkar, W. M. Dyab, M. N. Abdallah, M. Salazar-Palma, M. V. S. N. Prasad, and S. W. Ting, "Application of the Schelkunoff formulation to the Sommerfeld problem of a vertical electric dipole radiating over an imperfect ground," *IEEE Transactions on Antennas and Propagation*, Vol. 62, No. 8, Aug 2014, pp. 4162–4170.
6. "IEEE standard definitions of terms for radio wave propagation," *IEEE Std.* 211-1997, Publication Year: 1998. See also: http://ieeexplore.ieee.org/stamp/stamp.jsp?tp=&arnumber=705931&userType=inst&tag=1.
7. M. N. Abdallah, W. Dyab, T. K. Sarkar, M. V. S. N. Prasad, C. S. Misra, A. Garcia Lampérez, M. Salazar-Palma, and S. W. Ting, "Further validation of an electromagnetic macro model for analysis of propagation path loss in cellular networks using measured driving-test data," *IEEE Antennas and Propagation Magazine*, Vol. 56, No. 4, Aug. 2014, pp. 108–129.
8. T. K. Sarkar, W. Dyab, M. N. Abdallah, M. Salazar-Palma, M. V. S. N. Prasad, S. Barbin, and S. W. Ting, "Physics of propagation in a cellular wireless communication environment," *Radio Science Bulletin*, No. 343, (http://www.ursi.org/files/RSBissues/RSB_343_2012_12.pdf), Dec. 2012, pp. 5–21.
9. T. K. Sarkar, W. Dyab, M. N. Abdallah, M. Salazar-Palma, M. V. S. N. Prasad, S. W. Ting, and S. Barbin, "Electromagnetic macro modeling of propagation in mobile wireless communication: Theory and experiment," *IEEE Antennas and Propagation Magazine*, Vol. 54, No. 6, Dec. 2012, pp. 17–43.
10. R. E. Collin, "Hertzian dipole radiating over a lossy earth or sea: Some early and late 20th-century controversies," *IEEE Antennas and Propagation Magazine*, Vol. 46, No. 2, April 2004, pp. 64–79.
11. A. Ishimaru, *Electromagnetic Wave Propagation, Radiation, and Scattering*, Prentice Hall, Englewood Cliffs, NJ, 1991, Chapter 15 and Appendix to Chapter 15.
12. A. Baños, Jr., *Dipole Radiation in the Presence of a Conducting Half-Space*, Pergamon Press, Oxford, UK, 1966.

13. A. R. Djordjevic, M. B. Bazdar, T. K. Sarkar, and R. F. Harrington, *Analysis of Wire Antennas and Scatterers (AWAS) Software and User's Manual, Version 2.0,* Artech House, Norwood, MA, 2002.
14. A. De, T. K. Sarkar, and M. Salazar-Palma, "Characterization of the far field environment of antennas located over a ground plane and implications for cellular communication systems," *IEEE Antennas and Propagation Magazine*, Vol. 52, No. 6, Dec 2010, pp. 19–40.
15. M. Lazarus, "The great spectrum famine," *IEEE Spectrum*, Vol. 47, No. 10, October 2010, pp. 26–31.
16. R. W. McMillan, "Terahertz imaging, millimeter-wave radar," in *Advances in Sensing with Security Applications*, **2**, Springer, Netherlands, 2006, pp. 243–268.
17. T. Okumura, E. Ohmori, T. Kawano, and K. Fukuda, "Field strength and its variability in VHF and UHF land mobile service," *Review of the Electrical Communication Laboratory*, Vol. 16, No. 9–10, 1968, pp. 825–873.
18. K. Fujimoto, *Mobile Antenna Systems Handbook*, 3rd Edition, Artech House, Norwood, MA, 2008.

Index

C

D

E

F

G

H

I

K

L

M